TEACHING the
NATURE of SCIENCE

TEACHING THE NATURE OF SCIENCE
Perspectives & Resources

Douglas Allchin

SHiPS Education Press
Saint Paul

SHiPS Education Press
2005 Carroll Avenue
Saint Paul, MN 55104
shipspress.com

Distributed through Itasca Books
5120 Cedar Lake Road
Minneapolis, MN 55416
1.800.901.3480
www.itascabooks.com

Printed in the United States of America

ISBN 978-0-9892524-0-9

Contents

| Illustrations and Tables

| Preface

Science pervades our culture. It informs—or purports to inform—social policy, from climate change and clean water to the safety of food, drugs, and workplaces. Science may also potentially inform personal decision making, from nutrition and weight-loss diets to energy-efficient appliances or eco-friendly products. Science educators thus have an important responsibility.

How do we prepare students for the challenges that lie ahead? What does effective science education look like?

In recent years, a public sense of urgency about achieving high-quality science education has fueled increased accountability in schools. In the current cultural and economic contexts, that has yielded more standardized tests. Given their inevitable format of multiple-choice questions, education has become reduced to memorizing itemized tidbits of knowledge. Teachers, obliged by circumstances to teach to the test, focus ever more on piecemeal content.

In the cases central to the ultimate aim of science education, however, the primary concern is less content than understanding the practice of science. Which claims are reliable, and why? Which experts can you trust, especially when they seem to disagree? Do the circumstances reflect a warranted change in scientific consensus? What are the possible sources of error and how do they shape the certainty of the claims? What assumptions may have been made and how might they bias the conclusions? Who sponsored the research, and what are the affiliations and interests of the researchers? Where does verifiable information end and value judgment begin? Namely: how do scientists arrive at their conclusions, and when are they thus worthy of trust? Students need to learn foremost about the *nature of science*, or NOS: how science works—or doesn't work—and why.

This book is a roadmap and resource on how to think about the nature of science. What fundamental analytical skills contribute to fruitful ongoing reflection, which might be nurtured in an educational setting? It is a guide especially for emerging teachers, although the perspectives may equally inform veteran teachers, teacher educators, researchers, curriculum specialists, textbook writers, science education administrators, museum educators, and others.

The strategy is unlike that of other books on the nature of science itself.

My aim here is not to survey the vast knowledge of the history, philosophy, and sociology of science (HPSS), although such knowledge is, in a sense, foundational. Nor is my aim to profile basic scientific methodology, or the principles of "good science," particularly in contrast to pseudoscience or fraud. These approaches to learning NOS focus on content, too, but of a different kind. They are relatively unhelpful as an introduction.

Rather, my aim is how one can foster the development of thinking skills: ways to analyze and reflect fruitfully about NOS. The goal is for the teacher to develop a knack for *delving into NOS* and *asking important NOS questions*. Given a particular case, is one able to notice and highlight the relevant dimensions of science that affect the reliability of the claims? Can one interpret and articulate their significance? Can one identify further information that will deepen that understanding? The target is thus a repertoire of tools in NOS inquiry, not declarative knowledge. Ideally, a prepared teacher will be equipped to (1) serve as an explicit model for students, (2) guide others in their own emerging efforts, and (3) reflect further and continue to learn about the nature of science.

This strategy involves, in part, engaging the reader in new and different perspectives. To expand one's analytical repertoire, one needs to open new ways of seeing. Some things may have always been there to see. By recognizing how to notice them, one can appreciate their significance more fully. One needs to learn, in a sense, how to probe the nature of science and to pose the appropriate questions.

Accordingly, the style of presentation here is chiefly by illustration, or demonstration through cases. Perspectives are *exhibited* and *articulated*. There are no formal abstract principles to elucidate, no arguments to master or dissect, no lists to memorize. This book does not resemble a conventional textbook in that sense, even while it endeavors to guide deep and meaningful learning. In style, it follows basic findings in cognitive science. Learning, here, is fostered through anomalies and paradigmatic exemplars and generalizing from them.

Of course, NOS conceptions abound among teachers and students, even without any explicit instruction. So cases that illustrate commonplace notions or conceptions already entrenched in our culture are not, in general, addressed here. At the same time, many widespread NOS conceptions—including those entrenched in the culture of teaching—are ill informed. There is much to *un*learn about NOS. One must separate the wheat from the chaff. The focus here is thus selective, especially in targeting caricatures and naive conceptions. Further, in adopting a pedagogical constructivist perspective, most examples mindfully challenge the NOS preconceptions that are ill informed. Other examples consider teaching methods that seem intuitive to some but that, ironically, prove ineffective or even counterproductive. The overall posture may seem critical. Still, the intent is to motivate reflection and conceptual growth. Individually and collectively, the essays may seem to challenge many assumptions or practices. Yet this is precisely where one may expect genuine learning to occur.

The book is also incomplete in many ways. Learning involves engagement. Thus, ideally, the reader will couple reading of the text with an extended exploration of at least one case in the history of science. The perspectives introduced here become vivid when applied to and measured against real science. Appreciation of the many dimensions of NOS will be greatly enhanced by familiarity with the concrete details of one case. This, too, follows an educational approach that highlights the value of depth.

Some educators believe that the primary or best way to learn about scientific practice is to become a scientist oneself. Yet this may overstate the goal. Yes, every individual should develop some basic investigative skills that enable them to troubleshoot a lamp that does not light, say. They should be able to read simple graphs and evaluate simple data charts. But dealing with evidence at an elementary level is not sufficient for interpreting most science today: for example, the complexities of climate change models, or vast meta-analyses of the significance of mammograms at different ages. Students, like scientists themselves, must rely on the expertise of others. One must learn to be a prudent consumer of science: interpreting the difference between science that is well done and that which should be regarded skeptically or jettisoned outright. The relevant dimensions of scientific practice extend well beyond what one might expect individual students to perform themselves.

One may develop an understanding of the practice of science in many ways. The primary ways are through (1) a student's own labs or inquiry activities, (2) contemporary case studies, and (3) historical case studies. I will not have much to say about student investigations: this is familiar territory for most teachers (including from their experience as students). My primary focus is the role of historical cases (see Chapter 2). Indeed, I contend that deeper appreciation of historical case studies—when styled in an inquiry mode or problem-based format (Chapter 14)—provide deeper, more complete lessons. Indeed, they may well inform how a practicing teacher guides student reflection in the other two formats.

Another major emphasis in contemporary discussions of NOS education is profiling the nature of models and the process of model building. The theme has an important deflationary function, qualifying widely held popular views of the monumentality of scientific theories. Models have been and continue to be important tools in science. But not all science is model driven or explanatory in nature (see Chapter 1). The current enthusiasm for models carries the traces of a short-lived educational fad. One would do well to disregard the hype and focus just on the enduring features of models that have been acknowledged by philosophers of science for decades (see Chapter 8). Thus, while a model-based perspective complements and provides an important context for the themes discussed in this volume, I leave the articulation of that theme to others.

Nearly all the material in this book has been published elsewhere. It seemed appropriate, however, to bring it together in a single volume and present it as a coherent constellation. In addition, I have rewritten and updated every chapter to convey a more unified vision of teaching the nature of science. I have added cross-references and additional comments to underscore their integrated themes and enhance their integrity as an ensemble. In particular, I address topics that are overlooked elsewhere and stress themes that probe common yet misleading cultural conceptions. It is not a comprehensive, "textbook" introduction to the field. Still, one should find the basic tools for NOS reflection here, along with methods for empowering students with those tools, to promote self-guided NOS learning.

In assembling this volume, I have drawn on a unique cross section of experience in science teaching, scientific research, and advanced study in the history and philosophy of science. I taught high school biology for several years, both introductory and AP levels, and later introductory college biology at several institutions. One interdisciplinary science course was structured as an episodic history of science, supplemented with historically inspired labs. I have also participated in research at three biological field stations: looking at treegaps and forest succession in a mid-Atlantic forest; mapping long-term succession in a tropical rainforest; and measuring sexual selection in flowers through the differential transfer of pollen in meadows of the Rocky Mountains. I certainly recommend to any science teacher the exhilarating and enriching experience of living and working in a research community for a season. My own work for a master's degree in evolutionary biology focused on a mathematical model for information-center foraging, such as one finds in honeybee hives, ant societies, and many bird colonies. I couple this firsthand experience with a Ph.D. in the history and philosophy of science from the University of Chicago—and I continue to publish scholarly articles in these fields (from the *Dictionary of Scientific Biography* and an examination of eighteenth-century geologist James Hutton's views on coal, to analyses of disagreement in science, error types, and the conceptual dilemmas of Mendelian dominance). Many of my students in undergraduate history and philosophy-of-science classes have been en route to careers in science teaching, and it is rewarding to shepherd them to deeper understanding of, and reflection on, scientific practice. Through these diverse experiences, I have gained immense respect for science teachers, scientific researchers, historians, philosophers and sociologists of science, and, above all, students. I hope the present volume offers a fruitful synthesis, integrating and honoring these multiple contexts.

The contents of this book are presented in two sections. In the first, I provide perspectives for deepening an awareness and appreciation of NOS. The second section profiles some sample classroom case studies in several disciplines, each with supplemental pedagogical commentary.

In Chapter 1, I survey the territory ahead. In part, I address a large handful of NOS preconceptions, especially common among science teachers, hoping to clear the field of some entrenched notions that make deeper understanding more problematic. The project is expansive: tracking the assembly of scientific knowledge from laboratory or field observations to public communication of science. Reliability is at stake at every step along the way. And the scientifically literate citizen needs to be ready to cope with any of them. In Chapter 2, I describe the relevance of history and inquiry-type historical case studies to the task. In Chapters 3–5, I profile several challenges of using history effectively and, thus, portraying NOS faithfully. These include the problems of myth-conceptions, rational reconstructions, and ideologically shoehorned pseudohistory. In Chapters 6–8, I elaborate on important views of nature and science that are currently unduly peripheralized: the role of culture, the tendency for simplification to drift into oversimplification, and the dominating image of laws in science. In Chapter 9, I address the challenge of assessing NOS understanding, so critical in our current age of accountability.

In Chapters 10–13 (Part II), I present a sampling of case studies as exemplars—showing just what it means to use history to teach nature of science. These are both ready-to-use resources and models for developing such resources. Chapter 14 introduces a sample of other fine case studies, provides a guide for assessing yet others, and offers a framework for the novice case-study author. While these concrete resources are closer to classroom practice, the conceptual perspectives offered in Part I are critical to understanding how the NOS lessons are structured and how to use them fruitfully.

Those who wish to pursue the topics of this book further, who seek additional concrete resources for teaching as profiled here, or who want to continue the educational dialogue on NOS are invited to explore the SHiPS Resource Center, a website that I developed and have edited for the past two decades: http://ships.umn.edu.

Teaching the Nature of Science

Part I: Perspectives

1 | The Nature of Science: From Test Tubes to YouTube

Reliability as a benchmark • the demarcation project • falsifiability • scope and nature of NOS in science education • NOS for scientific literacy • mapping NOS • Whole Science

Emissions from a proposed local waste incinerator. Reported links between the measles vaccine and autism. Revised vitamin D recommendations. Underground seepage from a chemical waste site into the groundwater. A new Earth-like planet with the potential for life. Such cases in the news are striking because scientific knowledge will not help the typical citizen interpret the key issues about the reliability of the claims. One needs to understand instead the *nature of science* (NOS): Whose expertise can be trusted? What public presentations of scientific findings are credible? How do scientists reach conclusions about things they cannot see directly? When is a change in scientific consensus justified, if ever? How do emotions shape assessments of risk? How might scientists make honest mistakes, and how does one evaluate them? These cases all exemplify vividly the educational goal of scientific literacy. But sheer mastery of textbook concepts will not help. Rather, to inform real-life decisions, both personal and public, one needs knowledge *about how science works*. Knowledge of NOS may be as important as—if not more important than—knowledge of content.

Approaches to teaching about the nature of science have deep roots, now decades old. World War II seemed to demonstrate the public significance of science—even basic research—with the development of the atomic bomb, penicillin, sulfa drugs, radar, sonar, the pesticide DDT, the proximity fuse, cybernetics and early computers, cryptography, chemical warfare, and rockets. Scientist and political titan James Bryant Conant, who had helped shepherd the United States into applying science to the war, capitalized on the postwar spirit and began advocating teaching about the "tactics and strategy of science."[1] Efforts ensued for many decades, with varying effectiveness, generating scores of tests and surveys for NOS knowledge.[2] Science education reforms in the 1990s, however, buoyed the significance of NOS dramatically and placed it squarely among science curricular goals.[3] NOS remains prominent in major profiles of the science curriculum.[4]

But what is the 'nature of science'? Or what ideas about science, the process of science, and its cultural contexts are important to teach? One can easily imagine that there is opportunity for a wide range of professional judgment, even disagreement, among educators as well as historians, philosophers, and sociologists of science. And so there is. The practicing teacher negotiates through a sometimes contested territory. One fruitful approach, adopted here, is to be inclusive of multiple perspectives, rather than limiting or exclusive (see Chapters 3 and 5). At the same time, the goal of scientific literacy offers a valuable touchstone[5]:

> *Students should develop a broad understanding of how science works to interpret the reliability of scientific claims in personal and public decision making.*

Reliability, or trustworthiness, is a fundamental benchmark.[6] Any factor that significantly affects the reliability of scientific claims potentially merits our attention. That might range from the contamination of a Neanderthal DNA sample or the calibration of a gravity wave detector, to theoretical commitments that preclude a role for bacteria in ulcers, to fraudulent claims about stem cells or gender bias in research on heart disease. Analysis might extend equally to the sources communicating scientific results, from websites alleging that global warming is a hoax or television shows about mad cow disease, to news reports about cold fusion or testimony about the causes of an oil spill in a courtroom. That is, potential sources of error may arise in experimental materials and apparatus, in theorizing and reasoning about results, in social interactions or institutional politics, and in communication networks. Nature of science is about the whole of science: from the lab bench to the judicial bench (for a preview of the remainder of the chapter, see Figure 1.3).

Ultimately, students need to understand the whole of scientific practice, but not abstractly or philosophically. As citizens, they need a functional understanding that can guide analysis of scientific claims in particular contexts or cases.[7] General knowledge about testing hypotheses, theory-laden observations, or the tentativeness of scientific results will not suffice.

Learning the nature of science is a journey—perhaps even an adventure. As with any topic, the more experience one gains, the deeper one's understanding becomes. Indeed, the field of Science Studies, encompassing history, philosophy, and sociology of science, as well as other perspectives (rhetorical, cultural, experimentalist, feminist, Marxist, visual), continues to grow. Just as new insights appear in science, so do new insights emerge into how science works—or, sometimes, doesn't work. This book aims to launch you on the journey with a repertoire of questions for investigation and reflection, for students and teachers alike. This chapter surveys the NOS territory and presents an introductory conceptual framework and the basic tools for exploring scientific practice.

The next two sections address major conceptions of science entrenched in our culture and in the lore of scientists and science teachers. Historical examples and philosophical analysis can inform our understanding and help us rethink these widely accepted but misleading truisms about science. They help illustrate the importance of reflecting on the nature of science, even about commonplaces we may at first consider beyond question. Thus we may begin to clear the field and open fruitful exploration. Readers anxious for a more positively expressed view—where this preliminary skepticism leads us—may prefer to jump ahead to the subsequent three sections on characterizing the scope of NOS in science education, especially in the context of functional scientific literacy as profiled in this opening. Namely, what do we need to teach? The final section introduces the guiding notion of Whole Science and helps map out the issues addressed in the remaining chapters: how does one deepen an understanding of scientific practice and help students develop similar skills in analysis and reflection?

THE DEMARCATION PROJECT

The first impulse—common among students, at least—may be to rush to a dictionary for a clear definition of science. Of course, dictionaries define the use of *words* only. They do not articulate concepts. Still, the common urge is telling. It reflects a number of tendencies and unschooled beliefs. First, many assume that the world is organized so as to yield quick, easy, and clear answers. Second, problems are to be solved by external authorities. Third, science is an abstract concept that can be stipulated, rather than a human activity that is to be interpreted or understood. There is great faith that a simple factor distinguishes science as a special form of knowledge.

Philosophers of science have duly considered this challenge, known as *the demarcation problem*. But without clear success. Early efforts were largely motivated by ideological distaste for Marxism and psychoanalysis. Each claimed to be empirical and thus "scientific." Many prominent philosophers did not want to share the privileged status of science with such allegedly wrong-headed pursuits. They sought criteria by which to exclude them.

One effort focused on logic, which could be imbued with mathematical certainty. Yet while one might hope to express a theory and its derivative concepts in a rigorous logical framework, observations themselves have no logical structure. Linking them securely and unambiguously to linguistic or mathematical expressions proved difficult. The effort failed.

Another effort to demarcate science focused on verifiability. Only claims that could be verified by observation would count as science. Yet that seemed to exclude all kinds of unobservables and theoretical entities, such as atoms, magnetic fields, and genes. Reasoning about unrepeatable historical events, whether in geology, cosmology, or evolutionary biology, also seemed problematic. So that criterion fell by the wayside as well.

Yet another proposal focused on science as uniquely progressive. The cumulative growth of knowledge in science seemed intuitive. Even today, students tend to regard science as synonymous with progress. However, in many scientific revolutions, scientists seemed to jettison former knowledge in favor of new approaches that did not fully replace them. We no longer talk of phlogiston, caloric, electrical fluid, or worldwide floods. We no longer discuss chemical affinities, pangenes, bodily humors, or fixed continents. All of these were once widely accepted concepts. Even the apparently secure universality of Newton's laws was abandoned in favor of relativity; they did not apply to very light or very fast bodies. One more demarcation criterion for the creative wastebin.

Other ways of defining progress proved equally elusive. A later perspective focused on growth in the ability to solve problems. But clearly defining a problem or solution proved no easier. In addition, some acknowledged sciences, such as mineralogy, taxonomy, and astronomy, focused on documenting or collecting information about the natural world, not solving problems.

And so on.

The successive failures to demarcate science are now deeply informative. First, they help demonstrate that the reliability of knowledge so commonly associated with science is not simple. One needs to clarify the often arbitrary connections between phenomena and the terms one uses to discuss them. One needs to be mindful of theoretical context in interpreting observations. One needs to ascertain the limited scope of concepts. One needs to consider the value of theories in guiding further research as well as explaining existing results. Reflection on scientific reasoning has sharpened how scientists think and communicate.

With the privilege of retrospect, one may also wonder: what motivated the repeatedly unsuccessful efforts to define science? Ultimately, the project aimed not only to characterize science but also to distinguish it. Indeed, special distinction was central. Each proposed nature of science was imbued with value. A positive value. The role of demarcation was not neutral. It was *normative*.

Philosophers did not endeavor to describe science as it is. Rather, they rendered an *idealized version* of science. We *want* science to be logical. We *want* science to be progressive. We *want* observation to be independent of theoretical perspective. It makes the task of justifying and interpreting scientific knowledge much easier. It is a tribute to science, perhaps, that it has achieved so much without being reducible to any single identifiable principle or attribute. Science seems to work, even without adhering rigorously to some ideal. Normative and descriptive views of science differ. And this distinction becomes an important recurring theme in teaching the nature of science.

The demarcation project also tended to treat science as a paradigm (if not the exclusive domain) of rationality or objectivity. Scientific knowledge earned special authority. Authority, hence power. As reflected in the early aim to disenfranchise Marxism and psychoanalysis, the repeated efforts were inherently political. Such political overtones remain. Scientists—and science teachers—often enjoy unquestioned privilege. And many imagine they should be able to summon such power or prestige through a simple definition or appeal to a mere label.

The impetus to demarcate science persists. The target now is primarily pseudoscience: creationism, astrology, alchemy, telepathy, precognition, psychokinesis, aliens and UFOs, and New Age-ism, among other topics.[8] They are typically dismissed as inherently unscientific. Ironically, many of these "pseudoscientific" pursuits were once regarded as science or intimately related to it. As for astrology, Galileo, renowned advocate of the Copernican system, made a horoscope for his daughter, Virginia, at her birth! Johannes Kepler, who elucidated the elliptical orbits of planets, wrote more than eight hundred horoscopes. His views on the order of the cosmos and their effect on Earth guided his astronomical studies. (At the same time, he had little tolerance for *fraudulent* astrologers!) Alchemy was pursued by Robert Boyle (whose work with the vacuum led to the law that bears his name) and reputedly, a century later, by geologist James Hutton (known for opening the vast scope of Earth's age). Yet their work pales when compared with the archive of at least 131 alchemical manuscripts left by Isaac Newton, who devoted over three decades to alchemical researches. Historian of science William Newman has reproduced many of Newton's experiments: some on mineral growth—silica gardens and the star regulus of antimony—certainly give the immediate impression that, as alchemists contended, metals can grow like living things. Spiritualism, the belief that some aspect of human existence persists after death, was investigated by two great scientists of the nineteenth century: chemist William Crookes and co-discoverer of natural selection Alfred Russel Wallace. Both used their experimental skills to expose charlatans, while they considered other observations sufficiently controlled to warrant belief. Robert Boyle also wrote about the powers of gems, which presumably trapped vaporous corpuscles from their specific geographic location when they crystallized—not unlike the kind of unseen particles from human artifacts that dogs could smell and use to track their owner.

All we can say now about these various claims is that they are wrong. That does not mean the questions or proposed ideas were (or are now) inherently unscientific. Indeed, science has been integral to ascertaining the numerous errors in all these cases, apparently once worth entertaining by some great minds. The prudent teacher might exercise a bit of caution, therefore, before denouncing a student's naive wonderment on these topics. Today, such beliefs are simply ill informed. The most effective antidote may thus be information. The deeper challenge may be understanding how such claims continue to percolate through the culture, even after being discredited scientifically. That is a profound sociological challenge—but quite different from defining or understanding the nature of science.

Still, it is common to find that being wrong is often equated with being unscientific. Yet scientists can be—and, often enough, are—wrong. Error is almost a hallmark of science, properly conceived. For some critics, however, demonstrating that certain claims are wrong is apparently not enough. They must add the epithet "unscientific" or "pseudoscientific." It is more than just a rhetorical flourish. It is an appeal to the political authority of science, apart from the specific evidence for any error. And it is also, in many ways, a measure of the immense power accorded to science and the deference given to anything that can earn the label "scientific."

It is worth endeavoring, therefore, to tease apart the political image of science from the factors that contribute to the reliability of its claims. That is a significant challenge in teaching the nature of science.

FALSIFIABILITY

One concept from the demarcation era still lingers and holds wide currency: falsifiability. Even the name of Karl Popper, who introduced the idea, is widely known and celebrated among scientists and science teachers. They frequently appeal to falsifiability as a hallmark of science. Open the pages of the journal *Science* and you can find diverse scientists—from archaeology, the chemistry of bonding, climate change, and paleontology—expressing common sentiments about proper rigor in science[9]:

- "Science is based on the falsification of hypotheses."
- Scientists "work late into the night in order to destroy or falsify another scientist's hypothesis."
- Researchers who fail to present falsifiable theories are "not playing the game."
- A theory that cannot predict falsifiable hypotheses is not "sophisticated enough."

Falsifiability has played a legal role, too, in judicial decisions prohibiting the teaching of creationism and "intelligent design" in science classes. Yet the legendary virtues of falsifiability, like other demarcation criteria, are often overstated.

The gist is familiar and often introduced in science textbooks: you can never *prove* a theory, but you can *disprove* it. The basic reasoning does not require great philosophical sophistication—surely part of its enduring appeal. Namely: we can never exhaustively sample all cases. So, no matter how much evidence we may gather, we cannot rule out a potential exception. Framed in this way, falsifiability echoes the classic philosophical problem of induction. For example, no number of stable chemical elements entitles one, logically, to conclude that all atomic elements are immutable. Witness the eventual (and unanticipated) discovery of nuclear fission.

However, Popper's formulation cleverly added another dimension. He highlighted the role of negative evidence. A single exception could, he claimed, upend a theory. The logical pattern is easily grasped: any clear counterexample discredits the premise in a deductive argument. The corresponding formula for science is easy: "man proposes, nature disposes." Accordingly, many theories end up on the scrap heap. Van Helmont's willow-tree experiment, with its careful measurements of soil, falsified the Aristotelian notion that plants are primarily composed of earthy matter. Pascal's Puy de Dôme experiment with mercury barometers at different altitudes effectively falsified the doctrine that nature abhors a vacuum. Diatomic gases falsified Berzelius's theory of the electrical nature of molecular composition. Pasteur's swan-necked flasks, filled with clear lifeless broth, falsified (finally!) the resilient doctrine of spontaneous generation. Or so the lore goes. History seems littered with falsified theories: an ironic tribute, some might say, to scientific progress. The extraordinary leverage of falsification seems both simple and powerful.

The simplicity is deceptive, however. Further "simple" reflection can reveal the flaws and weaknesses in the principle of falsification, as it functions in practice. Most notably, the notion of a single exception can only be critical where theories are expressed as invariant, universal laws. A black swan does not mean much if you contend only that "*generally* swans are white." Thus, when one finds exceptions to Mendel's "laws" or Ohm's "law," one does not wholly discount their value as generalizations or as models describing a particular set of cases. No one abandons the notion that mammals have hair when a congenitally bald lemur is born at the zoo (although they may stop and stare in bewilderment). Many theories take the form of models. They describe how nature works, often in particular, specified contexts. They are not framed as universal statements (Chapter 8).

In other cases, scientists develop claims about what can or might be observed. Paleobotanists often find fossilized plants in pieces: a stem here, a leaf there. The pieces may always appear together in the same rocks, yet one cannot thereby reliably conclude that they are from the same plant. In this case, one well-articulated whole-plant fossil can be significant for *establishing* the fact that the parts represent the same species. This is the reverse of falsification. The history of subatomic physics, too, is filled with celebrated golden events: individual bubble-chamber images that persuasively established the existence of certain particles. In these cases, one example proves, rather than disproves, the hitherto uncertain theory.

Far more importantly, scientific reasoning is much more complex than simple deductive arguments. There are typically many assumptions, premises, and context (or boundary conditions). When an exception or counterexample occurs, one cannot be sure just which premise or assumption is being "falsified," even if one knows logically that something is amiss.

Historically, then, one finds that outright falsification is rare, except for claims of very small scope.[10] Of course, scientists respect the evidence. But they do not reject a major theory in the face of the first counterexample or anomaly. Instead, they typically revise it. They accommodate the theory to the new findings, rather than abandon it. Or they wait until the exception itself is found to be mistaken. All towards developing reliable knowledge.

For example, William Thomson contended on the basis of simple thermodynamic evidence that Darwinian evolution was wrong. By measuring thermal gradients at the Earth's surface and using known rates of heat dispersion, he calculated the age of the cooling Earth. His initial determinations of 40 to 200 million years did not seem to allow enough time for the gradual changes that Darwin proposed. The physical evidence, it seems, falsified Darwin's theory. Yet Darwin did not thereby abandon his conclusions. Nor did geologists. They maintained their own, quite different estimates of the age of the Earth, based primarily on interpreting sedimentary rocks. Decades later, the status of the evidence changed dramatically. The discovery of radioactivity introduced a new source of heat for the Earth's interior. Also, new ideas about convection currents in the mantle indicated that internal heat could be redistributed. The original cooling calculations had erred by at least an order of magnitude. The Earth was indeed very old. In the long run, disregarding Thomson's "falsification" seemed justified.

Another renowned episode of apparent falsification appears frequently in physics texts. In the late nineteenth century, Michelson and Morley tried to document the ether in space, by measuring the Earth's movement through it. They detected no ether wind: hence (apparently), no ether. With no medium for propagating light waves, Newtonian physics was (the oft-told story goes) falsified, generating a crisis that led to the theory of relativity. Yet physicists at the time experienced no such crisis. They found many ways to interpret the experimental results. Michelson acknowledged that an ether wind sweeping the surface of the globe seemed unlikely. But the Earth might very well be dragging the ether with it (as suggested earlier by Stokes). In addition, the apparatus, while expertly assembled, had limited sensitivity. The measurements only set an upper boundary to the density of the ether. Some physicists readily accepted that the ether might be very thin. Lorentz and others, by contrast, did rethink the basic physical laws. They postulated that bodies might contract while moving through the ether. If so, then the Michelson-Morley experiments would not have measured anything about the ether. All these alternatives reconciled the experimental results with theory, without abandoning Newtonian physics.[11] While falsification may contribute to a more dramatic story, it does not seem to describe how scientists actually work (also see Chapters 3 and 4).

The fate of Prout's hypothesis is another informative case.[12] William Prout proposed in 1815 that all atomic weights were whole-number multiples of hydrogen, viewed as a basic unit. However, there were many atomic weights—most notably, for chlorine—that did not follow this rule. Was Prout's hypothesis falsified and thus abandoned? Over the next century, chemists' views were mixed. Many were impressed with the theoretical match for many elements. Others focused on the exceptions. In the early twentieth century the role of isotopes became clear. Individual isolated isotopes tended to fit the hypothesis. Mixtures, typically found in nature, yielded non-whole-number measurements. In addition, hydrogen proved to be an approximate but inappropriate unit. One could think more precisely in terms of whole-number multiples of protons, neutrons, and electrons. (There are some exceptions even here, attributed to combinatorial interactions.) Ultimately, Prout's hypothesis was not strictly falsified; it was successively revised to accommodate the evidence.

Consider, finally, the case of a proposed fourth physical force—the weak interaction—to accompany gravity, electromagnetism, and the strong interaction in atomic nuclei. Murray Gell-Mann explained:

> You know, frequently a theorist will even *throw out* a lot of data on the grounds that if they don't fit an elegant scheme, they're wrong. That happened to me many times. The theory of the weak interaction: there were *nine* experiments that contradicted it—all wrong. Every one. When you have something simple that agrees with all the rest of physics and really seems to explain what's going on, a few experimental data against it are no objection whatever. Almost certain to be wrong.[13]

For Gell-Mann, as for other scientists, what mattered is the overall balance of the evidence, not individual results exclusive of others. The role of falsification is widely overstated.

As these historical examples illustrate, falsification is a romanticized ideal that mischaracterizes real, productive science. Researchers are eminently pragmatic. They typically finesse the evidence rather than regard theories as falsified. They redefine terms. They modify the theory or restrict its scope. They may even tolerate unresolved anomalies. Effective reasoning seems to integrate evidence and counterevidence both. Eventually, weaker theories do wane—but rarely because they are disproved. Philosopher of science Imre Lakatos, having profiled these flaws in what he called naive falsificationism, quite justly declared an "end of instant rationality." Science is not so simple or one-dimensional as it may at first seem. To teach the nature of science, therefore, one must endeavor to convey some of its subtlety.

The demarcation project and appeals to falsifiability each reflect widespread intuitions about science. However, simple consideration of actual episodes in science shows that these intuitions are mistaken. Historical cases can prove relevant in interpreting NOS. Here, then, is a primary strategy for teachers: *to render science through concrete examples, or case studies.* Aiming to understand scientific practice, one should delve into such historical or contemporary cases, rather than rely on platitudes or claims easily distorted by ill-informed preconceptions. One should observe authentic science in action. In this way, students address their preconceptions about the nature of science, just as they might address scientific misconceptions themselves through well-crafted observational activities. A remedy to the common tendency to reduce science to a simple programmatic definition is to explore concrete examples of scientific practice. (Chapter 2 details this strategy more fully.)

NOS IN SCIENCE EDUCATION

An invitation to explore scientific practice is potentially quite vast, more than one can expect of the typical student—or teacher. What is essential to understand? What is the appropriate scope of the nature of science for standard science education? What is the central focus or set of core principles? Namely, what ideas *about* science should be addressed in school?

As noted earlier, approaches to NOS in science education go back many decades.[14] As awareness deepened, educators' efforts to demarcate science with a simple definition, or to appeal to a single exclusive scientific method, were duly abandoned. By the mid-1990s, however, amid various reforms, a set of basic NOS principles seemed to emerge (Figure 1.1).[15] An analysis of eight major curriculum documents, including the influential Project 2061 of the American Association for the Advancement of Science (AAAS), the *National Science Education Standards* of the U.S. National Research Council, and other international counterparts, yielded a short list of shared tenets.[16] These formed the basis for a widely used NOS assessment instrument, Views of the Nature of Science, or VNOS.

Another group of educators adopted a more structured approach to developing consensus, using iterated rounds of discourse to resolve disagreement and articulate the reasons justifying each idea about science.[17] Their work, occurring several years later, benefited from some intellectual distance from the stormy debates over postmodernism and social constructivism that plagued the 1990s. Perspectives were well balanced. The analysis was neither radical nor reactionary. In addition, this group of experts was well versed in contemporary scholarship in the history, philosophy, and

---

FIGURE 1.1. NOS consensus items in U.S. National Science Education Standards (analysis by McComas & Olson, 1998).

- Scientific knowledge is tentative.
- Science relies on empirical evidence.
- Scientists require replicability and truthful reporting.
- Science is an attempt to explain phenomena.
- Scientists are creative.
- Science is part of social tradition.
- Science has played an important role in technology.
- Scientific ideas have been affected by their social and historical milieu.
- Changes in science occur gradually.
- Science has global implications.
- New knowledge must be reported clearly and openly.

sociology of science. At the same time, those specialized perspectives were filtered through the lens of educators. Discussion was insulated from the bias of both scientists (all too inclined to safeguard their authority) and science-studies scholars (sometimes too academic or rhetorically hyperbolic). The team developed as fine a characterization of NOS for the classroom as one is likely to find in this listing style. It largely paralleled and affirmed the earlier analysis.

The consensus list (Figure 1.1) presents a healthy corrective to common stereotypes about science. For example, "scientists are creative." Popular impressions depict scientists as ruled by logic and by strict methods, both associated with irrefutable conclusions. This item underscores that scientists make imaginative insights (perhaps even as reflected in another stereotype, the "eureka" form of discovery). Generation of plausible hypotheses, design of laboratory apparatus or decisive experiments, interpretation of anomalous results: all require creative thinking. Another NOS list item notes that "observations are theory-laden." That is, the data do not speak for themselves, a phrase one often hears in personal disputes that appeal to scientific authority. Experimental results require interpretation. Conclusions may differ with varying theoretical perspectives. Or: "science is affected by its social and cultural milieu." In the public arena, science seems autonomous, its independence contributing in part to its objectivity. Accordingly, scientists are sometimes caricatured as isolated, drifting into pathological personalities (whether as a mad scientist or more quaintly as an absent-minded one). Yet social needs often shape research problems and may affect the reception of scientific findings. Cultural ideas can bias concepts and conclusions on the basis of race, gender, economic class, or nationality. In all these cases, the roots of scientific objectivity and authority are commonly misconstrued. Science educators may thus regard the NOS consensus list as a fruitful benchmark. It is particularly useful as a concise reminder of major misconceptions about the nature of science, ideally to be addressed in the classroom.

At the same time, the consensus list is not without its problems. Most notably, some puzzling contradictions lurk beneath the surface. For example, the student must reconcile views that "science is empirical" and "science is affected by its social and cultural milieu." In one case, scientific conclusions are based on observation and experiment; in the other, on personal values and beliefs. How can it be both? If it is both, how can one tease apart their respective roles? Similar problems are encountered in trying to reconcile "investigation is theory-laden" with "scientists are creative." In one case, observations seem limited by a scientist's preconceptions; yet in the other, scientists are supposedly able to escape or transcend such limitations. Ultimately, NOS understanding is not well expressed in the kinds of general statements that typically appear on NOS lists. The various ideas about science need context and concrete particulars. As general statements, they are imprecise—and even potentially misleading.

Consider, for example, the most widely recurring theme in NOS discussions for the past half century: the provisional nature of scientific knowledge.[18] With further evidence, theories may change or be abandoned as wrong. Without the "test of time" and further scrutiny, science remains vulnerable. In some cases, we might acknowledge overtly that all the relevant information is not yet available or yields only statistical probabilities. The key word for the concept has become 'tentativeness'. The consensus list accordingly declares that "scientific knowledge is tentative."

Conceptual changes that are relevant to consumers and citizens occur frequently. Not long ago, for example, a change in recommended ages for mammograms generated considerable public controversy. New recommendations for levels of vitamin D have also been announced recently. Individual scientific studies—even if published, of course—may be flawed or incomplete, and their conclusions later invalidated. For example, a study linking the measles vaccine with autism was retracted as fraudulent, notably after many subsequent studies failed to replicate the original findings. In this case, however, thousands of individuals acted on the premature conclusions, leading to a significant risk of a measles epidemic in Britain. Understanding tentativeness is thus surely important to functional scientific literacy. As our knowledge grows, concepts are not only added; they may also be replaced or rejected, sometimes quite dramatically. But the question is whether mere recognition of this principle suffices.

Other cases indicate how the concept of tentativeness may be misinterpreted and even misappropriated. Consider the bane of biology educators: anti-evolution critics. When creationists advocate teaching the controversy (or affixing warning labels on textbooks), they implicitly appeal to a principle of critical distance, or tentativeness in science. In his creationist diatribe *Icons of Evolution*, Jonathan Wells opined that Darwinists are closed-minded, dismiss simple evidence, and thus fail the norm of skepticism in science. He derided Darwinists as "dogmatic" *twenty-three* times in the final chapter alone. Wells presented the concept of tentativeness as reason to question evolution, like any science. Consider also an ordinary person objecting to a newspaper treating evolution as a fact:

> Perhaps the wisest science teacher I know told his class that science proves nothing true; it can only prove things false. Until something is proven false, we can only assume it to be true until further notice. Science has proven wrong in the past. Remember Pluto? When I was in primary school, everyone knew it was a planet. Now, kids are taught that it's not. Science is constantly updating itself, and things that we knew for certain 20, 50, 100 years ago will eventually be refuted.[19]

Ironically, tentativeness has proved to be powerful rhetoric in promoting *misunderstanding* of the nature of evolutionary science. As an appeal, it is the last refuge for those who dislike the scientific consensus.

Consider also the case of global warming. Despite the scientific consensus expressed by the Intergovernmental Panel on Climate Change, skepticism dominated American politics for many years. Critics cited patchy data, questionable models based on numerous assumptions, the unpredictability of the daily weather, isolated results that contradicted general conclusions, the newness of climate science, the limitations of peer review, and so on. Note the telltale catchphrase of the former website, ClimateChangeFraud. com: "Because the debate is NOT over." The website seems to delight in quoting Mark Twain: "There is something fascinating about science. One gets such wholesale returns of conjecture from such a trifling investment of fact."[20] In the skeptics' rhetoric, climate science suffers from incautious overstatement and premature conclusions.

Of course, these are not the voices of reliable science. A study of scientists unconvinced about climate change confirmed that they are typically in peripheral fields and that their work is far less widely cited.[21] Indeed, as documented by historians Naomi Oreskes and Erik Conway, the public doubt has been deliberately orchestrated by just a handful of politically connected scientists.[22] Their strategy has been to generate an image of uncertain, still actively debated science. That has been enough to stall political action. That is, they did not need to present evidence for their own counterclaims. They were able to leverage the alleged status of tentativeness. The tactic is not new. Earlier, the same cadre of scientists planted seeds of doubt to mislead the public on secondhand smoke, acid rain, the ozone hole, and DDT. If all one learns is that "science is tentative," without learning how or why, mischief remains possible.

Finally, consider the case of a parent in Minnesota aligned with an anti-vaccine movement inspired by the fraudulent paper (noted above) that purportedly linked the measles vaccine to autism. In a letter to the local newspaper, he contended that

> Health professionals demonstrate great hubris when they claim to know all there is to know about the safety of a vaccine. Medical studies cannot prove that a vaccine is safe, only that it has not yet proven to be unsafe. Studies need to be continually performed, re-peated, and expanded to get us ever closer to the "truth" about a drug or vaccine's safety.[23]

Here, someone seems to have learned the consensus-list principle but does not appreciate the fabric of reliability in science. When combined with misconceptions about falsifiability, the superficial fragment of NOS knowledge can open the way to scientific nihilism. In this case, knowing about tentativeness, but only incompletely, proved counterproductive.

As illustrated in these three cases—creationists, climate-change naysayers, and vaccine critics—merely acknowledging science as tentative is insufficient. The concept can backfire if not understood fully. Understanding needs context. Again, students need to explore actual examples of scientific practice.

Ultimately, nature of science is poorly profiled by a list of general declarations. Understanding needs to be functional and concrete, as expressed in the principle of teaching through historical and contemporary cases, noted above.[24] Here, it is helpful to recall the broader goal of scientific literacy. Students should be able to interpret scientific practice in *particular* cases, not abstractly. A general level of understanding, as exhibited in the current consensus list (Figure 1.1), is not specific enough, say, for interpreting the safety of high-voltage power lines, waste incinerators, or pain-killing drugs. Memorizing or explaining a short list of principles is inadequate, even if they serve as convenient benchmarks for teachers. As AAAS noted in presenting its revised benchmarks in 2009, NOS is not diluted philosophy of science. Focusing on a prescribed set of stated concepts, then, misplaces the goal of NOS understanding. NOS understanding is best characterized functionally, towards supporting analytical skills in personal and public decision making.

NOS FOR SCIENTIFIC LITERACY

Although the NOS consensus list of the 1990s sketched an important core, when viewed in the context of scientific literacy, it also now seems significantly incomplete. That is, the limited set of principles—even if learned in fully functional terms—is insufficient to address the diverse and sometimes complex cases encountered by consumers and citizens in daily life and public discourse. If the primary aim of NOS understanding is to inform these cases, one must be aware of the spectrum of cases themselves and their relevant NOS dimensions. Ascertaining the appropriate scope and focus of effective NOS education is, in part, an empirical question. Apart from an intuitive NOS list, what is the concrete role of science in personal and social contexts?

Consider, for example, the case of Climategate. In November of 2009, someone anonymously released e-mails hacked from a university server, written by a leading member of the Intergovernmental Panel on Climate Change, the premier international scientific body on this topic. The messages included comments about scuttling efforts to release data under the Freedom of Information Act, a "trick" used to graph data, and ways to limit publications by critics. James Delingpole, in a blog for England's *Telegraph*, promptly dubbed it "Climategate." Within a week, the term 'Climategate' could be found more than nine million times on the Internet. Climate-change naysayers proclaimed vindication of their allegations of fraud and collusion.

Imagine a prospective scene at the lunch table: one coworker sighs how the case just proves that global warming is a hoax, while another contends that scientists don't do things like that and that the posted documents themselves are probably fraudulent. Mutual epithets fly across the table, and the person in the middle is asked to settle the matter. Here is a prime example of a role for scientific literacy.

What NOS concepts does one need to interpret this case effectively? The consensus list is relatively unhelpful. For example, the central issue here is credibility, not tentativeness. But credibility does not even appear on the list. Does it help to know the difference between a law and a theory? No. The nature of an experiment? Not really. "Science can be shaped by its social milieu"? Perhaps, but political bias could well influence both views. Has it? As noted above, one needs analytical tools, not general tenets.

To interpret Climategate, one needs to know instead about

- the spectrum of personalities in science
- the nature of graphs
- the norms of handling data
- how scientists communicate
- credibility and expertise
- robustness of evidential networks
- fraud or other forms of misconduct

That is, one needs to acknowledge that scientists are humans and that their activities are not immune to emotion and personal rivalries. Scientific discourse can get testy, especially behind the scenes and along informal networks (such as e-mail). At the same time, the system of formal publication helps filter arguments to relevant evidence. Graphs function to express data in meaningful formats. That may entail creative arrangement of results. (Here, data from two different studies and time scales were combined on the same graph.) Using "tricks" (in the researchers' jargon) is normal and does not indicate fraud. Original data are not always released publicly—although, in this case, public laws dictated the sharing of information. That was about legal misconduct, however, and hardly affected the scientific status of the data. Researchers may jockey politically for status and prominent publication venues. Yet past performance and expertise matter. Credible voices earn more profile. And here, especially, James Delingpole is a journalist with a strong ideological edge: he is not a reliable source on interpreting science or the nature of science. Ultimately, nothing in the e-mails provided grounds for challenging the evidence itself, which came from many sources and many converging lines of research. As subsequent investigations into this incident bore out, the violation of open access to data was a serious offense, but nothing weakened the scientific consensus, as alleged by so many online commentators.

NOS includes the whole spectrum of features that affect the reliability, or trustworthiness, of scientific claims. One cannot responsibly escape teaching any relevant factor. For example, credibility (even among scientists) has a central role, yet is missing from the NOS consensus list. Virtually all social interaction of scientists, especially the system of checks and balances through mutual criticism, are generally absent from various available NOS lists (see

Chapter 6). One also needs to consider the role of funding, motivations, peer review, inherent cognitive processes, fraud, and the validation of new methods—all features of scientific practice that become relevant at different times in public discussions. When one considers the diversity of cases that emerge in contemporary society, such as Climategate, one finds that the current NOS lists are severely truncated—and that, in a few items, emphasis is misplaced. Using science in daily life as the relevant context, what is an appropriate scope and focus of NOS in education?

Several approaches and studies may be worth noting. For example, Dankert Kolstø offered a prospective framework based on addressing socioscientific issues, especially controversies where one needs to resolve or address conflicting scientific claims. Such debates are often local and reported in the media: occasions where students as adults will likely participate in the community. The objective is "to empower the students as citizens" by describing "science as an institution and the processes by which scientific knowledge is produced." One aims to "increase students' competence in interpreting science-related statements."[25] One need not expect students to be scientists themselves. Rather, the student who is familiar with the methods of science, its social processes, and institutional norms is better equipped to interpret and assess claims made by experts in various fields. Kolstø outlined eight essential topics in four categories (Figure 1.2).[26]

First, science is a process. Science is not always complete. In public issues, especially, the science is often still at the frontier. Results are uncertain, meaning especially that different scientists may hold contrary views. Without instruction, students tend to attribute such disagreements to personal interests, opinions, or incompetence: that is, as pathology, rather than as normal in emerging science. One thus needs to understand science as a *social* process. Consensus is achieved through criticism, argumentation, and peer review. Students need to appreciate the difference between what

---

FIGURE 1.2. A general framework for analyzing the science dimension of socioscientific issues (Kolstø, 2001).

Science as a social process
1. "Science-in-the-making" and the role of consensus in science

Limitations of science
2. Science as one of several social domains
3. Descriptive and normative statements
4. Demands for underpinning evidence
5. Scientific models as context-bound

Values in science
6. Scientific evidence
7. "Suspension of belief"

Critical attitude
8. Scrutinize science-related knowledge claims

sociologist of science Bruno Latour dubbed "ready-made-science" (the stuff of textbooks) and "science-in-the-making" (still-active research).[27] This topic parallels earlier concerns about tentativeness but differs substantially by focusing concretely on why uncertainty arises at all and how it is resolved.

Second, science has particular limitations. Foremost, perhaps, students need to learn the distinction between descriptive statements and normative judgments. "Is" and "ought," facts and values, are validated by separate processes. Still, scientific information can be relevant in decision making—say, in assessing costs, benefits, and consequences of potential risks—without strictly dictating what one ought to do. Accordingly, one must differentiate the domain of science from politics and ethics and expose the fallacy of extreme scientism and technocratic postures. At the same time, one must underscore the rationale behind, and the role of, scientific demands for evidence, theoretical coherence, and underlying assumptions. In addition, students need to appreciate the ways in which scientific theories and models represent the world in selective ways and with particular contexts or scope. Scientific conclusions do not hold the same status as logical or mathematical truths, despite common impressions.

Third, science exercises its own epistemic values, relevant to the pursuit of reliable knowledge.[28] For example, scientists distinguish between anecdotal and systematic evidence. Also, there seems to be widespread misunderstanding of scientific conservatism, the circumspect "suspension of belief" until evidence is thoroughly secure. This differs from the widely reported norm of skepticism. Scientists tend to guard against error, rather than accept a best guess prematurely. Familiarity with these customary practices helps one interpret the statements of researchers as they move from a professional scientific forum to a public one, where precaution may indicate actions other than deferral of judgment.

Finally, scientific arguments follow certain patterns and standards, independent of the content or specific evidence involved. Although non-scientists have limited expertise, they can probe the epistemological foundation of claims and the social context(s) in which they are presented. Developing such skills in the classroom may also be coupled with fostering a critical attitude or habit of questioning.

Kolstø's framework for NOS differs noticeably from a list of philosophically oriented tenets, although it is still philosophically and sociologically informed. Most important, there is a thematic focus: diagnosing science-related claims encountered in personal and public life. NOS understanding is a tool, rather than an endpoint in itself. This perspective highlights, for example, the descriptive/normative distinction, the (ir)relevance of anecdotal knowledge, and the importance of respectful disagreement within science, not found on the NOS consensus list. The context clarifies how to envision the scope and content of NOS in the science classroom.

Jim Ryder reached similar conclusions. He drew on a set of 31 case studies (from the 1990s) where sociologists analyzed non-scientists dealing with scientific findings.[29]

While some conceptual knowledge, or traditional content, was needed, NOS features were again most prominent. His analysis highlighted the roles of theory and creativity in interpreting data; the frequency of disagreement and competing interpretations in active research; the role of professional credibility; and source of funding as a possible bias.[30] Scientific uncertainty, or indeterminacy, also emerged as important. "Individuals needed to appreciate that unequivocal findings are often unattainable, particularly in complex settings outside the laboratory."[31] For example, estimates of risks—of the likelihood that power lines cause leukemia, or of whether the meat supply will be free of disease-causing *E. coli*—may simply be out of reach given the complexity of the situation. The cases also indicated a role for understanding how science is communicated in the public domain, whether by scientists, government bodies, commercial organizations, or the media. Is the information complete? Might it be biased? In general, students need an overall understanding of the "ways in which knowledge claims in science are developed and justified": namely, an epistemic perspective on scientific practices (more below).[32] Ryder's analysis, like Kolstø's, helps significantly broaden the view of NOS beyond the topics typically addressed in conventional science education.

MAPPING NOS

The demarcation effort, rhetoric about falsifiability and tentativeness, and analyses of the components of scientific literacy all share a common theme. They all focus on the *reliability*, *trustworthiness*, or *authority* of scientific knowledge. In this common thread, one finds a unifying concept for NOS. What a student needs to learn above all is how to judge what (or whom) to trust—and why.

Indeed, former debates over what to teach about NOS—amplified to extremes during the so-called Science Wars of the 1990s—nearly always hinged on how to interpret the authority of scientific claims or of scientists as spokespersons for those claims. Without prejudicing the resolution to such problems, this is foremost what students should learn about science: tools for assessing whether any claim is reliable. To become well-informed adults and responsible citizens, they need to understand how evidence works—and, equally, where it can fail. Securing a definition of science—even a good one—will not solve the central challenge: discerning the reliable claims from the unreliable ones. Labels and formal definitions must yield to a practical and functional understanding. Philosophizing about the abstract nature of experiment or theories or the essence of science can be left to... well, philosophers.

How, then, should one map the nature of science around this theme?

What features are central to characterizing how science works?

One prospective strategy is to try to fully equip students to evaluate evidence on their own: to prepare everyone to make the same judgments scientists do. Such skills certainly seem appropriate where problems and evidence are simple. Few will dispute the goal of developing skills in recognizing relevant empirical findings, interpreting graphs and statistical measures, thinking about controls, considering alternative explanations, and so on.

However, there are limits. For example, in 2009, a U.S. Government task force issued new recommendations on appropriate ages for mammograms. A typical citizen, no matter how well informed, is simply unable to collect and evaluate all the evidence on the benefits, costs, and risks of the procedure at different ages. This was the rationale for a special expert task force. One relies on their expertise. Even scientists inevitably rely on other scientists. Robert Boyle saw that clearly even in the mid-1600s. Boyle helped establish the Royal Society, the first institution for exchanging scientific findings. He wrote at length about the criteria for trustworthy testimony.[33] In recent decades, sociologists of science have examined further the roles of expertise, "epistemic dependence" on others, and social epistemology: how knowledge develops in a community (also see Chapter 6).[34]

By comparison, a mistaken impression of one's abilities to evaluate evidence opens the way to mischief. For example, websites critical of global warming rely on readers' intuitions that their own common-sense judgment can trump the expertise of climate-change scientists. All they seem to need is a little evidence to judge for themselves. Just a few bits of selective counterevidence seem sufficient to persuade them. Ironically, perhaps, a scientifically literate individual needs to acknowledge the limits of his or her scientific knowledge. No one can legitimately pretend to train each person to always evaluate the evidence on their own or to participate in science in every instance. Teaching an understanding of the nature of expertise and systems of credibility seems essential in a modern society where technical knowledge is widely distributed among specialized experts.[35]

Yet understanding the role of expertise, while important, still falls short. Credibility may be challenged. Here, one needs to understand, more deeply, just how scientific practices contribute to credibility. For example, in the case of Climategate, using "tricks" with graphs or trying to limit publication by critics (discussed above)—while it sounds suspect on the surface—does not reflect fraud. In other cases, knowledge of how science works may help keep claims of credibility in check. For instance, Andrew Wakefield's study of autism and vaccines was funded by a legal group suing the vaccine manufacturers. That would have been a significant signal, if Wakefield had disclosed it, well before the journal formally retracted his claims.

Further, experts may be mistaken. For example, in late 2009, noted Belgian neurologist Steven Laureys announced that a patient who had been in a coma for twenty-three years following a car crash was apparently able to communicate using a special touchscreen and the assistance of an aide. Laureys linked this to his research, noting that people in noncommunicative states are misdiagnosed up to 40 percent of the time. Major news media, including National Public Radio, CNN, Fox News, and MSNBC, aired the remarkable story. Several months later, Laureys himself acknowledged that the method of "facilitated communication" proved to be bogus and that he had not exercised appropriate critical judgment of the method when it was first "demonstrated" to him. Even an expert can be an unwitting victim of fraud.

In addition, experts sometimes disagree. Credible claims may conflict. One needs additional resources to assess the nature of the disagreement and the relative status of alternative claims. Even credible claims may come with qualifications and caveats, whose meaning becomes clear only when one understands the various methods for ensuring reliability, as well as their limits. Trust should not be blind. Credibility merely signals responsible communication; it does not wholly substitute for it.

One needs to understand the nature of uncertainty and possible sources of error. As science on a particular topic matures, problems of debate and uncertainty tend to be resolved. In most contemporary decision-making cases, however, the science is young—still science-in-the-making, as Kolstø observed.[36] In such circumstances, uncertainty is high. Neither credible voices nor evidence can fully resolve the uncertain possibilities. At such times— those most typical of the challenge of scientific literacy—assessments of the nature and limits of reliable knowledge are especially important for guiding decisions and helping to plan for contingencies. Teaching an understanding of the uneasy status of scientific uncertainty, between ignorance and well-founded claims, seems just as important as understanding (the more familiar) tentativeness.

The informed citizen, then—the mature, well-educated student—will be able (at least) to interact with experts on topics they may know next to nothing about; recognize relevant evidence as well as presentations of bogus evidence; appreciate the limits as well as the foundations of emerging scientific claims; and negotiate through scientific uncertainty. One will be a competent interpreter, or critic, of science, even if not a practitioner of science (in the same way that film or music critics can effectively assess art without necessarily producing art themselves).[37] Interpreting the reliability of scientific claims requires a broad understanding of scientific practice, or how science works, from a simple laboratory or field setting to science journalism.

A simple yet synoptic approach tracks the genesis and movement of scientific claims. Namely, how are scientific claims generated and then transmitted? What ensures reliability at each step, as each may prove important in different cases?

One may conceptualize a path beginning with the most basic observations or measurements and leading to a public scientific claim. Simple measurements are first assembled into meaningful graphs. Observations are arranged into significant patterns. Order begins to emerge. But already there are many checkpoints for reliability. Are the samples and reagents free from contamination? Has the instrument been designed properly? Has it been properly calibrated? Has potential observer bias been prevented? Is the experiment designed to control for or monitor possibly confounding variables? If all is secure, one can next compare patterns in parallel sets of data, and apply statistical analysis. But are the statistical models and the corresponding statistical measures appropriate? Patterns and numerical trends can then be set in the context of models or theoretical explanations. But have alternative explanations been fully considered? Are there conceptual blind spots—from a researcher's theoretical framework or cultural context? When significant findings have been established, they may be published—and reach other investigators for criticism or novel development. Yet one may wonder whether peer review has been both suitably critical and fair. Even so, results are gauged against reputations. Has the system of credibility succeeded, or has charisma or a conflict of interest distorted the perceived significance of particular claims? At this stage, critical exchange, when functioning well, will isolate and remedy errors. There may be reviews of the literature or meta-studies, synthesizing, consolidating, and re-mapping information even more. When research becomes relevant to social issues, it then continues to travel even farther from the original inscriptions in lab and field notebooks. It may appear in the news, in legislative hearings or courtrooms, in marketing, or in the media. But here, too, reliable science depends on information and conclusions being faithfully conveyed and fairly represented. Scientific claims thus follow a vast trajectory of successive re-mappings, from disparate clusters of observations to the actions of citizens, customers, government agencies, or corporate leaders. As science unfolds, it forms longer and longer chains as raw evidence is assembled into larger and larger networks.[38] Science knits together local phenomena and observations into successively more global perspectives and concepts. Nature of science encompasses the mapping of processes through all these layers: from test tubes to YouTube.

By reducing this knowledge-generating process into individual steps or actions, one can parse the nature of science into its components. One version is provided in Figure 1.3. The resulting inventory may at first seem long and unwieldy. Yet (unlike the consensus list) it is unified. There is an arc to the history of scientific claims (also a structure of their implicit justification). Reliability is an important theme throughout. Still, for convenience, one can sort the NOS features into a handful of functional epistemic categories; one prospective taxonomy is presented in Figure 1.3.

Generating scientific knowledge requires care at each step of the process.

FIGURE 1.3. Partial inventory of dimensions of reliability in science.

Observational

Observations and measurements
- Accuracy, precision
- Role of systematic study (versus anecdote)
- Completeness of evidence
- Robustness (agreement among different types of data)

Experiments
- Controlled experiment (one variable)
- Blind and double-blind studies
- Statistical analysis of error
- Replication and sample size

Instruments
- New instruments and their validation
- Models and model organisms
- Ethics of experimentation on human subjects

Conceptual

Patterns of reasoning
- Evidential relevance (empiricism)
- Verifiable information versus values
- Role of probability in inference
- Alternative explanations
- Correlation versus causation

Historical dimensions
- Consilience with established evidence
- Role of analogy, interdisciplinary thinking
- Conceptual change
- Error and uncertainty
- Role of imagination and creative syntheses

Human dimensions
- Spectrum of motivations for doing science
- Spectrum of human personalities
- Confirmation bias/role of prior beliefs
- Emotional versus evidence-based perceptions of risk

Sociocultural

Institutions
- Collaboration and competition among scientists
- Forms of persuasion
- Credibility
- Peer review and response to criticism
- Resolving disagreement
- Academic freedom

Biases
- Role of cultural beliefs (ideology, religion, nationality, etc.)
- Role of gender bias
- Role of racial or class bias

Economics/funding
- Sources of funding
- Personal conflict of interest

Communication
- Norms for handling scientific data
- Nature of graphs
- Credibility of various scientific journals and news media
- Fraud or other forms of misconduct
- Social responsibility of scientists

Any one element can be a source of error if not addressed properly. Ironically, then, a complete profile of NOS also parallels potential sources of error, or error types, in scientific practice. Accordingly, in a science classroom, one may need to inform students about all the ways in which science can fail, so that they might understand how scientists prevent, mitigate, or accommodate potential errors. Paradoxically, perhaps, error can be a potent vehicle for teaching the process of science (see Chapter 5).[39] At least this would initiate the lesson, widely regarded as central, that scientists can—and sometimes do—err.

WHOLE SCIENCE

One might call this framing of NOS, sensitive to all the dimensions of reliability in scientific practice, *Whole Science*. Whole Science, like whole food, does not exclude essential ingredients. It supports healthier understanding. Metaphorically, educators must discourage a diet of highly processed, refined "School Science." Short or truncated lists of NOS features are simply unhealthy for understanding science.

The notion of Whole Science echoes and extends ongoing efforts to characterize NOS inclusively. In recent years, treatment of NOS in some places has yielded to discussions of "science as a way of knowing" (or "how scientific knowledge is constructed," "scientific inquiry," or "the scientific worldview"), "scientific practices" or the "scientific enterprise," and "how science works."[40] These labels tend to partition and treat as distinct experimental, conceptual, and social processes. They splinter material, cognitive, and cultural contexts. The label of Whole Science is a reminder that these components function together.

Many characterizations of the nature of science are incomplete. Targeting Whole Science helps restore the fullness to science. For example, some science educators profile science as fundamentally explanatory and focus almost exclusively on building theories and models. Yet science includes a variety of investigations, such as documenting, describing, and organizing natural phenomena; mapping causes (not always explaining them); and producing certain effects. Other educators advocate scientific arguments as the primary means for understanding what justifies scientific knowledge. Yet scientists exchange material demonstrations and samples as well as textual arguments. They assemble grant proposals and secure resources as well as presenting claims and evidence. In addition, knowledge-generating practices include not only cognitive and evidential methods, but social interactions. They find flaws in each other's work, adapt existing models to new domains, collaborate to bring together complementary skills, and so on.[41] Nor is science just a conceptual exercise: it includes lab skills and quasi-autonomous work on experimental systems.[42] Most important, perhaps, a Whole Science approach underscores the integrity of scientific practice, or how all the various NOS strands interact towards epistemic ends.

The Whole Science framework fosters a responsible balance between the foundations for reliability and the limits of science. Blind skepticism is no better than blind faith. As Henri Poincaré once reminded us, "Doubting everything or believing everything are two equally accommodating solutions, either of which saves us from reflection."[43] Neither incautious scientism nor anti-science cynicism should gain traction. Students need to develop analytical tools to assess both the promoters and critics of science. In this way, interpretations of the reliability of knowledge can inform our decisions, both as individuals and, collectively, as a society.

Ultimately, Whole Science provides a simple basic structure for teaching NOS. Engage students in samples of Whole Science, whether through student-initiated investigations or historical or contemporary case studies. Fill each case with reflective questioning about generating reliable knowledge. While the characterization of NOS may seem expansive, one need not teach everything all at once. A teacher who delves into *any* case of authentic scientific practice and discusses *whatever* details prove relevant to that particular case promotes a deeper understanding of NOS. Teachers may thus feel free to explore widely and reflect deeply. The inventory and taxonomy of NOS features (Figure 1.3) merely provides a reference for integrating and organizing such lessons and, over the long term, gauging the completeness of an ensemble of NOS lessons.

Framed in this way, the goal of teaching NOS may seem quite simple indeed. Equipped with a skeletal characterization for reflecting on NOS, the veteran teacher, or even the novice, may well feel primed to launch into the sample cases presented in Chapters 10–13, browse other historical cases (listed in Figures 14.1, 14.2, and 14.3), or, more ambitiously, assemble their own, as outlined in Chapter 14. This is certainly the ultimate aim. But patience is warranted. Perspective matters. Especially where assumptions about NOS already permeate our culture and conventional approaches to science education (as exemplified above in the cases of demarcation and falsification).

First, the project of delving into epistemics—the dimensions of reaching reliable conclusions—entails a shift from the familiar, final constructed form of science to the process of its construction. From finished ready-made science to uncertain, somewhat blind science-in-the-making. Ironically, the evidence and reasoning cited in retrospect as justifying a scientific conclusion are not always those that led us to that conclusion. Epistemics turn science inside-out, yielding a view quite different from prepackaged School Science. History, especially, can be valuable for seeing that process fully. The critical educational role of history in rendering science-in-the-making and its importance for understanding NOS are addressed in Chapter 2, providing the initial foundation for developing skills in NOS reflection.

Simply acknowledging a role for history, however, may not suffice. Even experienced educators can bring assumptions about NOS to history, biasing their interpretations. In some cases, the naive educator can unwittingly rewrite history, thereby subverting the desired NOS lessons. For example, ordinary habits of storytelling can themselves distort impressions of what scientists actually did. Strong tendencies to romanticize science, to reconstruct and idealize it, or to shoehorn it into prescribed norms can all confound honest NOS understanding. Strategies for safeguarding against various pitfalls in approaching history are profiled in Chapters 3–5.

Third, science in educational contexts is often construed simply as a "way of knowing," with a corresponding emphasis on its cognitive dimensions: scientific reasoning, or thinking skills only. A focus on Whole Science reminds us of the relevance of social interactions among scientists—such as response to criticism—as well as the social contexts of science, from sources of funding to the influence of conceptual metaphors and cultural perspectives. These components especially shape how we interpret science in contemporary cultural contexts—namely, the very aim of scientific literacy. The significance of the sociological dimensions of NOS, essential in broadening NOS awareness, is highlighted in Chapter 6.

A fourth set of challenges appears where authentic science intersects educational contexts. For many, a primary objective, epitomized in conventional School Science, is to simplify science to facilitate student understanding. Yet the science of modern decision- and policy-making is typically complex. Scientific literacy involves coping with uncertain, incomplete, or sometimes messy science. Notably, citizens need to understand the status of scientific models, including their contexts and limits. History, again, can inform our perspectives and expand the domain of NOS reflection, as sketched in Chapters 7–8.

Finally, any educational endeavor is generally sharpened by clarifying the concrete outcomes or objectives. What does the test look like? What should students ultimately be able to do? Transforming the aim of scientific literacy and NOS understanding into concrete forms of performance and assessment is the subject of Chapter 9.

With these perspectives from Part I in hand, the teacher is more fully prepared to appreciate the sample resources for classroom use in Part II. Chapter 10 presents a model historical case study, featuring a guided-inquiry format and NOS problems and reflection. Chapter 11 profiles a contrasting case style, where history informs largely student-driven inquiry. The NOS lessons here focus primarily on the virtues and limits of models. Chapters 12 and 13 present more complex role-playing simulations, showing how one can use history to model scientific-literacy contexts in the classroom. Chapter 14 provides some guidance on finding and reviewing— or writing—additional cases. By integrating the perspectives and examples presented throughout the book, the final chapter serves as a capstone in orienting and preparing the reader for teaching NOS in practice and for continuing fruitful analysis and reflection on the nature of science.

2 | History as a Tool

> *History, if viewed as a repository for more than anecdote or chronology, could produce a decisive transformation in the image of science by which we are now possessed.*
> —Thomas Kuhn, *The Structure of Scientific Revolutions*

History in science education • nine ways history can benefit the science teacher • case studies as samples of Whole Science • science-in-the-making and inquiry • summarizing history as a tool

When one contemplates aims and ideals in teaching science, another discipline, such as history, might seem unimportant or utterly irrelevant. Yet history is already a fixture in most science classrooms. What student does not learn of Darwin and his voyage on the *Beagle*, Mendel and his pea plants, Newton and his three laws, Mendeleev and his stunning predictions based on his periodic table, or Wegener and his underappreciated insights on continental drift? Vignettes of discovery and biographical sketches haunt the margins and feature-boxes of almost every textbook. It is strange, indeed, to find an experienced teacher who has not assembled a repertoire of anecdotes to amuse and entertain students—and to capture their sometimes elusive attention.

Some teachers certainly view these as occasions to teach about more than content. They may try to portray the human dimension of science, conveyed in personality quirks, foibles, or incredible life struggles. Or the stories may have implicit morals about the process of science or the nature of discovery or evidence. That is, in current teaching practice, fragments of history often function to convey the nature of science. Given the growing importance of NOS in the curriculum, one may find it valuable to consider these informal methods more thoughtfully. How might history inform science teaching, especially about the nature of science?

A role for history was entertained by Thomas Kuhn in the opening of his 1962 (now landmark) book, *The Structure of Scientific Revolutions* (epigraph above). Kuhn envisioned history as a gateway to understanding the nature of science. Moreover, historical knowledge, he claimed, could yield a "decisive transformation": perspectives dramatically different from popular beliefs. To achieve that, one needed first to set aside the caricature of history as mere chronology: a mind-numbing cascade of names and dates. By delving more deeply into discoveries and other episodes of the past, one could articulate how science worked. The history of science is a "how story" of science. One can understand science in action, its processes, its character. If one were to pursue a "science *of* science," history would be the essential data. Kuhn's analysis inspired a dramatic transformation at the time: guiding many philosophers into studying the history of science. In the decades that followed, the focus on history expanded to the sociology of science, then to laboratory ethnographies, rhetorical analyses, Marxist and feminist critiques, cognitive perspectives, cultural studies, and more. This collective enterprise, now known as Science Studies, functions as a valuable resource for the teacher of NOS. This chapter explores some of those opportunities, opened when one views history and case studies as a tool, not merely an endpoint, for teaching science and NOS more effectively.

NINE WAYS HISTORY CAN BENEFIT THE SCIENCE TEACHER

1. Contextualizing and Motivating Science

At first, history may seem peripheral. Yet, paradoxically perhaps, history's foremost value may be in making the science seem more relevant and, thereby, engaging students. History is a motivating tool. In that respect, history may address the most widespread challenge reported by science teachers.

Motivation marks the gap (sometimes the gulf!) between what a teacher teaches and what a learner learns. In a school context, of course, grades, getting ahead, and competing in the job market can be potential motivators. But these motivate students to memorize content for tests and pass courses, not to learn science. Students enjoy and even revel in learning on their own, when it is meaningful. The average young adult accumulates vast knowledge of music, sports, social networks, or communication technology almost effortlessly and rarely with any formal study. A major challenge for science teachers, then, is to find what triggers students to learn science.

In trying to make science meaningful, many teachers' first impulse may be to link the topic to students' lives today. For example, teachers may dutifully highlight events in the news (even though, ironically, most students rarely keep up with current events). These connections may indeed be important to *applying* concepts. One should not abandon them. However, they do not necessarily convey the reason for *doing* science. Instead, we should be asking: what problem or question led anyone historically into deeper investigation? Where did all those textbook ideas come from? Ask a student to learn the double circulation of the heart, and you get a memorized explanation as thin as the paper that the textbook is printed on. Ask a student instead to explain, as William Harvey did, why there are two sides to the heart (and perhaps give them a real heart to examine), and they are soon submerged in thinking and interpreting even "obvious" facts.[1] Ask a student to learn oxidation–reduction reactions, and you get complaints piled upon confusion. Take cues from eighteenth-century chemistry, instead, and you ask a student to explain how to make metals from ores; why metals burn, corrode, or rust; what happens when wood burns; and how all these processes are related (see Chapter 11). History allows teachers to shift from the alienation of prescribed answers to the wonder or unsolved problems that motivate learning. The original context makes the reasons for doing science "real."

Historical context inevitably highlights the human and cultural dimensions of science. The context connects the abstract—and, for students, often lifeless—scientific concepts to human concerns, values, and emotions. History contextualizes, and thereby motivates, the science. It matters little that the science may have happened at some other place or time. Good stories are compelling. Science is a human endeavor. It is conducted by and for real people. Research is fueled by sheer curiosity and the desire to improve the human condition—feelings that students share or readily appreciate. The human element is inherently engaging, even if not completely the same as the students' own lives. When well framed, history inspires students to appreciate scientific problems, experiments, debates, and concepts.

Historically, much scientific research and discovery has been motivated by sheer curiosity, what Horace Freeland Judson aptly called "the rage to know." Students rarely imagine that research can be emotional. Yet scientists experience the joy of investigative play, the frustration of malfunctioning equipment, the disappointment in anomalous results, and the thrill of discovery.[2] We can convey these feelings in many ways, but the experience of earlier scientists is perhaps the most vivid and situates even the most mundane concept in a human context. René Descartes, following many Medieval scholars, wondered at rainbows and wanted to know what caused them. Ulisse Aldrovandi was amazed by the hairy face of Petrus Gonzalus and by other human "monsters" and catalogued them, looking for patterns that might reveal the reason for their unusual form. One motivating tool, then, is to recover the original context of a discovery. That may come in the form of an unlikely phenomenon, a problem, or a puzzling observation. The teacher may borrow the original historical context or creatively adapt it to a corresponding contemporary scenario. Both cases may stimulate a desire to know or understand.

Other concepts emerged historically from vivid cultural contexts. Problems draining mines led to deeper consideration of the limits of "suction" pumps, water pressure, the existence of vacuums, and the weight of atmospheric air. Epidemics of beriberi among shipping crews in the Dutch colonies led to an investigation of the role of diet and, eventually, to an understanding of a new class of nutrients, vitamins (Chapter 10). Navigational challenges based on determining latitude and longitude and the problematic declination of the compass led to deeper studies of the nature of the Earth's magnetism. Science is rarely remote from human concerns. History offers an opportunity to re-motivate science for students.

2. Clarifying Concepts

Equally important, history can guide the core aim of teaching concepts. Students are not that different from their historical counterparts who encountered today's basic concepts for the first time. Indeed, the original historical context can certainly be a healthy reminder to teachers that these concepts are not so obvious to those not yet familiar with them! History is a kind of refreshing antidote for teachers who may forget how much they know.

Historical episodes outline a set of experiences and/or reasoning for students to follow. They can appreciate how each concept fills a particular need, emerges from particular observations, and, ultimately, is justified. Concepts emerge from context and evidence. Indeed, without this information, can a student really understand any concept fully or apply it meaningfully?

On a broader scale, the history of the emergence of a concept—or a family of related concepts—over many decades, or even centuries, may help structure or unify a series of lessons. The structure of the atom, electromagnetism, and Mendelian genetics are common examples already adopted in textbooks. The historical development is a framework for the student's own gradual conceptual growth: say, of genetics, the electrical atom, or the forces that build mountains.[3] History can help motivate and guide such a series of lessons by tracing a lineage of successive questions.[4]

In the simplest approach, one transforms each concept into a narrative of discovery. One begins with the problem (its cultural context as motivation) and proceeds through the investigative strategy, observations, and conclusions (see sample in Chapter 10). (There is opportunity, of course, for more discussion of the process of science—as profiled below in section no. 6.) The concept becomes an answer: a reason for existence beyond the test at the end of the unit. This strategy may well sound familiar. It is often adopted in upper-level college courses. Students encounter recent work in the field, study individual papers, and often interpret and discuss the results. Instruction is embodied in a recent historical review. (Such engagement with original work seems essential for developing skills in experimental design and critical thinking, as well as for conveying the limits of current concepts.) The implicit challenge here is to adapt and extend this method of conceptual development to all levels of science teaching.

Another approach follows the educational ideals of active and participatory learning. That is, one may immerse the student in history as an imaginary participant. In the 1950s, Harvard University, under the leadership of James Bryant Conant, developed a set of now landmark case histories in experimental science. Students followed the work of Louis

Pasteur on fermentation, Antoine Lavoisier on combustion, Joseph Black on latent heat, Jan Ingenhousz on plants' effects on the atmosphere, and others. Important case-study volumes have also been developed by Klopfer; Aikenhead; and Hagen, Allchin, and Singer.[5]

Another strategy is to invite students to read and analyze original papers.[6] Alternatively, students themselves may reconstruct historical dialogues[7] (also see Chapters 12 and 13).

The adventurous teacher can make the experience exceptionally vivid. Imagine a student arriving in class one day to meet a great scientist from history—who looks strangely like their regular teacher.[8] Historical role-playing is now a standard fixture in many museums. One of the great masters of the art was Richard Eakin, a zoology professor at the University of California, Berkeley, whose portrayals of Mendel, Pasteur, Darwin, and others were captured on film and are now available online.[9] From personal experience, I can report that the first time can be quite unnerving. Will students just laugh? But students (all too often starved for a break from routine) kindly buy in to the masquerade and usually have fun with the occasion.

Borrowing the historical context may also extend to laboratory equipment. Students may work with physical replicas of original apparatus or with other historically based equipment.[10] Such replication work is not necessarily "cookbook." Work in the physics department at Carl von Ossietzky University of Oldenburg in Germany, has—in the cases of James P. Joule and Charles Coulomb—revised our historical notions of the science itself.[11] Moreover, when students situate themselves historically to answer questions, they gain a deeper sense of "owning" the resulting concepts.

In some cases, one may elect to use the history as a guide, but disguise it. Darwin's name, for instance, can elicit stereotyped and misleading views of "survival of the fittest" or competition in society that obscure an appreciation of the role of "reproductive fitness" or divergence in "the origin of species." Instead, one may simply borrow the original puzzles from history—about biogeography, species, varieties and races, and variations produced by domestic breeding—that led collectively to our current concept—in this case, evolution by natural selection. Darwin's and Wallace's experiences provide a valuable framework for learning about evolution, even without Darwin or Wallace as central characters.[12]

3. Revealing Misconceptions

One unexpected pleasure of turning to history is finding how many scientists—even those we now recognize as having the greatest minds—once misconstrued, misinterpreted, or even rejected today's concepts. Their struggle reminds us how difficult the task can be for students today. More importantly, the misconceptions in the past often resonate with those of students today. Students frequently begin with an Aristotelian-like concept of motion or a Medieval concept of impetus, for example.[13] They resemble eighteenth-century chemists, Lamarckians, vitalists, etc.[14] History thus allows teachers to anticipate and be sensitive to common naive concepts or preconceptions.[15] The teacher can also learn from history to appreciate the specific contexts in which these potential misconceptions so often seem to make sense—a pedagogical step essential for transforming students' conceptions from within.

Historical data may likewise be useful in revealing all the likely conceptual hurdles that different students may encounter in conceptual development.[16] Students may be consoled, perhaps, knowing that their criticisms, reservations, or difficulties were shared by other great thinkers of the past. (On the other hand, the teacher may keep such views as hidden points of reference, clues for guiding students through the reconceptualization.)

Moreover, history can help identify the key arguments, critical experimental evidence, or ways of seeing that can lead to more sophisticated thinking.[17] Here, again, history may well be used clandestinely—as a hidden, but nonetheless powerful, tool for the teacher.

4. Celebrating Achievements

Perhaps the most widespread use of history already in the classroom is the celebration of landmark discoveries and great scientists. Historical episodes give human dimension to scientific knowledge, lending names, times, and places to ideas that often seem coldly objective or impersonal. More plainly, however, praising a discovery conveys the value of science in our culture. It may also illustrate the scientist's aesthetic—the appreciation of elegant ideas or insightfully designed experiments. From a cultural perspective, it is important to understand our historical conceptual triumphs over an Earth-centered universe and a human-centered organic history.

However, the teacher must also beware of the power of history—and of pseudohistory (Chapter 5). Lauding specific scientific achievements and scientists is not idle. It is value laden. Many implicit lessons can emerge and carry substantial unintentional influence. Too often, for example, scientists of the past are characterized in two distinct categories: those who were "right," lauded as heroes, and those who were proved "wrong," dismissed as fools. Such judgments fail to respect how all scientists work within certain historical contexts. When we ridicule nineteenth-century vitalists who believed in spontaneous generation, for example, we fail to see that they defended a fundamental (thermodynamic) principle: that ordered complexity cannot arise from simple homogeneity. Although these scientists were ultimately wrong in their conclusions, they exhibited exemplary scientific reasoning. Their work expressed the very virtues we want to highlight. Sorting scientists simply into heroes (to be celebrated) and fools (to be ridiculed or forgotten) fails to portray the nature of science faithfully.[18]

Simple tales of discovery can easily become simplistic. When they do, they no longer convey an accurate picture of science (see Chapter 7). Though scientific achievements are rarely attributable to one person alone, classroom stories tend to credit single individuals. Despite textbook portrayals, for instance, Le Chatelier was not the prominent author of the chemical equilibrium concept.[19] Revolutionary reconceptualizations, as well, are generally complex, compounded from many earlier, more modest contributions.[20] Likewise, emphasizing conceptual achievement rather than innovations in instrumentation or technology conveys another kind of bias: about the relative value of intellect versus labor.[21] Teachers who celebrate history need to reflect on the values and ideological overtones that their histories imply and promote. Venturing into history responsibly requires reflective mindfulness. See Chapter 3 for further elaboration on this important topic.

5. Promoting Scientific Careers

One reason for celebrating individual scientists is to promote science as a possible career. Scientists of the past may become role models. Selecting and portraying role models from history requires care, however. A widespread assumption is that such models must be larger than life and, hence, that only their achievements and virtues matter. Yet portraits of a Newton, a Darwin, a Curie, or a Carver as geniuses can create superhuman role models that are as hopelessly unattainable to some as they may be inviting to others.[22] Paradoxically, perhaps, favorably biased portraits may attract young adults precisely because they distort or dishonestly convey what it really means to be a scientist.[23] Such students later abandon their aspirations, sometimes bitterly. Overselling can lead to anti-science cynicism. A challenge for the teacher, then, is to render scientists in human scale—demystifying their achievements and acknowledging their flaws in personality—while still dramatizing the value of their contribution (also see Chapter 3).

History helps render science as human endeavor. That includes, of course, the human flaws. Historical portraits thus can help students address the stereotypes, so prevalent in our culture, of scientists as perfect, yet also impersonal and inhuman.[24] Thus, students benefit from hearing how sometimes scientists' personalities may affect their research style or even the content of their theories. Scientists' very human motives are visible through their acrimonious priority disputes, behind-the-scenes politics in publishing or getting grants, ambitions for Nobel prizes, and even reporting of fraudulent results.[25] Yet great scientific achievements emerge all the same. Students should see the human elements as a part of the scientific process, reconciled with the efficacy and general reliability of scientific conclusions.

Establishing role models in science may be especially important for able students who, inundated by images of science in popular culture, may not likely envision themselves becoming scientists.[26] This seems especially important where the concern is on recruiting more women and minorities into science and engineering fields. As important tools, classroom histories must be applied carefully. Presenting science (unreflectively) as primarily the result of white European males, for instance, carries with it an implicit message about who can participate in science. A teacher may thus adopt a selective cross section of historical cases. To ensure a more balanced prospect, teachers need to convey diversity. They may well use an unrepresentative sample—assuming that the history itself is not biased. As yet, we do not know how best to present this sample: should role models be explicit, as a way to expose and undermine current stereotypes; or implicit, as a way to convey the naturalness of diversity?[27] Research on role models is still significantly underdeveloped.

If the aim is for diverse representation in science, teachers may wish to be concerned about many categories: economic class, personality type (introverted vs. extroverted, competitive vs. cooperative), or thinking style (mathematical vs. verbal, visual vs. kinesthetic, abstract vs. concrete, speculative vs. conservative). The historical role models that the teacher selects or emphasizes will likely affect who pursues careers in science.

6. Developing Inquiry Skills

History may also be a prime context for developing inquiry and scientific reasoning skills. Of course, one need not turn to history to learn how to frame a problem, design an experiment, interpret its results, generate alternative explanations, and so forth. But history is a valuable resource for understanding how scenarios might be recreated in the classroom, especially in leading towards basic concepts (see sample in Chapter 11). The original efforts help show which experimental demonstrations will be most relevant and—equally important—which can lead to successful outcomes.

Echoing a theme introduced above, history can help by first contextualizing and thereby motivating problem-solving or investigatory efforts. The inquiry has human or cultural meaning. There is a perceived need to know or understand: to inspire and guide the work.

The role of the Harvard *Case Histories* was in part to provide such experience. James Bryant Conant, the co-editor, remarked that one vital way to learn science was by "recapturing the experience of those who once participated in [the] exciting events in scientific history."[28] In the case studies, students encountered the work of Boyle, Coulomb, and others, read their papers, replicated their experiments (as demonstrations), and interpreted the results in context. History became an explicit vehicle for learning scientific-thinking skills.

Historical scenarios can be especially valuable in consolidating this process. One can focus on the analysis and reasoning at various junctures without doing everything on one's own. Historical results can substitute for time-consuming investigations in the classroom. Years of research can be conducted vicariously in a single class period. One can also focus on reasoning skills, skipping from one relevant problem to the next, linked by historical narrative.[29]

History also provides information valuable to the instructor in guiding inquiry and the self-assessment of inquiry. One is the historical record of alternative approaches to solving a problem, designing an experiment, interpreting results, and assembling explanations. A teacher can anticipate the various possibilities or misconceptions that students may encounter in their work. At the same time, these alternatives serve as a baseline for assessing whether students are fully engaged in the creative process.[30]

Reasoning about evidence and alternative explanations generally leads the list of skills to teach in science. In the past few decades, however, historians and sociologists of science have documented more clearly the additional significance of technical or craft skills in the laboratory itself.[31] The program for physics teachers at Carl von Ossietzky University emphasized encounters with the actual equipment from historical designs. Students gained hands-on knowledge about building equipment, discovering hidden parameters, adjusting or de-bugging the apparatus, and replicating and refining procedures until reliable results can be achieved.[32] History is a basis for learning both cognitive and practical skills.

Historical scenarios offer a wealth of resources for teaching students skills in inquiry, or doing science. Not the least of these is that students may become aware that in their own work they achieved the same endpoint as some famous scientist.

7. Profiling Nature of Science

As noted in the introduction to this chapter, the history of science is a "how-story" of science. It allows students to appreciate how new ideas emerge: from confusion, ambiguity, surprise, or deliberate probing—sometimes dramatically in sudden insights, sometimes gradually and with great difficulty. For example, without historical examples, how can students understand the recurring role of serendipity or chance in science?[33] Without case studies of scientific disagreement and controversy in the past, how can students know how to approach or interpret current debates among professional scientists—on such topics as global climate change, genetic determinism, or the safety of cell phones?[34] History is a rich source of lessons on how science works.

For example, science is a human endeavor. History is replete with fascinating stories about how human details influenced the course of science. Charles Darwin, for example, almost did not go on his historic voyage on the *Beagle*, because the captain did not trust the shape of his nose! Teachers often already use such stories to entertain (or distract?) students. But when embedded in thoughtful reflection on the process of science, they may become significant lessons about the human dimension of science. Scientific knowledge is formed through human agency. Facts do not lie scattered on a beach like seashells, merely waiting, preformed, to be collected. This is a foundational lesson for students interpreting the claims of science. It also parallels what we know about how students, too, must participate in building their own knowledge.

Sometimes, the narrative of science is not merely about adding new concepts. In some cases, existing theories are wholly upended.[35] The dramatic shifts in worldview of the Copernican or Darwinian revolutions are most notable among them. Yet more modest episodes of conceptual change occur frequently. History provides ample illustrations. As noted in Chapter 1, scientists no longer conceive the world in terms of phlogiston, caloric, electrical fluid, or worldwide floods. Or chemical affinities, pangenes, bodily humors, or fixed continents. More than that, the history profiles how such changes occur: unexpectedly, yet still relying on evidence. There can be good reasons for abandoning one theory and adopting another incompatible with the former. Tentativeness has consistently been profiled as a central lesson on the nature of science (Chapter 1). Yet such lessons are empty without understanding why or how scientists change their minds. Historical examples help convey those reasons. The lessons are especially vivid when students, through inquiry mode, are shepherded through such episodes of conceptual change themselves (no. 6 above and sample case study in Chapter 10). Here, history can provide a depth of experience unavailable to students pursuing their own investigations.

Reviving the doing of science from history can also highlight many philosophical issues about the nature of scientific knowledge, proof, evidence, and so on. What grounds scientific theories? How should we characterize scientific knowledge, its permanence, reliability, or authority? How should one interpret scientific debates or uncertainty? Scientists themselves often turn to such philosophical questions in times of crisis. When chemists wanted to talk in terms of atoms, for example, they first had to debate and think through the role of unobservable, hypothetical entities.[36] At the time, those ideas took science beyond the acknowledged boundaries of concrete observables. Such reflection is equally relevant to an ordinary citizen today hearing about many theories beyond everyday experience. Even occasions of error or fraud can be exciting opportunities for students to reflect on how we know: for example, the notorious case of N-rays, which seemed to exist only in the laboratory of René Blondot and his colleagues.[37] Students need reflection to avoid both extremely naive views of progress and casually dismissive claims of complete relativism. Glimpses of science-in-the-making, in contrast to ready-made science, help reveal all the techniques that scientists use to triangulate and stabilize uncertain knowledge. History of science is a tool for opening discussion on the important methodological and philosophical principles that underlie scientific conclusions—and providing useful examples for interpreting their subtleties.[38]

8. Highlighting Science as Social

Historical perspectives invite an extremely broad view of the nature of science. Certainly, one is accustomed to thinking about the nature of observation and experiment and how to conceive the growth of scientific knowledge (such as the meaning of "progress"). Exploring the details of science in concrete cases, one also finds a role for social factors. Science is social in two ways. First, it is an activity of societies, and so reflects various social interests, politics, and economics (next section). But, more plainly, science is also practiced in scientific communities. The social dynamics of these communities can be important to science.[39]

The social dimension of research includes, for example, the significance of scientific communication: journal publications, correspondence, conference dialogue, e-mail, inter-lab visits, and so forth. Likewise, it underscores the role of persuading peers and the corresponding importance of organizing and presenting evidence effectively. Science is unique in that credit must come from peers, often from critics.[40]

Most important, perhaps, the ability to meet the evidential standards of a diverse population of scientists is one strong factor in establishing reliable knowledge.[41] Mutual scientific criticism functions like a system of checks and balances in isolating error or bias and leading the community to more well-founded conclusions. That is, there is a strong epistemic reason for promoting cultural and cognitive diversity in science, especially through educational settings (see no. 5 above). An emphasis on reliability through collective interaction also supports arguments for, and resonates with, recent trends in collaborative learning.[42]

Once again, history offers vivid illustrations. The social dimension can also be modeled for students experientially, just as case studies can model problem solving. Simulations of historical debates—whether it is retrying Galileo, debating Rachel Carson's *Silent Spring*, or trying to decide whether one can (or should) build a nuclear weapon—engage students in the social level of science (Chapters 12 and 13).[43]

9. Portraying the Cultural Contexts of Science

Seeing science through history reveals the poverty of most textbooks in conveying the richness of what science means. Science has economic, political, ethnic, ethical, rhetorical, and gendered dimensions, as well as conceptual. All of these can shape the reliability of scientific claims, as well as the public image of science. They are of special importance in science education because, unlike the other dimensions of science discussed above, they are typically absent from the education of science majors—the primary background of most science teachers.

First, history offers examples of how scientific ideas have realigned cultural attitudes, even worldviews, and how technologies have materially affected industry, labor, lifestyles, etc. Conversely, cultural values have entered science, sometimes substantially promoting productive research. Even religious views, sometimes considered antithetical to science, have proved productive for science on occasions. At other times, cultural views about race, sex, ethnicity, or power have biased scientific claims. History shows how culture and science are intertwined—both favorably and unfavorably—and helps students become sensitive to these factors in interpreting scientific claims.[44]

History can show, even more dramatically, how the whole concept of science and strategies for developing reliable knowledge have been viewed differently at various times and in various cultural contexts. Especially when one abandons a narrow Western or Eurocentric perspective of history, one learns about the scientific contributions and ways of science in China, India, Africa, pre-Columbian America, Australia, Asia, and the Pacific.[45] The important themes of the social and cultural dimensions of science are addressed further in Chapter 6. History is a tool for the teacher and student alike to think more deeply about what science itself means.

FROM HISTORY TO CASE STUDIES

As noted at the outset of this chapter, history is certainly no stranger to the science classroom. Yet, as profiled above, its role can be far deeper than is commonly acknowledged. Merely celebrating scientific discovery on occasions or bringing humor or brief anecdotes into a lecture hardly plumbs the potential for teaching science as a whole. Indeed, such brief glances might well mislead students or convey distorted images antithetical to scientific literacy (also see Chapter 3). The scope of history as a tool is a reminder of the aim of introducing students to the nature of science in its entirety: Whole Science (Chapter 1).

As described in the many applications above, history is also a versatile resource for teaching science, not just NOS. A teacher can find clues about how students develop concepts and find occasions to exercise emerging scientific-thinking skills. Indeed, in a historical scenario, these three elements—content, process skills, and NOS—are fully integrated. Using history as a benchmark, then, one may well conceive Whole Science in an educational context as a synthesis: teaching scientific concepts while nurturing process-of-science skills and fostering NOS reflection. History is a vehicle and guide for teaching Whole Science (see Chapter 10 as a model example).

The multiple aims of Whole Science may at first seem quite ambitious. But they are easily addressed in classroom case studies—a format encountered repeatedly in discussions above. Case studies reduce the potentially unwieldy vastness of science to manageable units. They serve as small samples or cross sections of Whole Science. Although fragments, they function holographically, embodying a representation of the whole. That is, they limit the scope of study without sacrificing the integrity or complexity of scientific practice. Most importantly, perhaps, they show how the many components or dimensions of scientific practice interact.

Case-study teaching has many virtues.[46] Most notably, the concrete context helps to motivate student engagement and learning. Accordingly, a case-based approach has been adopted in some courses and textbooks: at the secondary level, the American Chemical Society's *ChemComm*, and it biology counterpart, *BioComm;* and at the college level, *Explore Life, Chemistry in a Community Context,* and, in England, "Science for Public Understanding."[47] The prepared case-based programs currently available, however, tend not to emphasize explicit NOS reflection. One of the primary virtues of case studies from the perspective of Whole Science is thus lost. Nor need one adopt an entire curriculum to benefit from case studies. Periodic case studies, especially when they revisit dominant themes, can effectively convey NOS understanding.

Case studies may be either historical or contemporary, of course. Given a goal of scientific literacy, or interpreting science in today's world, case studies drawn from current events or the media might seem optimal.[48] Such cases have many advantages, of course. Yet modern cases may also be problematic in instruction. Emotion-laden content can distort perceptions and, thus, learning about NOS. In addition, the science is not yet resolved. This can thus foster disillusionment and confusion.[49] Moreover, the key information on the internal mechanisms of science is largely unavailable through popular media.[50] The portrait of science will be incomplete just where one wants it to be most informative. Most important, contemporary cases lack a clear solution by which to judge problem-solving or interpretive efforts, critical in an instructional context. How can one learn what leads to reliable knowledge if one only addresses unresolved controversies (a problem exemplified in Collins and Pinch's provocative set of case studies, *The Golem*[51])? Understanding how an episode ultimately unfolded provides a benchmark for evaluating the methods of science. One may thus want to resist the temptation to rely exclusively on contemporary cases.

Indeed, some NOS lessons may be learned well only through historical cases. For example, consider 'tentativeness', or the provisional nature of scientific knowledge. Errors are only really recognized in retrospect. One needs examples of real and profound conceptual change that were wholly unanticipated. To compare a reasonable "before" with an unexpected "after," the episode must be past. Cases of historical error are good opportunities for showing honest mistakes, as well as inferring the methods for avoiding missteps.

Historical cases prove useful in another way. Students need the freedom to fail while they practice applying their skills. They also need to evaluate and adjust their emerging sense of NOS judgment. History provides clear solutions for assessing one's growing analytical skills. Contemporary cases, still in process, cannot. By understanding the ultimate historical outcome, one can calibrate one's developing NOS thinking.

In a similar way, appreciating how social or cultural perspectives influence science requires a relatively remote vantage point. One must be able to see the culture as culture. For example, we no longer share Victorian England's views of competitive society and racial hierarchy. One can thus see how they influenced Darwin's conceptions of natural selection and the evolution of morality. We can also see how nineteenth-century views of women once shaped theories about the female skeleton, mammals, and even flowers. Cultural beliefs enter science without conscious awareness. Only by studying such historical cases might one be able to see today's cultural perspective of, say, biological determinism and how it shapes concepts of genetic identity, cloning, and genetically modified organisms.[52]

History can thus inform contemporary cases. For example, a student's interpretation of the 2009 Climategate case (Chapter 1) would be greatly informed by earlier study of Millikan's oil-drop data, or Mendel's data on inheritance in peas; the "tricks" used to map chromosomes in the 1910s or a Mercator-style map; or the publication politics of Newton, Lavoisier, or geologist Roderick Murchison.[53] The case of revised mammogram recommendations (like many cases of revised assessments in public health) would be informed by learning about how more data once dramatically altered theories about the causes of pellagra or beriberi or the assessments of the risks of thalidomide or genetic engineering. The case of facilitated communication of coma patients (Chapter 1) brings to mind the earlier cases of Clever Hans and psychic-showman Uri Geller, or the recurring pitfalls of research on spiritualism and the paranormal.[54] Historical case studies may adopt a variety of forms, as demonstrated in Part II of this book.

HISTORY, SCIENCE-IN-THE-MAKING, AND INQUIRY

History finds a welcome home in classrooms oriented to inquiry or active learning. From a Whole Science perspective, case studies become an occasion for active engagement: say, working in the footsteps of historical scientists. That is, the history of a particular concept may provide a clear framework or underlying narrative to structure inquiry. History maps a trajectory and potential activities, reasoning exercises, and decision points. Along the way, the student can exercise and develop process skills and reflect on the nature of science. All engage students in scientific practice at different levels.

One reason for turning to history is an analogy between students and their counterparts in the past. Both are engaged in learning and discovery. Learning in the past (by individuals or a scientific community) can be a model for learning in the present. While fruitful, however, this "cognitive recapitulation" analogy can easily be overstated. For example, students do not learn the basic concepts by merely replicating or redoing historical experiments. Nor do they learn anything about the process of science. Consider the little volume, *Famous Experiments You Can Do*.[55] Portraits of

Newton, Galileo, and Faraday adorn the cover, along with drawings of their apparatus—suggesting that the reader will get a glimpse of real science. But the exercises are all cookbook. There are no unknowns. So how can the reader "share in the discoveries"? Imagine someone advising Michael Faraday, for instance, to "use adult supervision." Scientists do not fill in the blanks of an equation. Historically, the formula was the outcome of his investigation, not the means guiding it. These "experiments" for the home are not experiments. They are demonstrations privileged by hindsight. They turn science inside out. While the projects may seem like fragments of history for the classroom, they differ profoundly from real investigation—and thus from real science.

The problem with such approaches to history is that they begin at the end. They rely on knowing the answer in advance. The process of science is then reconstructed to lead surely and directly to this outcome—frequently portrayed as the obvious outcome. This upends the whole context of investigation, based on unknowns and uncertainties. In a sense, the whole process of science is short-circuited. By contrast, the great virtue of history for education is that it can portray the process as it moves blindly forward, exploring and gradually eliminating alternative ideas. In the real world, scientists work from a position of ignorance and persuade themselves, through empirical questioning and observation alone, how the world, in fact, works. Understanding these two complementary perspectives is critical for the science teacher using history. One looks backward from ready-made science and uses the "right" answer as the benchmark. The other looks forward, from the perspective of science-in-the-making, and (like a working scientist) relies solely on evidence and reasoning as a benchmark.[56] History unfolds from the perspective of science-in-the-making, and in this respect history has great affinity with inquiry learning.

The irony of using history effectively in the classroom, then, is that the past must be made "present" again. One must forsake ready-made science and restore science-in-the-making. In case-study work, students must experience a historically situated perspective, blind to the outcome, akin to the uncertainty in modern cases that one hopes to inform. For example, a student will learn about conceptual change personally by unexpectedly reorienting *their own* ideas. They might adopt the position of a famous scientist in a historical scenario—before a revolutionary discovery is made—and address the same problem. They struggle with the new observations or data and the apparent contradictions with earlier theories. Ultimately, reasoning from the available evidence, they reject old concepts and accept a new one (see sample in Chapter 10). Knowing the answer in advance makes this lesson impossible, just as a spoiler ruins a mystery or suspense thriller. It is essential to frame a historical case with its problems open ended (Chapter 4).[57]

The challenge for teachers, then, is to revive historical context, and with it the sense of scientific uncertainty and excitement. Case studies should, variously, challenge students to ask questions, to think critically about what information is relevant, and to design experiments that specifically collect that information. Those tasks might be accentuated with further problems, say, about sample size, sources of experimental error, or controls. Students should examine results critically (even original data from a famous historical study), imagine alternative explanations, and suggest investigations to resolve the alternatives. Again, effective problems are open ended. They do not expect single targeted "right" answers. The teacher must adopt a status of uncertainty alongside the students.[58] Along the way, one may reflect on the ethical use of animals in experiment, or how to secure funding, or what social decisions are appropriate given the findings—all highlighting NOS contexts. Even if there is a central story or central character (helpful for preserving a fruitful focus), there must be opportunities for envisioning alternative trajectories and pursuing diverging inquiries, at least on a limited scale. In this way, history becomes inquiry learning.

Inquiry learning does not leave the whole task of discovery to the student. So, for example, questions or problems will surely be selected to highlight particular target lessons. They will be adapted to the cognitive level of the students. The teacher may also initially *model* effective analysis—for example, showing how to tease apart empirical questions from ideological values. Or how to find and assess assumptions in an argument. Maybe an occasional sample of cautious, mature judgment. History is a guiding structure or scaffolding, not a script.

Recapturing historical uncertainty, or science-in-the-making, is integral especially to applying historically based NOS lessons to the present. Most contemporary decision-making cases involve scientific claims that are young and uncertain. The science has not yet benefited from the proverbial "test of time." Debate is often still active—unlike the voice of absolutism that haunts textbooks. The evidence for evolution, plate tectonics, or atomic theory is already well established. Of course one can recite the evidence. But rationalizing an answer already known differs markedly from reasoning blindly towards a yet unknown solution. Problems in today's culture rely on the second, more demanding skill. Respect for historical context is thus not a concern isolated to some fussy historians; it is central to the NOS aims of using history in science (Chapters 3–5).

History transformed into inquiry (or creative simulation) applies to learning the nature of science as much as scientific concepts or skills. That is, NOS lessons (like all lessons) are more vivid when students actively experience them first hand. For example, one cannot expect prepared lessons that merely illustrate NOS tenets with historical cases to be effective. One must engage the students in *explicit reflection.*[59]

Here, however, the aim is to pose problems specifically *about the nature of science.* The level of problem is different. Science teachers thus need to *problematize* NOS in their case studies. Accordingly, a teacher may pose questions not just about content, but about the context: how science works. *How* does it generate reliable claims? Does it always? The teacher might develop a habit of asking

- Why should anyone believe this?
- Can you see any potential for bias or error? How would you remedy it?

In historical case studies, the questions may be more focused:

- How did the rivalry between Pasteur and Koch influence their research?
- In what ways did Newton's religious views shape his scientific ideas?
- How did new techniques for isolating gases alter chemical knowledge of combustion, heat, the elements, and the nature of chemical reactions?
- How did the proximity of the offices of Walter Pitman and Neil Opdyke affect the interpretation of the Eltanin-19 data on sea-floor spreading?
- What do each of these events indicate about the nature of science in general?

Widespread educational strategies encourage teachers to first bring prior conceptions to mind and then engage them with discrepant events, anomalies, or other new information. This helps each student situate new ideas into his or her own way of thinking. These challenges can become part of learning NOS in historical case studies, too.

Paradoxically perhaps, motivating and framing questions may be far more important than providing ready-made answers. The aim is for the learner to learn, not for the teacher to teach. Working with case studies may thus involve a profound, often unrecognized gestalt switch in professional ethics. In the lecture model, one respects students by being a reservoir of knowledge and generously sharing one's expertise. In a case-based, problem-based, or inquiry approach, respect is based on supporting student autonomy and scaffolding self-directed skill development. For some teachers, the shift in self-image may be quite challenging—even if ultimately rewarding.

The great benefit of framing questions about nature of science is that the teacher can engage in answering them, too. The teacher can learn through active reflection, as well. No one starts off as an expert on NOS. But knowledge quickly deepens with more examples and reflection on each. A habit of posing questions and reflecting explicitly helps both student and teacher grow. That's where an inventory of questions or potential issues can be helpful (Figure 1.3). History is a playground for NOS inquiry.

SUMMARIZING HISTORY AS A TOOL

Historical perspective is an indispensable tool for effective science teaching. It is critical, in particular, for complete NOS lessons. Accordingly, one typically finds history *and* nature of science coupled together in curriculum standards: as HNOS.[60]

Those eager to see concrete examples of historically informed science teaching can browse Part II for a sampling of classroom case studies. These demonstrate the aims and perspectives sketched in the first two chapters.

Yet while history is a potentially quite valuable tool for learning science, teachers must learn to use the history—like any tool—appropriately, lest they betray the very subject they hope to enlighten. The next three chapters address some of the challenges: a sort of *User's Guide for History*.

First, historical narratives may at first seem to fit into familiar habits of telling stories. Yet the context of storytelling, so common to human experience, tends to shape the very way that stories are told. One can easily distort the historical information, even while trying to be informative. Histories of science are not just stories of science. They are implicit accounts of the nature of science. Here, the teacher must be aware of the potential for misleading "myth-conceptions": Chapter 3.

Second, in a context of science teaching, one can easily let the science overshadow the history. As hinted above, there is a tendency to reconstruct the science according to what is known now, even to "correct" the history. Of course, omitting or changing historical details can distort NOS understanding, inadvertently subverting the very lessons one hopes to convey. The teacher will also want to know how *not* to teach history in science: Chapter 4.

Next, preconceptions about the nature of science can be as powerful as preconceptions about scientific concepts. Through them, one can easily (that is, subconsciously) interpret history selectively. One can find facts that confirm one's prior perspectives, disregard others that challenge them, and blindly shoehorn history into one's own way of thinking. It seems to be an inherent cognitive tendency. One must learn to recognize how our cognitive blind spots work and "listen" to history, lest one succumb to pseudohistory: Chapter 5.

Of course, few science teachers can afford the luxury of full-time historical study, as fascinating as it may be. Thus, the next three chapters offer some rough-and-ready rules for sorting reliable history from ideologically mangled junk. Foremost, teachers need to respect historical context. That is, they need to adopt the perspective of science-in-the-making (Chapter 1) and be open to what history has to tell them. With that foundational respect, history becomes a gateway to understanding the development of scientific concepts and the nature of science—able to transform the image of science by which we are now possessed.

3 | Myth-Conceptions

Images, icons, and caricatures of scientists • rhetorical dimension of scientific narratives • Gregor Mendel • Alexander Fleming • William Harvey • Joseph Priestley • the architecture of scientific myths • strategies for teaching

Scientists, in popular imagery, seem epitomized in the puppet characters of Dr. Bunsen Honeydew and his hapless assistant, Beaker (Figure 3.1). They are whimsical caricatures. Yet they illustrate common impressions. Perhaps they even contribute to creating them. An earnest study of the nature of science begins here, with such widespread cultural images—and their unacknowledged potency. By exploring the roots of such preconceptions, educators may find essential first clues for guiding the development of more informed perspectives.

Consider Dr. Bunsen Honeydew, aptly named, perhaps, after the ultimate symbol of science in school: the Bunsen burner. No matter that students could probably not tell you who Robert Bunsen was, nor how the burner developed, nor why it was important.[1] Images, names, and impressions can easily shape thinking more than substantive, yet more remote, facts and ideas.

The Doctor's appearance is equally significant. First, he wears a white lab coat. That uniform is emblematic of experimental life.[2] He also wears glasses. That signifies keen observation, the principal activity of science. Of course, the dear Doctor has no eyes. But that may indicate the metaphorical blindness of scientists, who (in the popular image) care more about the insular world of research than about humanistic pursuits. Honeydew is stout,

FIGURE 3.1. Stereotypical scientists Dr. Bunsen Honeydew and Beaker. (Image reproduced courtesy of Disney and Muppet Studios.)

with a round head like...well, a melon. It holds a big brain: for memorizing all those formulae and facts. He is bald, and certainly not charismatic: suitable for someone whose life excludes common human emotion or romance. Nor is he athletically fit, having spent too much time confined in the lab. And, finally, he is male.

Beaker is quite different. Ironically, he looks more like a test tube than a beaker. As Honeydew's counterpart, he also exhibits features widely viewed as characteristic of science as a profession. For example, he also wears a white lab coat. He has hair, but it sprouts up, completely unkempt: preoccupation with work seems to eclipse even simple grooming. Beaker's oversized eyes bulge, all a-goggle. In context, they signal the dangers that Dr. Honeydew always seems eager to introduce, completely absent-mindedly. Yet Beaker's comic pleas (inarticulate stuttered beeps of alarm) never succeed in forestalling impending catastrophe (which usually claims him as a victim).

Finally, the two indomitable scientists are surrounded by impressive equipment, whirring and flashing away. They always have new discoveries to announce. The motto of their lab, "where the future is made today," reflects how science is generally synonymous with progress and typically conflated with technology. For Dr. Bunsen Honeydew and Beaker, science always begins with great promise. But it also inevitably ends in pyrotechnic disaster. Dramatic justice, with a hefty dose of humor, seems to keep the apparent hubris of scientists in check.

One may well be tempted to dismiss these puppet characters as irrelevant fictions. Yet their features are exactly what students typically depict when asked to "draw a scientist." Several decades ago, this simple exercise was crafted into a diagnostic test, designed to probe student preconceptions. Applied repeatedly over the years, the test results have been remarkably consistent across various groups and ages.[3] Dr. Bunsen Honeydew and Beaker reflect the archetypal preconceptions through which all subsequent information about science is measured. From the perspective of teaching the nature of science, the whimsical characters are ultimately quite serious and informative.

A solution to the popular preconceptions seems simple enough: depict real scientists. Delve into history. Render authentic science from famous examples. Yet such stories are already common, even in the science classroom. Alas, such icons as Newton, Darwin, Mendeleev, Watson and Crick, or Galileo all become caricatures, too. Popular histories romanticize scientists, inflate the drama of their discoveries, and cast the process of science in monumental proportion. They foster unwarranted stereotypes about the nature of science—notably, all for the sake of telling a good story. Although based on authentic events, the histories are deeply misleading. They subvert the goal of teaching the nature of science.

The challenge, then, is to change *how* scientists are depicted and *how* the science stories are told. It is first and foremost about the framing and the rhetoric.

A first step for the teacher of NOS is simply to recognize the caricatured icons and their import. For inspiration, one might consider Andy Warhol's eye-popping images of Marilyn Monroe, Mao Tse Tung, or Elvis Presley. Warhol cleverly rendered these popular icons so as to comment on their larger-than-life status. He amplified them to monumental scale, emblazoned them with bold psychedelic colors, and alluded to their omnipresence through multiple side-by-side copies. By making the ordinary perceptions extraordinary, he invited the viewer to see the icons *as icons*. Through such awareness one could reflect on the nature of the icons and the culture that generated them. A similar strategy guided Jasper Johns in painting the American flag on top of the American flag, on top of the American flag. The visual message was to notice the flag as a potent symbol and design itself, rather than as what it normally symbolized transparently. These artists tried to awaken awareness of cultural assumptions lost in plain view.

In this chapter, I, too, want to raise awareness and invite reflection. First, I examine four familiar historical cases in science: Gregor Mendel and genetics, Alexander Fleming and penicillin, William Harvey and circulation of the blood, and Joseph Priestley and the "goodness" of air. All include historical errors and distortions. They thus mislead students about the nature of science and how science derives its authority. Yet what is significant is the striking sameness of the errors and *how* they arise. Ultimately, the stories' errors indicate certain shared storytelling patterns.

I focus specifically on the histories' rhetorical architecture. *How* do the narrative devices and formats generate the mischaracterizations? The key errors result from rendering science in a mythic form, in a literary sense. These are not just ordinary misconceptions of science (now so much a part of science educators' discourse). They are *myth*-conceptions.[4]

Ironically, science educators themselves tend to perpetuate such myth-conceptions. But educators are also ideally positioned to remedy them. The goal of the analysis here is to inform classroom practice on how to teach nature of science more effectively. After analyzing the problem, I profile some appropriate remedies and NOS teaching strategies. Fortunately for those with limited expertise in the history, philosophy, and sociology of science, some simple diagnostic tools are available to avoid succumbing to and perpetuating the myth-conceptions.

CASE 1: GREGOR MENDEL AND GENETICS

Consider first a quintessential icon of science: Gregor Mendel. Few biology textbooks fail to mention Mendel and his work on pea plants. He is, as historian-biologist Jan Sapp notes, "an ideal type of scientist wrapped in monastic and vocational virtues."[5] That is, he is an exemplary scientist. Stories about him implicitly contain morals about the nature of science. For example, Mendel worked alone in an Austrian monastery. Lesson:

scientists modestly seek the truth, not ambition. Mendel used peas. Moral: scientists design studies using appropriate materials. He counted his peas: scientists are quantitative. He counted his peas for many generations over many years: scientists are patient. He counted thousands and thousands of peas: scientists are hard working. After all this, Mendel was neglected by his peers, who failed to appreciate the significance of his work, but he was later and justly rediscovered: scientific truth triumphs over social prejudice. Above all, Mendel was right: scientists do not err. His figures, in fact, were too good to be true, statistically speaking. But he was ultimately correct. Any hint of fraud should only confirm the depth of his theoretical insight. The common stories about Mendel are simultaneously a lesson in the nature of science. And it is mythic in dimension. Although he has not been canonized by the Church, biologists certainly honor and revere him so.[6]

Visual images echo this lesson. Texts sometimes use photos to illustrate scientists from history, and Mendel is no exception. Some texts, however, use illustrations instead. For example, one popular biology text abandoned the stuffy posed photo in favor of a painting of the scientist at work.[7] Mendel is portrayed observing: that's how we know he's a scientist. He wears his eyeglasses and a white apron (instead of a lab coat). Moreover, a soft-edged medium and pastel color palette epitomizes the idealized, romantic tone of the standard narrative. Such illustrations, too—even with no caption—convey a lesson about the nature of science.

On the surface, nothing seems flawed with the conventional story of Mendel. But historians' more complete and informed interpretations differ in significant ways.[8] The discrepancies, however apparently subtle, valuably reflect common tendencies and pitfalls in adapting history for teaching the nature of science.

A typical approach is illustrated by Westerlund and Fairbanks.[9] Because they know modern genetics, they see modern concepts in Mendel's original paper and thus explore no further. They do not consider historical context: for example, work that may have preceded Mendel, meanings that may not be the same as today, or concepts that may not yet have been fully articulated. For example, knowing that *Mendel presented his paper orally* matters. His text thus included one statement, variously modified, repeated like a mantra to a listening audience and suggesting Mendel's central thesis. Mendel refers to a "law" of hybrids (as announced in his title): one-half the offspring of hybrids are hybrid again, the other half pure-breeding; of these, one-half are one original type, one-half the other. The mathematical regularity is presented (repeatedly) as a phenomenological law, like Snell's law of refraction or Boyle's law of gases. Here, if anything, is Mendel's law *in Mendel's own terms.*[10] Westerlund and Fairbanks, however, add more. Having referred to Mendel's garden studies, they say:

From the results of these experiments and his interpretations of them, he developed a theory to describe the fundamental mechanisms of heredity.[11]

A contextual analysis indicates that Mendel's claims were not so ambitious. Mendel focused on a rather narrow problem: trying to create pure-breeding hybrids and to characterize the identity of species. His paper presents only a law of hybridization and a mathematical formalism that describes it. He did not advance any general theory of heredity. That was for others who later revived Mendel's work. Mendel also did not regard his findings as universal. (Here, it is helpful to know the historical context of *Mendel's other publications*.) His subsequent 1869 work on another species, hawkweed, made that almost painfully clear: there were no clear 3:1 ratios to be found at all.[12] The effect of these apparently minor elisions is more profound than one might imagine. As a result, one credits Mendel with more than he did. One inflates the image of individual scientists and concentrates the gradual process of science into moments of genius. Good for drama, misleading for teaching about the nature of science. The apparent clarity of Mendel's original 1865 paper to modern readers—even perhaps high school students—can be deceptive.

The same pattern of inflation is reflected in other treatments of Mendel. For example, textbooks typically elucidate "Mendel's" two laws. Paradoxically, though, in his classic 1865 paper, Mendel did not explicitly formulate a second law, or principle of independent assortment. He certainly performed dihybrid crosses and reported 9:3:3:1 results. Yet there was no formal recognition that the independent behavior of two character states should be regarded as lawlike. In fact, geneticists did not distinguish Mendel's first and second principles until several years after the revival of his work, when they encountered anomalous ratios in offspring. Bateson, for example, found a 12:1:1:3 ratio in sweet peas for flower color and pollen shape: alleles segregated, but the genes did not assort. Ironically, then, while Mendel's second law bears Mendel's name, he himself did not state it. Again, accounts credit Mendel with more than he did and so distort the nature of science.

Some educators have even referred to Mendel's law of dominance.[13] While Mendel referred to his traits as dominant and recessive, the notion of prepotency—that one parental trait determines the trait of the offspring—was common among breeders at the time. Mendel followed a few earlier biologists in merely attributing it to the trait, rather than to one parent or the other. Texts also present dominance as foundational, its exceptions as "non-Mendelian." Mendel himself, however, seemed aware that dominance was not the norm. Just before introducing dominant traits, he noted that

> with some of the more striking characters, those, for instance, which relate to the form and size of the leaves, the pubescence of the several parts, etc., the intermediate, indeed, is nearly always to be seen.(§4)[14]

He noted other exceptions: stem length (the hybrids were actually *longer*, §4), seed coat color (hybrids were more frequently spotted, §4), flowering time and peduncle length (§8). For Mendel, his law applied only to "those differentiating characters, which admit of easy and certain recognition" (§8). Other characters followed another, different rule or law. This seemed the case, for example, with hawkweed. Mendel's concept of dominance presents several dilemmas, mainly due to an NOS perspective that Mendel is the only proper benchmark for "Mendelian" genetics.[15]

Mendel is also credited with the essence of the phenotype/genotype distinction. Yet Mendel worked at the level of observable characters. He did not distinguish clearly between traits and material units of heredity. Mendel discussed *'elementes'* (regarded as today's genes), but not as occurring in pairs in each organism. Mendel's notation shows that an A × A cross yielded A + 2Aa + a: the homozygous form was 'A', not a diploid 'AA'.[16] Again, Mendel's achievement is profiled as monumental, unambiguous, and unerring—with a corresponding moral about the nature of science.

The misrepresentation of Mendel applies to his methods as well. Introductory textbooks often report that Mendel studied seven character pairs in peas—tall/dwarf, smooth/wrinkled, green/yellow, etc. Each exhibits a clear dominant–recessive relationship, and in pairs the traits seemed to function independently. Textbooks often boast of Mendel's insight in experimental design. However, historian Frederico Di Trocchio contextualized Mendel's work *in the investigative practices of his own time*.[17] Standards for reporting were quite different. Mendel did not present his preliminary work. Mendel's records (versus his published paper) indicate that he first studied *fifteen* traits in *twenty-two* different varieties that could be regarded as breeding true (or pure). That is, Mendel seems to have abandoned traits that did not exhibit a clear pattern. His method was apparently selective. If so, then Mendel was subject to trial and error just as much as today's scientists. And that is a powerful lesson in the nature of uncertainty in science, easily obscured by an idealized account.

Not all things Mendelian are Mendel's.[18] Thus, Mendelian genetics— its concepts and methodologies—is not the result of just one person. Textbooks widely dub Mendel the founder of modern genetics. But this reveals more about the apparent need for a founder figure than about the history. Mendel, as reconstituted hero, is idealized and monumentalized. Indeed, the aura of Mendel and his achievement is often more important than what Mendel actually did or did not do. Biologists and historians can thus present Mendel as supporting contradictory claims, while all appealing to Mendel's monumental authority to bolster *their* claims.[19] The problem is not simplification or abridged history, suitable for a science classroom. In many instances, Mendel has been recreated historically to meet mythic expectations.

CASE 2: ALEXANDER FLEMING AND PENICILLIN

Another case is perhaps the most celebrated example of chance, or accident, in science: the discovery of penicillin by Alexander Fleming. In the conventional story, a stray mold spore borne through an open window landed on an exposed bacterial culture.[20] Then, as David Ho reported for *Time* magazine's *100 Persons of the Century*:

> Staphylococcus bacteria grew like a lawn, covering the entire plate—except for the area surrounding the moldy contaminant. Seeing that halo was Fleming's "Eureka" moment, an instant of great personal insight and deductive reasoning.... It was a discovery that would change the course of history. The active ingredient in that mold, which Fleming named penicillin, turned out to be an infection-fighting agent of enormous potency. When it was finally recognized for what it was—the most efficacious life-saving drug in the world—penicillin would alter forever the treatment of bacterial infections. By the middle of the century, Fleming's discovery had spawned a huge pharmaceutical industry, churning out synthetic penicillins that would conquer some of mankind's most ancient scourges, including syphilis, gangrene and tuberculosis.

Fleming himself often underscored the role of chance in his work. In receiving numerous honors, he was fond of reminding others, "I did not invent penicillin. Nature did that. I only discovered it by accident." Fleming, as hero, is a role model: someone who had the insight to capitalize on a chance observation, consequently giving health, even lives, to millions.

This story is deeply misleading, even where not demonstrably false. It excludes relevant details, mischaracterizes others, and arranges the narrative suggestively.[21] Undoing the rhetoric, I hope, shows how it creates its lesson about the nature of science, especially about the roles of context and contingency.

First, consider the phrase "when it was finally recognized for what it was." Because originally, in 1928, Fleming hardly envisioned penicillin as the great drug it later became. He did not strongly advocate treating humans with it until 1940. What happened in those twelve years? Initially, Fleming had indeed been searching for antibacterial agents. But he was not impressed with penicillin's therapeutic potential. It was not absorbed if taken orally. Taken by injection instead, it was excreted in a matter of hours. For Fleming penicillin was limited, perhaps to topical antisepsis. Hardly momentous. In the ensuing years Fleming used penicillin, but as a bacteriological tool. It suppressed the growth of certain bacterial species and allowed him to culture certain others. That became valuable for manufacturing vaccines—a major task Fleming managed at St. Mary's Hospital in the 1930s. Meanwhile, Fleming's research had turned to another group of chemicals, the sulphonamides. Without further work, Fleming's

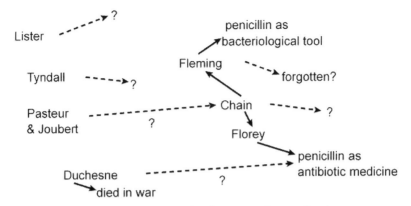

FIGURE 3.2. Alternative histories in the discovery of penicillin: history as a web, rather than a timeline.

discovery would have languished, another relatively mundane scientific finding (Figure 3.2). Chance is reserved for Fleming's first observation, not its subsequent development. It sparks the plot but does not let it wander without direction.

The ultimate pursuit of penicillin in treating human infections was due entirely to another lab, led by Howard Florey in Oxford. In 1938, Ernst Chain, Florey's associate, began searching for natural antibacterial agents, endeavoring to elucidate their mechanisms more fully. He chose three to study, penicillin just one among them. Chain used Fleming's 1929 paper, but with his own, quite different, purpose (Figure 3.2). By early 1939, Chain and Florey began to suspect the medical potential of penicillin. But Florey had problems securing funds for testing because of the war effort. They also faced several technical challenges. They needed to improve production and purification methods, refine an assay to determine the strength of their extracts, and scale up production. After five months of work, all with no guarantee of success, they had enough brown powder to test on a few mice, which yielded promising results. While this work reflects the bulk of the scientific process, the traditional story consolidates it as uninteresting drudgery.

Now, the popular story sometimes notes, "As the world took notice, they swiftly demonstrated that injections of penicillin caused miraculous recoveries in patients with a variety of infections."[22] But the work was hardly minimal. Or swift. For tests on humans, they needed substantially more penicillin. The Oxford labs culturing the mold scaled up from flasks and biscuit tins to hundreds of bedpan-like vessels stored on bookshelves. Purification turned from the laboratory to dairy equipment. After the first test, they had to find ways to remove impurities that caused side effects. The tests eventually went quite well, but it had required two professors, five graduate students, and ten assistants working almost every day of the week for several months to produce enough penicillin to treat six patients. While a narrative of science might well celebrate hard work, here the "swift" pace linking insight to triumph seems primary. Also, emphasis on the downstream work would reduce the dramatic role of the chance event as pivotal.

Fleming noticed Florey and Chain's striking results. Yet he did not disturb his research agenda. He knew that penicillin's value still lay in economical mass production. Thus, the research—and, in a sense, the discovery—was *still* not complete, and certainly not Fleming's alone. One can now imagine the details of three more years of work before the United States could produce enough penicillin to treat a quarter-million patients per month. The ultimate achievement was indeed monumental and worth celebrating in the classroom. However, the story exaggerates the scale of Fleming's role, thereby creating a distorted image of genius in science (also true for Mendel, Case no. 1). Fleming, Florey, *and* Chain all shared the Nobel Prize in 1945. If Fleming "changed the course of history," it was not without the help of Florey, Chain, and dozens, even hundreds, of technicians.[23] An aura surrounds Fleming, like an inspiring tale of a scientist winning the lottery: vicariously, we thrill in his good fortune. But the story inflates the role of one scientist at the expense of representing how science happened.

While this episode exemplifies the role of chance, popularizers nevertheless credit Fleming, as hero, with *noticing* the antibacterial properties: the "eureka" moment that Ho described. Others, however, besides Fleming had noticed the antibacterial properties of *Penicillium* (Figure 3.2). In 1871, Joseph Lister (noted for introducing antiseptic practice into surgery) found that a mold in a sample of urine seemed to be inhibiting bacterial growth. In 1875, John Tyndall reported to the Royal Society in London that a species of *Penicillium* had caused some of his bacteria to burst. In 1877, Louis Pasteur and Jules Joubert observed that airborne microorganisms could inhibit the growth of anthrax bacilli in urine that had been previously sterilized. Most dramatically, Ernest Duchesne had completed a doctoral dissertation in 1897 on the evolutionary competition among microorganisms, focusing on the interaction between *E. coli* and *Penicillium glaucum*. Duchesne reported how the mold had eliminated the bacteria in culture. He had also inoculated animals with both the mold and a lethal dose of typhoid bacilli, showing that the mold prevented the animals from contracting typhoid. He urged more research. After earning his degree, he went into the army and died of tuberculosis before pursuing that research. Chance, here, ironically worked *against* his discovery bearing fruit.[24] That is, Fleming was not as uniquely perceptive as the popular story suggests. Moments of mythic insight may involve large doses of opportunity, context, and contingency, not just intellectual prowess. Given other circumstances, the history might not have included Fleming at all. But this history is harder to package into a compelling narrative.

Classroom histories tend to follow only a single linear plot. The narrative connects Fleming directly to the status of penicillin today. Other plot lines and scientists become invisible. The outcome thus seems inevitable. But to understand the process of science as it moves forward, the alternative futures and potential alternative discoveries are essential (Figure 3.2). Educators must portray science-in-the-making, advancing blindly, not science-made unfolding predictably. Contingency does not define just the moment in 1928 when Fleming turned his attention to the discarded, now famous culture in the tray of Lysol. It permeates the whole process. It may not fit a standard plot trajectory conveniently.

In celebrating Fleming, therefore, one might focus instead on his habits: the *context* that fostered the moment so often depicted as critical. Fleming was not known for running a tidy lab. Abandoned cultures heaped unattended in a basin would not have been at all unusual—less chance than the story suggests. Such messy circumstances invite the unexpected. For molecular biologist Max Delbrück, this promotes discovery. He labeled it the "principle of limited sloppiness."[25] In addition, Fleming was accustomed to play and pursuit of idle curiosities. At first, he simply found the halo of inhibited bacterial growth *interesting*. Later in the day, he toured the building trying to interest his colleagues—who were largely unimpressed: no promise of miracle cures yet. There was no "instant of great personal insight and deductive reasoning," as dictated by the heroic plot template. Rather, personal amusement. Later, Fleming drew pictures with *Penicillium* on culture plates and watched them develop (like photographs) over several days as the bacteria grew in the negative spaces. A teacher might have complained that Fleming was frequently "not on task." A very different image of science emerges when one sees sloppiness and play as contributing significantly to Fleming's "chance" moment. The plot becomes less algorithmic. But also more human.

In the traditional history, science appears to rely on an exceptional individual and a rare moment of insight to propel it forward. The fuller story reminds us, dramatically, how it might have been otherwise. The conventional story is narratively cozy. It celebrates modesty and good luck. Science appears to be formulaic and sure, even when the critical event is portrayed as chance. A more authentic history, however, reveals the many contingencies and contextual factors that shape the scientific enterprise through multitude potential pathways (Figure 3.2)

CASE 3: WILLIAM HARVEY AND THE CIRCULATION OF THE BLOOD

Next, consider the case of William Harvey and the discovery of the circulation of the blood.[26] Harvey, physician to royalty, claimed that the blood did not move to its natural place on its own but was propelled by the action of the heart. Moreover, blood is not merely used up in the extremities. Rather, it continues to flow as in a natural cycle. Harvey's conceptual achievement was certainly recognized by his peers (although not without some dispute). He also epitomized the emergence of experimental investigation in the early 1600s and inspired a new generation of English physiologists.[27]

Harvey's work provides an excellent opportunity to convey the nature of science to students. Science educator Anton Lawson offers one such account, but it is riddled with error—and it is instructive to understand just how it fails.[28] Most notably (and the first signal of trouble), the author announces that he is *reconstructing* Harvey's reasoning. Harvey's "thinking can be cast in the form of hypothetico-deductive arguments," he notes. "Can be." This was not Harvey's own method. The *idealized* method of science takes precedence and forms an explicit context for shoehorning the history. Harvey's work is thus distorted and so, too, is the nature of science.

The rhetorical strategy is to first establish Harvey as a desirable role model by dramatizing his discovery:

> Galen's theory of blood flow was virtually unquestioned for near-ly fifteen hundred years until 1628 when the English physician William Harvey...published a book.[29]

Here, the monumental time scale—over a millennium—functions narratively to impress us with the scale and singularity of the discovery. The statement may seem harmless enough. But it casually discounts others who, before Harvey, had introduced new ideas about circulating blood flow. For example, the circuit of blood to the lungs had been recognized by Michael Servetus in 1553, Realdus Columbus six years later, and Andreas Cesalpino in 1603 (although each for different reasons). All three questioned Galen's authority. Moreover, Ibn al-Nafis discussed pulmonary blood flow in the 1200s, during the Golden Age of Arabic science. Cultural sleight and historical details aside, the magnitude of Harvey's achievement has been grossly exaggerated. Contributions from several physicians over at least a century are collapsed into just one person: William Harvey. The implicit lesson for the reader? Harvey possessed some special form of reasoning, which his peers did not, that was critical to his success in science. Inflating genius here has a persuasive narrative role.

This sample "history" emphasizes Harvey's reasoning against Galen. Galen believed that blood must flow from the heart to the lungs, and that some blood flowed back to the other side of the heart, but he also reasoned that blood might permeate the septum of the heart directly. This is treated as somewhat astonishing, even outlandish. Harvey, we are told, put the mistake right with hypothetico-deductive reasoning. (Never mind that Andreas Vesalius had criticized Galen on this very point, on the basis of his observations decades earlier.) The uninformed reader never learns that Galen was a pioneer in dissection. He hardly would have advanced such a claim foolishly, absent any observation whatsoever. Here, Galen, as straw man, fills the narrative role of adversary, or villain. We never learn how *Galen* might have reasoned, nor why *his* ideas were respected for so long. Indeed, the question doesn't even arise, although the purported aim is to illustrate scientific reasoning.

Later, the reader learns about what was supposedly Harvey's greatest triumph: the prediction of capillaries.[30] Although no one could observe them at the time, Harvey apparently saw the bold implications of his theory:

> *If*...the blood flows away from the heart in the arteries, and
> *If*...the bloods flows towards the heart in the veins,
> *Then*...the arteries and the veins *must* be connected.

Moreover, this impressive if–then reasoning was apparently dramatically confirmed fourteen years after Harvey's death by Marcello Malpighi, having been guided by this prediction.[31] However, both these historical "facts" are fabrications. Indeed, one can only reconstruct Harvey's reasoning this way if one already knows that capillaries exist. When I began teaching, I encountered this story about Harvey's prediction—and naively I believed it. Harvey's original work, however, reveals a different story. In his classic *De motu cordis*, Harvey describes how blood *percolates* in the lungs "the way water percolating the earth produces springs and rivulets."[32] For Harvey, blood permeates the flesh, which is porous like a sponge. The rate of blood flow even varied through the organs, depending on their "denseness or sponginess."[33] Harvey did not reason blindly, of course. He had dissected many "lesser" animals—"crabs, shrimps, snails and shell-fish"—that have hearts but no blood vessels (open circulatory systems, in our terminology). He had observed directly that connections were not needed. Ultimately, *Harvey did not predict capillaries.*[34] This misattribution comes from trying to fit Harvey, as scientific hero, into a predetermined, idealized reasoning pattern.

Ironically, perhaps, Harvey even argued *against* capillaries. Galen had reasoned (deductively) that anastomoses must exist to allow blood to travel from the veins to the arteries. Harvey proposed the opposite direction and, thus, was ready to discount Galen. Harvey reported many years later:

> I myself have pursued this subject of the anastomosis with all the diligence I could command, and have given not a little both of time and labour to the inquiry; but I have never succeeded in tracing any connexion between the arteries and veins by a direct anastomosis of the orifices.[35]

He boiled organs—the liver, lungs, spleen, and kidneys—until they were so brittle that their dust could be shaken from the fibers and he could trace every blood vessel distinctly. But no anastomoses. Harvey not only did not postulate capillaries, he looked for them and concluded that none were there.

Nor was Malpighi's discovery of capillaries guided by Harvey's theory.[36] He made his now landmark observations while focusing (instead) on the structure of the lungs. In a letter to his mentor Giovanni Borelli, Malpighi noted the limits of unaided observation of the lungs:

[T]he blood, much divided, puts off its red color, and, carried round in a winding way, is poured out on all sides till at length it may reach the walls, the angles, and the absorbing branches of the veins.

The power of the eye could not be extended further in the opened living animal, hence I had believed that this body of the blood breaks into the empty space, and is collected again by a gaping vessel and by the structure of the walls.

Echoing Harvey, Malpighi revealed his initial belief in the "empty space" between the observable blood vessels, where blood "poured out" and was "collected again by a gaping vessel." Using his microscope, Malpighi then observed "that the blood flows away through the tortuous vessels, that it is not poured into spaces but always works through tubules...."[37] The observation of capillaries was quite unexpected, not part of any planned test about circulation. Malpighi did not even refer to Harvey at all. *Malpighi did not address Harvey's or any other prediction about capillaries.*[38] Here, false history is created to fit a narrative role. Vindication, dramatically delayed, helps provide the intended moral that hypothetico-deductive thinking is the route to successful discovery.

To understand the nature of science, one wants to understand how Harvey actually reasoned, not how he *might* have reasoned according to some idealized scheme. Anything less trivializes what contributed to his very real achievement. The reconstruction distorts Harvey's work by making deductive tests seem central. In his 1628 book, Harvey reported numerous demonstrations whereby one might observe his claims. He encouraged the would-be skeptic to trust his own observations, rather than the authority of a text. In a famous diagram, he portrayed a ligatured arm, by which one might *demonstrate* the presence of valves in the veins. Harvey exhorted his reader to see first hand the evidence for one-way flow of blood: "But that this truth may be made more apparent...." This is presented as *Harvey's own* test *based on a theoretical prediction.* However, Harvey reported to Robert Boyle that the observation of valves in the veins was one of the observations that *led to* his concept of circulation. It was not a test derived through deduction or prediction. It was a clue to a creative synthesis.[39] At the same time, the role of Harvey's many years of observations and vivisections, well before he developed any explicit theory, is discounted. Also omitted are criticisms of Harvey's early ideas, based on the amount of blood flowing through the heart, which also contributed to his thinking. Harvey's discovery was not driven primarily by deductive logic.

Consider, finally, the treatment of one of Harvey's central ideas:

Harvey's guiding analogy was...circular planetary orbits and the belief that large-scale planetary patterns should be echoed in smaller-scale physiological systems.[40]

This is the microcosm–macrocosm concept of the chemical philosophy, shared by Robert Fludd, a close friend of Harvey's. The analogy also extended to chemical reactions, strengthening the analogical resonance. Harvey used this image throughout his book, sometimes explicitly as an argument. He described the heart as the sun of the microcosm, giving warmth and life to the body. That seems very strange to us today—and decidedly unscientific. Yet this analogy was *integral* to Harvey's very *reasoning* as well as to the reception of his idea.[41] Some may want to discount that this microcosm worldview could lead Harvey to discover something valuable. No doubt because the analogy is unfounded by today's standards. Attributing it to Harvey appears to lessen his status as a scientist. But it is coupled with Harvey's achievement. It is essential if we want to understand scientific reasoning and portray it faithfully to students. But in this particular treatment, the analogy is only curtly acknowledged, then abandoned as peripheral. The historical facts, even about reasoning, seem secondary to the persuasive aims of the "historical" narrative.

Errors pervade Lawson's two-page treatment of history. But the errors themselves are not as important as what generates the errors. Harvey is repeatedly shoehorned into a particular method of scientific reasoning for rhetorical purposes. History becomes a tactic of persuasion. By inscribing a philosophical perspective into the work of a renowned scientist, an author gives it the semblance of naturalness or authority. An imaginary example simply does not carry the same cachet. However, when one delves into history to prove a point, rather than to listen to what it has to say, one can easily err. Historians talk about *respecting* history—that is, regarding history as an end, not an instrumental means towards some other goal. Historians ideally endeavor to decipher the details and complexities of historical context, and not just re-map their own views onto the past.[42] By contrast, this account by Lawson tries to monumentalize Harvey for specific persuasive purposes, leading to telltale historical errors. Mythic grandeur and misconception are intimately related. That is what constitutes a *myth-conception*.

CASE 4: JOSEPH PRIESTLEY AND THE "GOODNESS OF AIR"

In recent years, there has been a resurgence of interest in eighteenth-century polymath Joseph Priestley.[43] He was, even by modest accounts, an extraordinary individual. Among other things, he founded the Unitarian faith, defended the French Revolution in class-stratified England, pioneered pneumatic chemistry, isolated oxygen, and vigorously defended the phlogistic doctrine (see Chapter 11). His investigations into various "airs" led him to recognize that vegetation helps restore air that has been "fouled" by combustion or breathing animals. (In today's terms, plants use carbon dioxide and generate excess oxygen.) Priestley's discovery appears in various educational contexts, most notably in a 1957 historical case study on "Plants and the Atmosphere."[44] Of special interest here, however, is a recent treatment for science educators by Michael Matthews.[45]

In this account, Priestley is an exemplary scientist—and that is paramount, as in the Harvey case. One finds first an adulation of Priestley, in the form of an endorsement of an earlier 1892 assessment (providing a patina of age-old wisdom):

> If we choose one man as a type of the intellectual energy of the eighteenth century, we could hardly find a better than Joseph Priestley…the hero of the eighteenth century. [46]

Priestley, as hero, is presented as a role model—and a public figure due all proper deference. The student is to identify with Priestley and adopt his valued traits. Yet Matthews overly romanticizes Priestley's work, distorting the history and, with it, the nature of science.

Priestley's reputation is burnished with hyperbole and emotion. Thus, Priestley did not just have wide interests. Rather, he had "*staggeringly* wide interests." His role in understanding plants was not just important. It was "*pivotal.*" Moreover, Priestley gains heroic status by overcoming imagined villains. For example, Aristotle "*thwarted* the provision of a correct account of the Restoration of Air." His ideas were not merely rough hewn. They were an "*obstacle.*" Further, we are to admire Priestley as a person, not just a scientist. Through his political beliefs, he apparently helped champion "*at great personal cost*" a victory over "*the most corrupt, venal and incompetent* government in all Europe."[47] We learn how Priestley was persecuted for his political and religious dissent, not just to add human dimension, but to evoke sympathy for Priestley. Ultimately, we are asked to feel "pity" that Priestley's bicentennial was not more widely celebrated. The emotion is all about aggrandizing Priestley, not just engaging the reader in the historical episode.

Priestley is presented as epitomizing certain *values*: "the tenets of the European Enlightenment tradition…whose essential features are manifest in Priestley's life and writings." The emphasis on promoting values may alert the critical reader. Indeed, the history here does not function to inform an understanding of the nature of science. Rather, it tries to leverage Priestley's greatness to promote an ideology. The explicit goal is for students

> to appreciate and understand key elements of the scientific tradi-
> tion: hard work, experimentation, independence of mind, a respect
> for evidence, a preparedness to bring scientific modes of thought to
> the analysis and understanding of more general social and cultural
> problems, a deep suspicion of authoritarianism and dogmatism,
> and the concern for promotion of an open society as the condition
> for the advance of knowledge.

Here, epistemic methods of science are confounded with cultural values and personal beliefs, implying that they are all of a kind, with little to distinguish them. The conflation further implies that science requires a particular political ideology or worldview. Priestley is quoted (approvingly) from his 1767 book:

> A philosopher [scientist] ought to be something greater, and better than another man. The contemplation of the works of God should give sublimity to his virtue, should expand his benevolence, extinguish everything mean, base, and selfish in his nature, give dignity to all his sentiments, and teach him to aspire to the moral perfections of the great author of all things…. Hence he is able to venerate and rejoice in God.[48]

That is a fascinating position by an individual scientist. But it is not fully contextualized. It thus becomes a general statement about science, apparently justified by history. Ultimately, the moralistic use of the case leads to a grossly misleading view of the nature of science.

First, Priestley's discovery about plants and the "goodness of air" is attributed to his character and Enlightenment values. The implication is that given the "right" values, the "right" methods and thus the "right" answers inevitably follow. But Priestley's discovery was far less straightforward. Historian Leonard Nash, for example, notes that Priestley's achievement was, "like so many of Priestley's other important discoveries, purely a matter of fortunate accident."[49] That is, it did not result from certain values. According to Stephen Johnson, Priestley's investigatory style relied in general on improvisation, hunches, "chaotic method," randomness, and "bricolage." His "inspired—and somewhat chaotic—explorations" typically involved "endless series of experimental variations."[50] His discoveries thus resulted from brute trial and error, not any particular method or set of intellectual principles. Even Priestley himself was rather modest: "I flatter myself that I have accidentally hit upon a method of restoring air which has been injured by the burning of candles." Many years later he commented, "In looking for one thing, I have generally found another."[51] Priestley succeeded, then, "because he was an indefatigable experimentalist."[52] Ultimately, he owed much to his leisure time, provided by his wealthy patrons.[53] In his essay for science educators, Michael Matthews trims the facts and, thus, misrepresents the nature of science—all guided by his narrative strategy. Namely, he wants to show that science advances only by adopting certain political values (such as an open society and suspicion of authorities), not by circumstance, good fortune, or sustained labor.

Priestley's work is also presented as free from error. This contributes to a positive-role-model image. For example, Priestley's discovery of carbonated (soda) water is celebrated. Yet one does not learn that Priestley promoted it, incorrectly, as a cure for scurvy. Instead, one is advised that while Priestley "recognised the enormous wealth that could be made, [he] nevertheless turned down the opportunity of commercial bottling of his Pyrmont water, saying he preferred to search for truth, not money."[54] The sentiment may be noble, but the implicit lesson is about moral precepts, not the nature of science.

Priestley's reports about the restoration of air also met with problems.[55] He returned to his experiments a half decade later (having moved to a new locale). Like others, he could not consistently obtain the same results. In subsequent investigations, he concluded (incorrectly) that sunlight alone could produce "pure" air in water. His earlier claims now seemed uncertain, yet he did not retract them. He had already received the Copley Medal, and his findings had been praised by the president of the Royal Society. Priestley's religious worldview, too, embraced a rational design of nature. Originally it had helped him adopt the notion that plants restore the quality of the air for human respiration. Now, that same view proved to limit his ability to acknowledge possible error. He disproportionately discounted the new negative results:

> [O]ne clear instance of the melioration of air in these circumstances should weigh against a hundred cases in which the air is made worse by it.[56]

Only later would Ingenhousz and others help solve the puzzle, by showing that plants purify the air only in light.

Most notably, Priestley characterized the restored air as "dephlogisticated air": drawing on the notion of phlogiston, which others soon abandoned. Phlogiston, the matter of fire, helped Priestley interpret the role of plants in the atmosphere. Indeed, his defense of the doctrine was warranted at some level (Chapter 11). Yet as a philosophical materialist, he could not envision phlogiston as an energy apart from matter.[57] That led to conceptual conflicts. Priestley's insights and blind spots were ironically related. Matthews lauds Priestley's materialism as integral to his science—when it leads to the "right" answers. Here, where it leads to error, it is disavowed. Indeed, one finds a complaint that even mentioning Priestley's adherence to phlogistic doctrine serves only to unjustly "blacken" Priestley's reputation.[58] Science, apparently, has no room for error. By defending a pristine legacy for Priestley, one effectively shuts out discussion of the role of fallibility or tentativeness in science (see Chapter 1).

Portraying Priestley's work as error free is important, of course, to the narrative aim of fashioning a heroic character. If Priestley appears less than perfect, the implicit lesson about his values is critically weakened. Here again, the *ideological* moral of the story eclipses an honest and accurate portrayal of the nature of science.

In addition to idealizing Priestley, this account inflates his historical role by discounting relevant work by others. For example, historian Leonard Nash underscores that our understanding of photosynthesis was a "progressive discovery…over a period of almost two centuries."[59] This account, by contrast, presents Priestley's role as "pivotal."[60] He thus equates awareness of the restoration of the "goodness of air" with the discovery of photosynthesis itself—wholly neglecting the significance of both light ("photo") and the fixation of atmospheric carbon ("synthesis"). It thereby peripheralizes numerous researchers whose contributions were integral to a complete understanding of the atmospheric consequences of plant chemistry. A half century earlier, Stephen Hales established the system for collecting gases or "airs" and also developed a setup for investigating plants in isolated systems.[61] Joseph Black first identified a chemically distinct component of the air, fixed air (our carbon dioxide).[62] Henry Cavendish figured out how to collect water-soluble gases over mercury.[63] Charles Bonnet helped establish the function of leaves.[64] Benjamin Franklin, in a letter to Priestley, critically articulated the implications of plants for the global atmosphere.[65] Jan Ingenhousz identified the role of light.[66] De Saussure clarified the respective roles of carbon dioxide and oxygen using the new chemical nomenclature.[67] Priestley himself acknowledged his limited role in identifying only the effect of plants in the "goodness of air," not the underlying causes:

> In what manner this process in nature operates, to produce so remarkable an effect, I do not pretend to have discovered.[68]

Of course, with all these major contributors, one can hardly regard Priestley's Enlightenment values as the sole source of scientific achievement. Still, Matthews claims repeatedly that Priestley *discovered* photosynthesis.[69] That outlandish error can plainly be traced to the narrative aim to credit Priestley alone—and thus to privilege his set of values.

By focusing narrowly on Priestley's worldview, the account preempts the relevance of cultural context and contingency. Priestley himself commented:

> More is owing to what we call *chance*, that is philosophically speaking, to the observation of *events arising from unknown causes*, than to any proper *design*, or preconceived theory in this business.[70]

For example, one may be impressed that Priestley began work on fixed air (carbon dioxide) because, when he initially moved to Leeds, his house was not yet ready. His temporary residence was near a brewery: the source of the fermentation gases he investigated. Recall, too, Priestley's research style: he was a "hacker, not a theoretician." Indeed, he seemed to exhibit an "aversion to theorizing."[71] Also, an apparatus for collecting and manipulating gases, or "airs," was a prerequisite to investigating them effectively. Indeed, Priestley himself helped improve the pneumatic trough.[72] That was not merely the product of some Enlightenment intellectual values. As Johnson notes:

> The question of air was "in the air" not for any vague, spirit-of-the-age reasons, nor because a solitary genius had experienced a heroic epiphany. Air had become an interesting problem in large part because a handful of technologies had shed light on that most invisible of substances.[73]

This account, however, discounts the critical role of laboratory instrumentation (relegating it to a footnote). Priestley also received plentiful encouragement from Benjamin Franklin and generous patronage from Lord Shelbourne and, later, friends and members of the Lunar Society in Birmingham. Johnson recently analyzed what allowed Priestley to make his discoveries. By casting a wider net, he was able to see a role for many factors:

> [Priestley's] brain and biography, his method, the technology of the day, the information networks, the scientific paradigms.[74]

In this view, Priestley's many successes were a convergence of many contextual factors, from his life experiences and the material culture to the ideas and social practices of the time. Priestley's work emerged from a convergence of interests, motivations, and opportunities, not a few personal dispositions. Scientific discovery is a complex achievement, not the exclusive province of a few specially endowed individuals.

There is no doubt that Priestley's discoveries were important and that he contributed significantly to the intellectual life of his time. For this reason, ensuring an accurate portrayal of the nature of science would seem very important indeed. Priestley's life also included some compelling stories. They can help to illustrate how scientists are human and pursue activities beyond science. The burning of Priestley's house and laboratory by an angry mob certainly seems tragic, whether Priestley was a scientist or not. But Priestley's political and religious views (his crusade against conventional Christianity, for example) did not logically flow from his science. Applying scientific analysis to *values*, whether political or religious, is a category mistake. As discussed above, Priestley's worldview interacted with his science in fascinating ways. A thorough analysis and discussion of his case can certainly contribute to deeper understanding of the nature of science. But one cannot, at the same time, responsibly eclipse other relevant factors prominent in this case: patronage, instrument technology, experimental tinkering, contingency or accident, error, and the contributions from multiple researchers. When framed ideologically, an account of Priestley's worldview becomes instead a misleading and counterproductive myth-conception.[75]

THE ARCHITECTURE OF SCIENTIFIC MYTHS

By now, one may well see the common narrative pattern shared by these various accounts of scientific discovery—whether about Mendel, Fleming, Harvey, or Priestley. While they all exhibit inaccuracies in historical detail, the infelicities are particular and relatively unimportant themselves. Rather, what matters is their consequences for understanding the nature of science. The effect is all due to the rhetorical framing. That is, the *narrative architecture* generates myth-conceptions in science. And one can find the syndrome of features just as easily in other popular stories—of Galileo and his observations with the telescope (Chapters 5 and 12), Bernard Kettlewell and the peppered moths (Chapter 7), Antoine Lavoisier and combustion (Chapter 11), or any number of accounts on the Internet, in television programs, or in textbook sidebars. The aim in this section is to characterize that narrative structure clearly enough for someone who does not study the history of science to diagnose and tease apart the mythic histories.

The pattern—and, thus, the myth-conceptions—are not arbitrary. They are rooted in the cognitive dimensions of storytelling. That is, one should interpret these stories contextually, as a form of human behavior. Particularly important, notes psychologist Thomas Gilovich, is the relationship between the storyteller and the listener/reader. That is, the social context itself can shape *how* the story is told. Stories fulfill at least two major social functions: information and entertainment. Each role contributes to scientific myth-conceptions (as elaborated below). Further, the intent to inform guides the storyteller in shaping the content:

> What the speaker construes to be the gist of the message is empha-sized or "sharpened," whereas details thought to be less essential are de-emphasized or "leveled."[76]

These inherent cognitive tendencies—*sharpening* and *leveling*—are the same techniques used in drawing caricatures. Dr. Bunsen Honeydew and Beaker—and all the scientist stereotypes they implicitly represent—are likewise products of storytelling craft. Sharpening and leveling help transform a plain story into a myth. Nevertheless, the resulting stories also typically bear signs of their transformation. A keen observer can use these features to help unravel the myth. What, then, in a rhetorical or literary context, is the telltale architecture of scientific myths?

1. Monumentality.

The first typical consequence of narrative *sharpening* is a focus on one central character: the scientific discoverer. The role of the scientific community is thereby *leveled*. So, too, is the causal role of any social context or critical convergence of separate events. Causation is easier to manage cognitively if reduced to just a single, simple cause. Omitted, too, are the roles of chance and contingency. Accordingly, the focal scientist has a lot to do to yield a discovery. In addition, outcomes tend to be sharpened as due to inherent human dispositions, rather than to a series of external events. No wonder, then, that scientists tend to exhibit extraordinary abilities. The time frame, too, can also be abridged, often into one momentous insight. The roles of alternative concepts, dead ends, and errors are leveled. Indeed, any trait that does not reflect favorably on the character would seem irrelevant to the ultimate achievement. Mistakes and character flaws thus seem superfluous and are discarded as well. Information critical to the nature of science is discarded, ironically for the sake of telling a memorable, "informative" story.

As the cases above vividly illustrate, scientists (as literary characters) accordingly become larger than life. They are heroic.[77] Their personality exudes virtue. As scientists, especially, they never err. As noted repeatedly, their personal achievements are inflated. Discoveries that, historically, were gradual and distributed among several individuals are concentrated into a single person, sometimes in a "eureka" moment. Historians have long criticized adulatory biographies that seem to willfully omit negative traits. They even have a special label: hagiography. But the cases here go beyond merely sanitizing history. They introduce historical error and generate superhuman feats or abilities. Indeed, they transform scientists into storybook characters. They thus share with their wholly fictional counterparts the monumentality of heroes, legends, and sometimes even gods.

For some, these mythic, superhuman characters function as role models that inspire students. Paradoxically, this seems to subvert the widespread goal of portraying science as a *human* endeavor.[78] Historian Stephen Brush once famously asked whether the history of science should be rated "X" in the classroom. He wondered if students exposed to real, non-mythic scientists might not want to become scientists themselves. More recently, Brush has observed that *some* scientists would seem to be good role models. Still, Brush noted, a primary focus on recruitment fails to consider the consequences for nonscientists of misrepresenting scientists or exaggerating their achievements.[79] In any event, educational goals now address science for *all* students. Moreover, they incorporate the history of science in roles other than as a vehicle for recruitment (Chapter 2). The importance of history— or *mythic* history—in providing role models needs further examination.

First, the assumption that role models must be universally positive has yet to be fully studied. Certainly, *some* anecdotal evidence suggests that *some* scientists have been inspired by such myths. But we do not know whether the mythic image alone was causally significant (such individuals may already have been oriented towards science for other reasons). We do not know whether great, but less hyperbolic, figures may serve the same role. Nor do we know whether the myths are significant across the entire population of scientists. Many, or even most, scientists may well be inspired by other factors. For example, other historical evidence (also anecdotal) indicates the importance of simple encouragement, even where no role model existed (for example, in the cases of women and minorities). One science student of mine who learned of Newton's conceptual blind spots and somewhat unsavory personality ironically felt encouraged. She could now see herself contributing meaningfully to science, without being perfect. Other students concurred. At the very least, educators need substantially more research on the role of mythic characters as role models before asserting their importance in recruitment.

Further, we do not know how many potential scientists are alienated by such myths. In what ways (that may now go unnoticed) do these mythic figures *discourage* students from pursuing science? They may be *negative* role models. That is, a student with a keen interest in science (but perhaps unproven ability) may infer that she or he cannot make a meaningful contribution. Experienced classroom teachers will surely recall students who opined that "science is only for geniuses" in justifying their lack of effort. Research results on attitudes towards science and recruitment should certainly be interpreted in terms of different types of history.[80] Further, the assumption that heroes must be perfect to be role models may well be questioned. Indeed, they are much more human and real—and more accessible to students—if educators acknowledge their flaws and limits as well as their triumphs.[81]

In addition, focusing exclusively on recruitment upstages questions of establishing role models in other contexts. That is, these mythic figures also generate expectations for how scientists should perform in society. They are role models of a very different kind. How is public sentiment shaped by expecting scientists to meet the standards of the myth? How much cynicism emerges when scientists fail to meet those unrealistic standards? More research would certainly be welcome. In any event, teachers ought not to regard mere engagement with larger-than-life characters as personally inspiring. The importance of mythic scientists as role models in education remains unclear.

One may, nonetheless, interpret their role within the mythic narratives themselves. Not just the characters are monumental. Everything is grand in scale. For example, Fleming did not just discover the antibacterial properties of some fungus. No, he *conquered some of mankind's most ancient scourges.* Harvey did not discover circulation merely. Rather, he rescued us from a mistake that had persisted *for fifteen hundred years.* Priestley did not just reveal how plants help restore the "goodness of air." Rather, he provided an account that had been *thwarted since Aristotle.* Mendel founded the whole field of modern genetics, as Priestley did for pneumatic chemistry. A single scientific study seems to have immense significance. Nor did Priestley just advocate liberal values. No, he helped champion a victory over *the most corrupt, venal, and incompetent government in all Europe.* Again, all this has strong implications for the relationship between the storyteller and the listener/reader. The monumentality impresses the listener. The storyteller, too, thus feels more valued—and more powerful. The mutually favorable emotions seem to validate the truth of the story—and foster similar storytelling in the future.

2. Idealization

Sharpening and leveling also occur in depicting the process of science and scientific methodology. Of course, any narrative of science is inherently limited or simplified. Every historical account must be selective in some respect (and hence can be viewed as biased). But the problems visible in the cases here are not due to truncation itself. Simplification does not inherently yield *myths*. It is the nature of the leveling. The second architectural feature of myths in science, then, is *idealization*. Qualifications are lost. *Uninformative* decontextualized generalizations emerge. For example, Mendel's initial fifteen traits, and the preliminary work they represent, are leveled. In the case of Priestley, the intellectual and social contexts are not just leveled, they are *razed*. The combined roles of Harvey's comparative anatomy, quantitative reasoning, and interventions are eclipsed. *All for the sake of telling a good story.* An idealized, decontextualized form of science, especially as witnessed in the conventionalized "scientific method," is sadly the predictable outcome.

One consequence of simplified narratives is a streamlined plot line. The multiple lineages of thought and action that characterize a *web* of history (Figure 3.2) are reduced to a single *timeline*. Stories, like those about penicillin, are reduced to the essentials linking the past to the present. Fleming's and Priestley's roles become sharpened. Duchesne's, Ingenhousz's, and others' become leveled. Although many will acknowledge that science involves trial and error, stories of science rarely include blind alleys (except for dramatic tension). The resulting sequence of events, leading item-by-item to the discovery and then to its meaning to us today, is all too easily interpreted as inevitable. Mendel established Mendelian genetics single handed. Hence, the narratives understate the uncertainty and misportray science-in-the-making (Chapter 2). Lessons about the nature of science suffer accordingly.

Historians have long criticized interpreting (or rewriting) history as leading to—and justifying—current states of affairs. It earns a special label, based on a classic example from Britain in the 1800s. There had been much controversy over voting rights, traditionally granted only to landowners. The Whig party, which promoted reform, finally gained power in the House of Commons. The House of Lords still opposed reform, however, and after one vote in 1831, Bristol saw the worst riots in British history. Troops had to quell the violence. Through shrewd negotiations, Earl Grey (of tea fame) persuaded the king to exert pressure on many lords, and the reform passed. With the Whig party in control, however, earlier history started being *rewritten*. Thomas Macaulay, in particular, wrote histories as though a constitutional government had been inevitable from the outset. Who was a hero changed. Macaulay's history was not a minor academic revision. The new history essentially justified the power of the Whig party by excluding prior disputes and erasing any contrary thoughts. The modified history was *political*. Historian Herbert Butterfield labeled this bias *Whig history*.[82] Whiggism is conventionally construed as a political bias. Here, instead, it seems to be due to *rhetorical* bias. The minimized plot lines and exclusion of alternatives can result merely from the tendency to sharpen and level in telling stories.

As the apparent informativeness of stories is sharpened, certain types of details will tend to be lost. The particulars of the discovery—details of time, place and culture, contingencies of personality, biographical background, coincident meetings, etc.—become secondary. Concreteness gives way to vague generality. Errors or failures, too, are eclipsed. Mendel's abandoned pea traits and the work on hawkweed are forgotten. Priestley's exploratory strategy is backgrounded. Harvey's microcosm analogy is discounted. All these details seem to drag down the plot. In a good story, the pace is exhilarating. As a result, stories tend to preserve the minimal elements needed just to *justify* the outcome narratively.

The stories of science thereby become idealized and universalized, in accord with their monumental status (above). The scientist's methods are cast as transcending their particular contexts.[83] They exemplify a Method of Science writ large. Here, the details or contingencies cannot be too important, lest they subvert the general lesson. Consequently, the idealized *narratives* promote the conventional *school philosophy* of a scientific method, in the sense of an algorithm guaranteed to find the truth. The well-known deficits in the standard account of "the scientific method" were nicely profiled by Henry Bauer.[84] He even labeled it a "myth." Yet Bauer and others have not considered how narrative leveling fosters this view. The attractiveness of the myths helps sustain perceptions of a universal scientific method. Storytelling tendencies themselves may be part of the problem.

The power of idealization in narrative is especially evident in the various errors it fosters. While the stories are all about history—events that happened—they sometimes drift into stories of what "should" have happened. Witness Harvey's imagined prediction of capillaries. Or Mendel's second law. Or Fleming's posture on penicillin's use for humans. Or Priestley understanding photosynthesis. Sometimes, the desire for a coherent story may overtake the historical facts (Chapters 4 and 5). Simplification may seem inevitable in education. Simple concepts, even if flawed by overgeneralization, seem essential. However, this leaves educators with the additional responsibility of articulating how such simple concepts can lie in more complex situations, like those encountered in social decision making (Chapters 7 and 8).

In failing to understand the work of scientists, people sometimes expect too much of science (Chapter 1). Not long ago, several individuals sued scientists for mistakes in their published papers.[85] What fostered such a stark frame of mind that expects science never to err? Did textbook accounts of famous discoveries help shape their thinking? For scientific literacy and addressing science in social decision making, the issues are quite complex. They often deal with scientific uncertainty and incomplete or conflicting results. Yet, in political contexts, partisans appeal to science (the science for *their* cause) as though it could resolve debate in simple black-and-white terms (Chapter 7). Their expectations seem to echo the idealized classroom histories. Again, there is opportunity for more research on how public sentiment about science is related to the implicit promise of scientific myths.

While method is unquestionably important in science, the mythic structure oversimplifies the process. It seems flawless. When coupled with the monumental scale, it overstates the guarantee on the claims it generates. With no place for error, except as pathology, the process appears more efficient than it actually is. No wonder that students fail to understand the nature of uncertainty or provisional acceptance of theories in science. One may turn to stories of science to teach NOS, but without attention to the architecture of scientific myths, such stories may ironically make the problem worse. In seeking to remedy misimpressions about science, applications of history of science should be a solution, not a source of the problem.

3. Affective Drama

A third element of the myth-conception architecture is literary techniques whose primary purpose is entertainment. One may enhance a story's power to engage (and persuade) with many *rhetorical devices*—that is, literary constructs or familiar plot patterns. They intensify images, heighten drama, and deepen the aesthetic response. They make a story more compelling, possibly even more persuasive or believable. This is also why histories are more potent cognitively than mere descriptions of science and its methods. Through their emotional effect, the stories also become more *memorable*. This may be one reason why the culture perpetuates the myths, even though they are false or misleading: simply because people remember them and enjoy telling them—and retelling them.[86]

This tendency of the storyteller was vividly rendered on stage by Peter Shaffer in the endearing character of Lettice Douffet.[87] Lettice is a guide at a historic house in Britain. Her scripted lecture of dry mundane facts is dreadfully dull, despite her best efforts. The tourists shuffle and yawn. One day, frustrated with the routine, Lettice goes off script, adding a few historical flourishes. The visitors are aroused from their indifference and now thank her sincerely for the tour. In successive scenes, we see the rendition become more adventurous. Soon Lettice is spinning fantastic tales and dramatizing the action of historic events, all richly fabricated. The tourists are enthralled. Both the guide and the visitors are happy, even as historical accuracy is richly and comically abused. Ultimately, Lettice's supervisor denounces her "farrago of rubbish." Lettice's defense? She appeals to her family motto, "Enlarge! Enliven! Enlighten!" Lettice's storytelling is a cautionary tale for the science teacher. Lettice neither sharpened nor leveled the story. Rather, she embellished it. And the known facts were obviously sacrificed for one reason: *entertainment*.

Science teachers may thus attend to the role of various rhetorical devices that artificially inflate the emotional drama of science stories. In addition to assessing scientific evidence, one must learn to analyze the literary craft: the style, the plot construction, relationships among characters, word choice, etc.

Among the familiar patterns in science, one might be alert to

- the thrill of the moment of discovery (the stereotypical light-bulb cartoon)
- struggle against (unwarranted) critics and opponents, with ultimate vindication
- the pleasant surprise of chance
- the reward of professional integrity (loyalty to evidence, resistance to social prejudice)
- shame (for example, challenging an ultimately correct idea)
- tragic irony of knowledge gained through some other human loss

In science stories, truth always triumphs. But typically only after dramatic conflict. As in melodrama, the hero conquers the villain. The pattern recurs frequently: scientist versus suppressor-of-the-truth, Harvey versus Galen, Priestley versus Aristotle, Darwin versus Lamarck, Lavoisier versus the phlogistonists, Wegener versus the stabilists, Galileo versus the Church. Students will never learn that scientists can legitimately disagree if dramatic emotion trumps history. The pervasive "aha!" phenomenon also deserves special note. Nothing drives a discovery plot quite like a well-framed "eureka!" For added effect, it may come in the wake of despair. Ultimately, the listener gets to share emotionally, even if vicariously, in the discoverer's glory. The monumental features of the heroes are also important here. They amaze and engage the reader. Hence the storyteller's ego is enhanced. Good teachers will understand what elements make stories entertaining—and manage them responsibly in their own storytelling.

4. Naturalized Explanation

The final element that makes these histories mythic involves their role in explaining and justifying the outcome. They are not just stories of science. They are *just-so* stories of science. Like Kipling's fables—"How the Leopard Got His Spots," "How the Camel Got His Hump," etc.—they explain a certain result, using the narrative itself as a form of justification.[88] Every history, every story, has an implicit lesson, or moral. Historical tales of science inherently model the scientific process by showing how a series of events *leads to* a certain conclusion. The narrative couples process and product. The story of a discovery *explains*, narratively, the methods of science and, hence, the authority of science.

The moral of the story is the capstone that fulfills the informative function of storytelling. The more profound the moral, the greater the perceived value to the listener—and the greater the status of the storyteller. One can appreciate even more deeply the significance of the monumentality, which amplifies the importance of the story and whatever moral it conveys.

Yet in the architecture of scientific myths, the narrative is also idealized. The consequence is an all-too-familiar just-so story of "How Science Finds the Truth":

- Science unfolds by a special method, independent of contingencies, context, or values.
- All experiments are well designed and forestall any mistakes.
- Interpreting evidence is unproblematic and yields yes-or-no answers.
- Achievement relies on privileged intellect. (Scientists are special, extraordinary people, whose authority is beyond question.)

Thus:

- Science leads surely and inevitably (and uniquely) to the truth, without uncertainty or error.

This simplistic account is already widely critiqued. Notably, it is the antithesis of the enduring goal to teach the tentativeness of science. It has also often been labeled a "myth." But this common label is based on construing a myth as a set of widespread and unquestioned false beliefs (see note 4). Here, the emphasis is on the *literary genre* and its distinctive features. One should note especially how the canonical features of the Method gain legitimacy through the mythic architecture. Because of the *idealization*, the simple method is deemed sufficient to account for all scientific achievement. Because of the *monumentality*, the resulting authority is monolithic. Because of the *rhetorically crafted affective drama,* the features, however misleading, become emotionally compelling.

Ultimately, then, the myths are forms of persuasion. They are value laden. The history is a way to legitimize the process of the discovery as told. That is, the historical tale is not just an illustration. It is a political action. The myth's lessons seem to derive status from the accepted value of the historical scientific achievement. This is especially clear in the cases of Harvey and Priestley: their purported reasoning and worldview are intended to reflect universal values about doing science. These narratives of science are arguments in disguise. Moreover, when cast in a grand scale, the importance of the story—and the storyteller, as well—appears amplified. In these cases, the authors essentially hijack the fame of great discoveries to promote a particular ideology of science and teaching (also see Chapter 5).

The context of amalgamated explanation, justification, and persuasion leads to the most important element of the mythic architecture: the near invisibility of its own mythic construction. In this respect, myths of science exhibit perhaps the central feature of classical mythology. Traditional myths typically explain natural phenomena—the movement of the Sun, the seasons, the rainbow, and so on—through the actions of human-like gods. While appearing to interpret nature, the myths also, conversely, inscribe human behavior in nature. Thus, the myths function to legitimize certain actions or forms of human conduct by *naturalizing* them.[89] They apparently describe an objective feature of the natural world. In a similar way, particular views of science benefit when inscribed into history. The author's view of the nature of science seems to emerge, unmediated, from the facts of history.

The explanation is apparently *naturalized* by history. The rhetorical work—all the many assumptions about what was important to include in or exclude from the story and why—is thereby hidden.

The reader focused on the story—with its drama, emotion, entertainment, and apparent informativeness—sees only the history. Unless sensitized to the architecture of the myths, one rarely sees the framework for composing and editing the story. The story format, by naturalizing the caricatured version of NOS, elegantly conceals its own rhetorical work. The architecture itself is invisible. Sometimes even to the storytellers themselves.

Myths of science are myths, not just idle stories, because their architectures—through rhetorically crafted drama—tend to naturalize the monumental, idealized, sharpened, and leveled view of science. It is the *myth*-conceptions, not merely the historical and philosophical *mis*conceptions that follow from them, that deserve our critical attention.

STRATEGIES FOR TEACHING

In opening this chapter, I noted that educators need to change *how* history is told and *how* scientists are depicted. Mythic classroom histories, like School Science, distort the nature of science and mislead students, even as they purport to show how science works. They become *pseudohistory* of science (discussed more fully in Chapter 5). Like pseudoscience, they foster false beliefs about science—here, about the sources and limits of scientific authority. One might imagine that the only solution would be to purge science textbooks—and the culture at large—of all historical error. However, such a utopian goal is hardly necessary. Nor does every teacher need expert credentials in history. Rather, one can leverage substantive change with just a little knowledge about how history is told. Educators might begin with a few simple strategies, profiled below.

First, teachers need to be careful in gathering their historical information. Of course, the credibility of sources is as important in history as it is in science. One should check the credentials of authors and/or scan the types of sources the author uses. Professional historians draw from the original sources and work in an intellectual culture attuned to the dangers of hagiography and Whiggism: their work can generally be trusted. Science journalists are typically well informed through research (no surprise), but they are not always as reliable as one would wish: their business often calls on them to entertain, and they are accustomed to reducing the nature of science to sound bites. Their work is variable. Scientists are well informed about how to do science themselves, but they are not always able to articulate it fully. They tend especially to downplay the cultural contexts of science. Although they deeply appreciate the historic scientific achievements, they also tend to reconstruct them in the abstracted, idealized fashion of myths. They also tend to celebrate their own field and romanticize its heroes.[90] In general, one should look for *textured* histories: rich in detail and balanced in critical perspective. Informative histories will illuminate the process of science by showing its possibilities, rather than shoehorn it into a particular, narrow philosophy of science (shoehorning is discussed more fully in Chapter 5).

Credibility of sources is only a rough guide. Ideally, teachers (or students) will learn to recognize the rhetoric in myths for what it is. The cases above may help vividly illustrate the dangers for the unwary. They also indicate the diagnostic features of suspect or marginal history, even where one does not know the historical background itself. A few brief maxims can help evaluate any history or apply it in the classroom. For example: Suspect simplicity. Beware vignettes. Embrace complexity and controversy. Discard romanticized images. Do not inflate genius. Mix celebration with critique. Scrutinize retrospective science-made. Revive science-in-the-making. Explain error without excusing it. And above all: Respect historical context. For those who like mnemonic devices, one may express the "SOURCE" of the problem and the "SOURCE" of the solution as summarized in Figure 3.3. Thus, while we may never eradicate mythic narratives about science, we might nevertheless be able to neutralize them in the science classroom. Analytical tools empower teachers to recognize myths and limit their effect, even without vast historical expertise.

Once one is aware of the tendencies of mythic storytelling, a supplemental strategy is to actively counterbalance them. For example, one might highlight the complexity of a case study (rather than level it) or underscore the role of multiple scientists (rather than sharpen a single hero). Alternatively, one might delve into a particular case and trace the role of particulars and context, resisting the urge to idealize methods into a preordained schema. There is ample drama and excitement in such approaches, although not always in the familiar patterns. The teacher might also try to collect and teach in part from a repertoire of original materials: original publications (including alternative theories), quotes from letters or diaries, interviews, personal recollections or autobiographies, and the like. Such resources are increasingly available on the Internet, sometimes on science education websites. Revisiting the historical documents also keeps the original history fresh and avoids the successive retelling that leads to layers of sharpening

Figure 3.3. Comparison of mythic narrative and authentic NOS history.

Mythic Narrative	Authentic NOS History
Science-made	**S**cience-in-the-making
Overinflated genius	**O**pportunities
Unqualified Universality	**U**ncertainties
Retrospect, Romanticism	**R**espect for historical context
Caricatures	**C**ontingency, Complexity, Controversy
Expected results & Excuses	**E**rror Explained

and leveling—as Lettice did comically with her description of the English manor house. In lieu of prepared stories (with designated NOS lessons), one might strive for direct engagement with episodes of science, equipped with a broad spectrum of NOS probes or questions (Figure 1.3; cases in Part II).

Another major strategy to dispel the myth-conceptions is to cultivate a repertoire of alternative stories of science that violate the norms of the mythic structure and, like Warhol's portraits, begin to expose the rhetorical conventions at work. That is, teachers may benefit from *discrepant NOS events*. By contradicting expectations or conventions, they may, like any anomaly, provoke rethinking and lead to deeper understanding. One core of idealized mythic narrative is a simple formula for science:

right methods ⇒ right conclusion
wrong methods ⇒ wrong conclusion

The view of scientific methods as foolproof is a caricature, of course. Akin to the characters of Dr. Bunsen Honeydew and Beaker. Science proceeds by trial and error, one often hears. Everyone seems willing to acknowledge that science is fallible. This means, of course, acknowledging that scientists can err. It means acknowledging honestly that *good* scientists can err. Even Nobel Prize winners. We cannot gently excuse the errors or explain them away, as in the myths. Nor can we cast all mistakes in science as fraud or as aberrant pathology. It is not just that scientists lapse from some ideal method. The "right" methods do not always yield right ideas. Sometimes:

right methods ⇒ wrong conclusion

or:

haphazard methods and chance ⇒ right conclusion

Hence, even a single case of fallibility, well articulated, may serve as a corrective to the mythic caricature. Ideally, educators should introduce *some* histories that chronicle how evidence at one time led reasonably to conclusions that were only later regarded as incorrect (see case examples in Chapters 10, 11, and 12). This is how teachers can *explain* the limits of science without vague handwaving about skepticism or tentativeness. They must help undo the cultural myth-conception and show how doing science can, on occasions, lead to error. Ideally, educators will also show what allowed scientists later to recognize a mistake and remedy it. Narratives of error and recovery from error can convey *both* what justifies *and* what limits scientific conclusions.

Myths of science are unquestionably seductive. They tempt the teacher eager to engage students. They entertain. On the surface, they seem to inform. This is why the mythic forms of the history of science already haunt the classroom and our culture at large. But they are misleading. They do not promote understanding of the process of science or nature of science. Thus, in contrast to some current science education reforms (Chapter 1), having *more* such histories solves nothing.

Instead, we need *different* histories. Teachers need to promote less mythic narrative frameworks. While less monumental, they may still be equally dramatic and humanly inspiring. While less idealized, they may still serve as exemplars (in concert with others) for understanding the process of science. Other rhetorical devices may evoke responses: the excitement of opportunities, the suspense of persistent uncertainty, the reward of hard work, the surprising significance of "trivial" events, the tension of even-handed debates, the tragic consequences of human limitations, and the aesthetic of resolving error. The new stories will celebrate insight achieved through perseverance, creative interpretation of evidence, and shrewd insights enabled by depth of experience, as well as through happenstance and blind luck. They will reflect how scientific conclusions are assembled, how they are challenged, how error can occur, and how knowledge is sometimes revised. Alternative narratives of science need not reduce the greatness of scientific achievement. But, ideally, they will portray equally both the foundations and the limits of scientific authority, and foster deep understanding of the nature of science. Effective histories of science will avoid engendering myth-conceptions.

4 | How *Not* to Teach History in Science

Characteristically, textbooks of science contain just a bit of history, either in an introductory chapter or, more often, in scattered references to the great heroes of an earlier age. From such references both students and professionals come to feel like participants in a long-standing historical tradition. Yet the textbook-derived tradition in which scientists come to sense their participation is one that, in fact, never existed.

—Thomas Kuhn, *The Structure of Scientific Revolutions*

Using history, abusing historical context • historical simulations and rational reconstructions • recapitulating history, upending history • virtues of "wrong" ideas • summarizing how not to teach history

History is a valuable tool for teaching the nature of science (Chapter 2). Yet teachers must learn to use history, like any tool, appropriately. For example, some renderings of history can generate unintended myth-conceptions, both about scientists and about the foundations and limits of scientific knowledge (Chapter 3). Others embody undue philosophical bias (Chapter 5). This chapter, aiming to further clarify effective NOS lessons, discusses another potential pitfall when history is applied unreflectively. Ultimately, one must *respect the historical context of science, adopting a perspective of science-in-the-making*, if one wants to convey the nature of science authentically.

The general strategy for teaching NOS through history is simple enough. Lead students through cases of scientific investigation and discovery (Chapters 10 and 14). Through a narrative presentation, they may follow the historical scientist's reasoning and investigative decisions, all in a human context. Ideally, they might also experience vicariously the thrill of discovery. Through explicit reflection on the episode, students begin to integrate NOS understanding into their native thinking.

Of course, from psychological research, educators have learned the significance of students' initial perspectives and motivations. Active learning, whereby students are engaged in developing the concepts themselves, is far more effective. Thus, one can adapt historical cases to inquiry learning (Chapters 2 and 14). The history offers a motivational context and a possible trajectory (whether followed closely or loosely). Students begin to participate in the narrative, which may be interrupted frequently to allow students to address questions, solve problems, and discuss the process. Of course, the teacher will want to chart a path between cookbook exercises (no different than most conventional labs, even if they are inscribed in history) and aimless search (the overly optimistic goal once espoused as "discovery learning"). Local judgment is required, then, to find the appropriate level of challenge to offer students and the degree of scaffolding they need. Regardless, explicit reflection is important here, too, in consolidating NOS lessons.

So: the general framework is simple. Yet whether one presents a narrative or guides participation in it, one must adapt history in assembling the central narrative. How? One approach focuses exclusively on the NOS lessons and selectively reconstructs the history accordingly. Another begins, in a sense, with the history and highlights and articulates the latent lessons. The latter tries to re-present the original history as closely as possible, sensitive to the historical context. The former edits the history into a more "rational" form, reflecting how science *should* function. One strategy is fundamentally descriptive, the other ostensibly normative. Teachers often aim to facilitate learning by reducing the process of science to its basics— and so the idealized reconstruction may seem the most appropriate at first. Yet, ironically, this form of framing history ultimately threatens the very NOS lessons most important to scientific literacy (Chapter 1). Educators need to think carefully about how history yields its NOS lessons and, thus, what form of history to adopt.

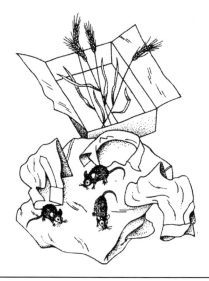

FIGURE 4.1. Textbook depiction of Helmont's spontaneous-generation experiment (Nason and Goldstein, 1969, p. 15).

USING HISTORY, ABUSING HISTORICAL CONTEXT

One potential problem in applying history in science education is strikingly demonstrated in an otherwise unremarkable high school biology textbook from 1965.[1] The authors give two accounts of work by Joan Baptista van Helmont, "an outstanding Belgian scientist, physician, and chemist" from the early seventeenth century. One lauds his work. The other disparages it. Yet both convey the same assumption about proper scientific method.

In the first example, the authors describe an "experiment" devised by van Helmont on the spontaneous generation of mice (Figure 4.1). Van Helmont "wrapped some grains of wheat in a sweat-soaked shirt." He returned several weeks later to find the kernels of wheat transformed into baby mice. The lesson, of course, is clear, even to the naive science student: van Helmont was another ignorant scientist of the past who

> did not consider whether his experiment had any loopholes which might cause him to draw a false conclusion from the results. Perhaps if you think about it, you can realize what was wrong with his experiment. Perhaps you can suggest an improved setup which would give a more valid conclusion. Perhaps you can even explain why van Helmont's experiment gave him the results that were just described.

The simple experiments on spontaneous generation by Francesco Redi (comparing covered and uncovered vessels of rotting meat) handily remedied the error. Now we know better. Science progresses from credulousness to certainty.

In the second example, the reader learns about van Helmont's renowned willow-tree experiment, sometimes hailed as the origin of experimental plant physiology (Figure 4.2). Van Helmont first weighed a willow branch along with a 200-pound potful of soil. Then he planted the branch, watered it, and five years later weighed the two again. The tree had grown an impressive 164 pounds, while virtually all the soil remained. Students are to see how elegantly van Helmont showed how the matter of a tree does not come from the soil. Here, van Helmont is the hero, not the fool. From his example, we might draw the deeper lesson: construct a test, quantify and measure. And be prepared in some cases to be patient for the data. Modern students sometimes repeat this lesson for themselves, albeit on a smaller scale, using radish plants, whose weight change can be observed in weeks rather than years. Moral? Identify the correct procedure, and the right answer inevitably follows. That's *how* science progresses.[2]

Ironically—even paradoxically, perhaps—van Helmont is both praised and ridiculed as a scientist in the same text. He is both hero and fool. It seems hard to reconcile the two images of the same individual. But the focus in

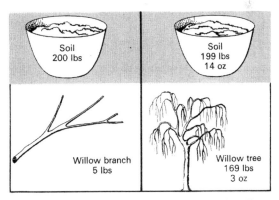

FIGURE 4.2. Textbook depiction of Helmont's willow-tree experiment (Nason and Goldstein, 1969, p. 276).

each case is not van Helmont, but whether he was "right" or "wrong." From that knowledge, one can apparently reason backward to judge the method. In one case, the method was surely misguided. In the other, it was necessarily exemplary. The text embodies the NOS notion that scientific method is algorithmic and leads inevitably to the truth, regardless of context. Therefore, the right answer will always reflect the right method.

The dichotomy of hero and fool in science was likewise noted by Stephen Jay Gould in accounts of a famous debate in developmental biology: the preformationists versus the epigeneticists. Preformationists claimed that the body grew from a tiny homunculus in the fertilized egg. Epigeneticists, on the other hand, claimed that structures and organs formed only gradually from undifferentiated tissue. Today, the improbable historical image of an adult human curled up inside the head of a sperm (with untold successive generations inside that) seems almost laughable: the preformationists were fools. By contrast, the epigeneticists were the heroes: they defended principles so plainly observable with any microscope. Yet Gould reconsidered the debate in historical context. The preformationists, he noted, themselves defended a fundamental principle: that complexity cannot arise out of homogeneity. The egg, therefore, must have some original structure (although not necessarily a fully formed human, as is so often caricatured). As a scientific posture, we might want to endorse that, even today. By contrast, the epigeneticists seemed to take observations at face value, without fully appreciating the theoretical consequences. Today we know about genes and how they work. But in the early 1800s, one would have had to appeal to rather mystical or vitalistic forces to explain the origin of form: straining the empirical boundaries of science. Reviving historical context supports a different image of science than prejudiced retrospect.[3]

So the case of van Helmont can be very instructive—not about the methods of science so much as how the nature of science is rendered through historical cases. Taking a cue from Gould, one can profitably delve into the historical context of van Helmont's experiments. Indeed, one finds that the two textbook stories abuse the historical context and so misrepresent the nature of science.[4]

The wheat-to-mouse experiment is especially striking in historical context. Van Helmont was fascinated with the problem of transformation, from the chemist's basic procedures to organismal development and metabolism. For example, he studied digestion, the remarkable transformation of food into flesh. Like others in the early 1600s, he saw similarities with the fermentation of grapes, as well as to bread rising and the formation of honey from nectar. Van Helmont viewed form as emerging from a specific agent: what he called at times a ferment, a seed, a leaven, or an archeus (miniature workman). Today, one might think vaguely in terms of enzymes. For van Helmont, such causal agents were not wholly spiritual, but material. They could be released and captured when things burn: what he called a *gas*. Disease in van

Helmont's view was similarly due to external agents that invaded specific organs and altered them. As a physician, he sought to remove the cause of the illness rather than merely treat its symptoms—a revolutionary view at the time. For van Helmont, then (and perhaps for us, too), the transformation of matter from wheat to mouse body was real and raised questions about the process. An effort to draw an elementary lesson about van Helmont *from our current scientific understanding alone* seems misguided.[5]

Consider also the willow experiment in historical context. Van Helmont concluded that "all Vegetables do immediately, and materially proceed out of the Element of Water only." His rather elaborate and fascinating world view matters here. For van Helmont, there was only one primal element, water. All other matter was derived through ferments (just described above). In this, he challenged prevailing Aristotelian doctrines that there were four elements: water, earth, fire, and air. The willow experiment was thus essentially designed to show that a tree was not earth mixed with some fire. Rather, it was transformed water. (At the same time, the experiment also showed the transformative ferments at work.) Thus, what many now consider van Helmont's great experiment was coupled to an error. Still, in its intended role, the demonstration was dramatically successful—especially in provoking others to think about the problem of elements and, in some cases, to repeat the experiment.

Some educational histories add that the "real" answer was readily available to van Helmont, implying that he must have been blind not to notice it. "It was ironic that van Helmont, the discoverer of gas, never guessed that something in the air might also have contributed to the growth of his willow tree."[6] Indeed, we are told that van Helmont "discovered carbon dioxide." And "had he not been so dogmatic," David Hershey alleges, "Helmont might have used his data to conclude that fresh plant matter consisted largely of water but that dry plant matter consisted mainly of carbon dioxide gas and a small amount of soil minerals."[7] The explicit cautionary lesson for students: do not ignore your own data when making conclusions. Here, again, the myopic fool.

Van Helmont's prospects look very different in historical context, however. In this case, delving into history becomes quite an adventure. First, while van Helmont did introduce the *term* 'gas', his *meaning* (as noted above) was hardly ours. For him, a gas was an essence specific to each substance, which would be given off when it was heated, burned, or dissolved in acid. Van Helmont compared gas to *blas* (a complementary word that, by contrast, has drifted into obscurity): an astral, non-material power of movement (akin to the Hippocratic *enhormon*). Gas, by contrast, was material. Yet although material, a gas was imbued with *spirit*. It was also the vital essence of the blood. It was converted from water by ferments. But a gas was not water vapor. Nor was it air. Nor a component of air. Nor one of three states of matter. Accordingly, neither Robert Boyle nor Joseph Priestley, who each worked with different types of "airs" (our gases), adopted van Helmont's term.[8] Van Helmont's 'gas' was not our 'gas'. A story that erases the historical context distorts the NOS lesson.

So, too, for carbon dioxide. Van Helmont did collect a substance that *we* recognize (in our terms) as carbon dioxide. But oxygen was not identified as an element until the 1770s, so van Helmont could not have conceived it as *carbon dioxide*. Rather, it was the spiritual essence (*gas*) of coal or wood, what van Helmont called *spiritus sylvestris*.[9] Air was viewed as an element for many years to come, not a mixture—and certainly not a mixture of Helmontian *gases*. Even a century later, chemist Joseph Black would refer to (today's) carbon dioxide as *fixed air*, not a distinct compound containing carbon and oxygen as elements. Conceptually, van Helmont's *spiritus sylvestris* is not our carbon dioxide.

Van Helmont's experiment looks very different to us because of our modern perspective and collateral knowledge—our own historical context. Unraveling these differences in meaning is part of the exciting, revelatory work of history. Van Helmont lived in a very different world conceptually: one where the notion of plants absorbing something from the air obviously did not seem to fit. Historical context makes a difference to interpreting NOS.[10]

Ultimately, if the lesson is to be about the nature of science, one needs to understand the process of actual science, including the role of historical context.

All too often, it seems, we delve into history to judge rather than to interpret science of the past. We seek values, not facts. One may see it as an oblique form of confirmation bias. Confirmation bias is perhaps the most fundamental cognitive error documented by psychologists (also known as the congruence heuristic, positivity bias, and belief persistence).[11] That is, our minds select or highlight confirming examples, while discounting or peripheralizing counterexamples. Brains tend to reinforce pre-existing beliefs. Here, we view ideas of the past as either matching or contradicting our current views. In celebrating those persons who concur with us—namely, the heroes—we ultimately seek to validate *our own* ideas. We readily assume that the knowledge (even if only vaguely expressed) was fully justified even then. Where the evidence might have seemed insufficient, we fill in the gap with the hero's insight, intuition, or genius. The methods, too, must have been sound (else, impossibly, we might be wrong now). By contrast, scientists in the past who advocated alternative ideas, we tend now to cast as fools, thereby disenfranchising their evidence or reasoning. Ultimately, history can become an exercise in evaluation and self-justification, not understanding. The sorting of heroes and fools becomes another form of *naturalizing* one's beliefs in history (Chapter 3). And it can evoke deep emotions.[12]

Accordingly, the heroes of science and their methods tend to become pristine and error free. Thus, if Isaac Newton or Robert Boyle practiced alchemy,[13] or Charles Darwin espoused non-Mendelian heredity or exhibited racist assumptions, or Galileo attributed the tides to the Earth's motions, or James Hutton reasoned theologically about the age of the planet, we kindly excuse our heroes. We hide the flaws or brush them aside because they were ultimately right about the *important* things. We willingly explain away or excuse the errors, thereby maintaining the integrity of scientists as infallible. Faulty conclusions, even if well reasoned, are not allowed into science. (And in this way, teachers can inadvertently perpetuate notions that science is not uncertain or tentative at all.)

A historical case study can thus be counterproductive if hindsight eclipses historical context. Consider another proposed exercise about van Helmont that asks students to interpret his willow experiment *on the basis of what we know today.* Students are expected to learn by "correcting" the history and making it come out "right." For example, biology educator David Hershey claims that students should understand how van Helmont performed the "wrong" experiment. To assess his hypothesis about the role of water, he *should* have grown the willow hydroponically in mineral-free water. Moreover, he *should* have used many replicates. He *should* have added a control: a soil-filled pot without a willow. He *should* have considered experimental (measurement) error. He *should* have explained why he chose a willow and why he sunk the pot in the ground. He *should* have cited Nicholas of Cusa, who earlier had suggested just such an experiment. He *should* have given more details about his method. He *should* have been more skeptical and critical in his conclusions. He *should* have recognized the role of carbon dioxide.[14] Alas, most of these standards are applied out of historical context, leading to an idealistic NOS lesson that van Helmont simply *should not have erred.*[15] All scientists work with limited knowledge—and what is important for students to understand is how they deal with those limits and extend their own knowledge. While van Helmont "might" have done many things, he did not. It behooves us to understand, fairly, why. Only then can we begin to understand his very real achievement and how science works *in human and cultural contexts.* Here, actual scientific practice is eclipsed by a series of anachronistic what-ifs. While the individual *methodological* lessons can be valuable, the overarching *NOS* lesson here—that science is error free— is wildly unrealistic and inappropriate. Good scientists can make mistakes. Ultimately, the history is irrelevant to Hershey's intended lessons. It has been hijacked to naturalize an NOS image of science as methodologically pure and yielding certainty (Chapter 3). For NOS lessons to be authentic, however, educators need to listen to history, not judge it.

In the view that Hershey merely epitomizes, one assesses historical scientific work on the basis of some current or idealized model of science. In a view that is sensitive to historical context and science-in-the-making— and thus to science as scientists practice it—one tries to understand van Helmont's pioneering new methods *in context.*[16] Still, some of his conclusions could nonetheless later be construed as "wrong." A well-designed or valuable experiment need not always yield an answer that is completely "right." Indeed, how much more would a student learn about the nature of reliability and fallibility in science by seeing how someone can sometimes get "wrong" answers using the "right" methods?

HISTORICAL SIMULATIONS AND RATIONAL RECONSTRUCTIONS

Mystery novels and suspense films lose much of their thrill when spoilers intervene. Uncertain prospects differ substantially from the confident clarity of retrospect, once all the clues are assembled and the ambiguities resolved. So, too, in science and history of science. The difference in perspective between science-in-the-making and ready-made science is profound (Chapter 1). And it is fundamental in NOS education. In this section, I profile how the distinction is critical to historical case studies.

For clarity, I focus on one exercise as a sample: on the "Causes of Earthquakes."[17] It draws on history and uses inquiry. Yet in configuring the history for the lesson, the historical context was lost. That process critically affected the NOS lessons. Because the effect was surely unintended, the analysis can highlight some concepts essential for the NOS educator already attuned to history.

The exercise aims to help college students understand, first, how evidence and reasoning can support different explanations at the same time. Further, students develop skills in comparing and evaluating theories. To begin, students encounter five different explanations for earthquakes. Each has a historical source, and original documents are introduced where possible. Next, students receive some original seismic data, which they transfer onto world maps and analyze. (Note that this parallels a now classic 1959 mapping by Bruce Heezen and Marie Tharp, which gave more precision to the earlier work of Beno Gutenberg and Charles Richter.) Finally, they construct a decision table, assessing the evidence for each theory according to several key questions, paralleling the challenge of geologists and geophysicists in the early 1960s.

Now, the value in this exercise should be obvious. Students come to appreciate that even with the same evidence, different explanations of the same phenomenon are possible. They see how each theory is rooted in certain assumptions. The perceptive student will recognize how theoretical evidence is built through field work and the creative act of interpreting data. And students gain "experiences with the higher cognitive reasoning skills associated with the evaluation and interpretation of knowledge claims."[18] These exemplify paradigmatic NOS benefits from a historical approach. One should not lose sight of them.

At first, the earthquake exercise seems like a fragment of history for the classroom, allowing students to work in the footsteps of famous historical scientists. Yet it differs from real history in a few ostensibly minor, but ultimately quite profound, respects. First, and most important, students probably already know the "right" answer. That is, in this case, the activity is biased by knowledge about plate tectonics, which has penetrated our culture. When students know the answer, they typically shape their reasons accordingly. They learn how to conform reasoning to a conclusion, rather than how to reason towards an unknown conclusion. That is, they rationalize. Ironically, conventional schooling trains them very well in this. Of course, scientists don't know the answer in advance. They must persuade themselves on the basis of the evidence at hand, not cherry-pick evidence to support the known outcome. The pivotal NOS lesson about historical uncertainty and science-in-the-making is not just absent, it is undermined. Students can gain the impression that judging evidence is relatively easy for scientists because they are just incredibly smart or, like the mythic geniuses, intuit the answers. The irony is that a primary objective of the exercise was to underscore how evidence helps resolve explanatory uncertainty. In this way, the exercise does not simulate science. It *reconstructs* it.

The exercise is further reconstructed—or artificially contrived—in its pre-established data set (Part 2). The evidence simply appears, *deus ex machina*, from the teacher. Students never have occasion to appreciate why these data were selected or what motivated collecting them. (Ostensibly, the data were gathered to answer the very question the students are addressing. But this was not the case historically and, hence, misleads about the nature of science.) How easy it would be to challenge students to imagine what information would be relevant to assess their alternative explanations. Also, the students receive only a small part of the evidence that was (or might be) relevant. Moreover, the data are largely significant only from the current theoretical perspective. Students may thus tend to regard reality as presenting itself preformed. They may perhaps be habituated to passively accepting whatever evidence is presented to them. A more contextual approach, by contrast, would underscore the role of creative thinking in experimental design and data collection, a major NOS concept.

Finally, students are asked to choose between the theories according to prestructured standards (Part 3). The exercise organizes each idea about earthquakes into comparable 'Giere frameworks'.[19] The hypotheses, background assumptions, initial conditions, and premises are all clearly stated. It conveys an NOS impression of strict logic governing science. Yet scientists rarely work with such formal reasoning. Their thinking—effective, as history demonstrates—is more flexible and adaptive than that. So instead one could work from students' native approaches, with the *teacher* using the formal structure to help inform, articulate, or deepen their thinking. In addition, no one in history sat down to consider the five theories all at once. Each explanation was a response to extant data and concurrent theories in its own time. The exercise thus models NOS again as some idealized reasoning process. The contrast NOS lesson is to understand *actual* scientific judgment in context.

In all these ways, the loss of historical context is significant. A *rational reconstruction* is not a historical simulation.

The problem with teaching through rational reconstructions is that the history—and thus the process of science—is backwards. One works *from* the final answer to trace the relevant evidence from history. The path forward is determined by a single endpoint. That differs from the historically situated view, with its uncertain paths and possibly diverging branches. Historical context is the educator's resource for modeling the blind forward-moving process of science-in-the-making. One cannot take for granted the generation of hypotheses, the search for relevant information, the design and critique of experiments, the elaboration of alternative explanations, the struggle with experimental anomalies, and conflicts between multiple interpretations. These are the elements of scientific discovery. This is where student simulation, rather than reconstruction, informs NOS understanding. There is more to how science works than just justifying the final outcome— or assuming that it is correct.

Historical simulations and rational reconstructions each seem to be historical in nature. But the rational reconstruction is not concerned with the process of history, only its product. The names, dates, and crucial experiments that are assigned to specific ideas are merely incidental to its purpose. The core structure is a constellation of arguments, not an unfolding process of true *discovery*. Rational reconstructions (like Hershey's "corrections" of van Helmont, above) are ultimately ahistorical in nature. They foster hero–fool dichotomies and associated unhealthy NOS views. Where the aim is to convey process of science, by contrast, teachers should highlight the historical context of a case. Ideally, students should have the sense that in peering through the eyes of working scientists, their decisions have several degrees of freedom, including the healthy opportunity to fail.

RECAPITULATING HISTORY, UPENDING HISTORY

When well framed, historical case studies can provide effective models for learning the process of science. It may seem, then, that the whole history of science reflects a pattern or outline for modern classroom learning: history as a syllabus. Advocates of this view, Jean Piaget famously among them, analogize from Ernst Haeckel's biogenetic law that "ontogeny recapitulates phylogeny," suggesting that "learning recapitulates history." Call it, perhaps, the model of *cognitive recapitulation*. Although the parallel is intuitively attractive, it is also misleading. It, too, erases historical context.

Cognitive learning theory now emphasizes that an individual builds cognitive structures on earlier ones.[20] Where history shows how complex concepts emerged from simpler ones, therefore, the cognitive recapitulation model can surely be an effective guide. But this can apply only within a family of related concepts. It can only suggest a format for addressing a series of atomic models, say, or a set of variations on basic Mendelian or electromagnetic themes: one 'constructs' the complex on (or from) the basic.

The reasoning behind the cognitive-recapitulation analogy, however, does not allow one to apply it more broadly. The so-called Chemical Revolution, for example, seems to offer a model opportunity to teach about 'revolutionary' conceptual change. Historically, Lavoisier's discovery of oxygen and its role in combustion dramatically replaced earlier explanations using phlogiston. Yet in one set of high school chemistry classes, I assisted in teaching about phlogiston after students had already learned about oxygen and the modern system of elements (Chapter 11). The unit focused on metals and the (oxidation and reduction) reactions that transform them. By observing phenomena highlighted by late phlogistonists, students found that the concepts of phlogiston and oxygen each addressed important, though different, aspects of combustion. One accounted for the material changes, while the other accounted for energetic changes (in today's terms). By approaching the topic "backwards" historically, they did not encounter a revolution. Instead, they found that the two concepts were complementary. As illustrated by this case, history is not an infallible standard for conceptual development.

The important element of history is its context. Thus, sometimes it may even be appropriate to disturb or modify actual history to preserve the sense of its context. In reaching his conclusions about the circulation of the blood, for example, William Harvey dissected many animals. Yet he never examined a squid, so far as is known. But the organization of a squid's blood vessels is excellent for such an investigation. Squids have three hearts: two gill hearts (pumping just to the gills) and one body heart. The separation of the hearts and their arrangement make it easier to conceive of the human heart as two hearts, joined in a circuit. Thus, examining a squid can be a good example for reasoning in the context of Harvey's other observations, though Harvey himself never did so.[21]

In a similar way, the lesson on metals and oxidation–reduction reactions (above) drew anachronistically on the pyrotechnic thermite reaction. In this reaction, iron oxide reacts with aluminum (once ignited) to yield molten iron and aluminum oxide—and an impressive display of heat and light. It was (and still is) used in welding railway tracks. Although the reaction was discovered more than a century after phlogistic concepts had been abandoned, we confidently interpreted it from a phlogistonist's historical perspective. The phlogiston in the aluminum was transferred to iron. The concurrent heat and light were the same as the release of phlogiston in combustion. Here, the simple concept of phlogiston helped underscore the notion that the phlogiston leaving one metal was paired with its appearance somewhere else. For us today, oxidation and reduction reactions are coupled. Historical authenticity was subservient, here, to the historical context of the phlogistic way of thinking.

On another occasion, I was challenged to teach about the Copernican Revolution to some eighth-graders. But how can one convey the magnitude of this reconceptualization to modern students so fully indoctrinated into thinking (or knowing?) that the Earth travels around the Sun? It is hard for them to imagine otherwise—which was exactly the core of the lesson. My strategy—designed to revive a sense of the historical controversy—was literally to upend history. I offered evidence, adapted from early-seventeenth-century arguments, that the Earth did not move, as the students insisted. For example, a ball dropped from a high tower falls straight down. The Earth does not move away underneath it as it falls. The Sun clearly rises in the morning and sets in the evening, moving across the sky during the day. We do not feel ourselves moving. Should not a good scientist rely on observational evidence? The students were adept at *describing* the current interpretation as an alternative, but they had trouble justifying it. Their frustration and annoyance that someone would question common-sense knowledge, I could point out, paralleled feelings about Copernicus's views in the 1600s. The lesson was about NOS, not history itself. The method thus aimed to appreciate the history, not to repeat it exactly. (For a variation set in historical context, see the historical simulation of Galileo's trial in Chapter 12).

These examples might show that applying the history of science to teaching science and NOS is not the same as teaching the history of science per se. Historical fidelity is not necessarily paramount. Still, faithfulness to historical context *is*, lest one lose the intended NOS lesson.

There are other ways in which the cognitive-recapitulation model is limited.[22] Fundamentally, the analogy is weak. It assumes that the growth of scientific knowledge is linear—and that the sequence of history is one sequence, not a collection of multiple overlapping and interacting sequences. The context of contemporaneous theories in other fields, though, is an important element of science.[23] The context of geophysics, for example, greatly affected early judgments about continental drift, which had emerged from the contexts of biogeography and structural geomorphology. Likewise, William Thomson's argument about the impossibility of biological evolution, based on thermodynamics, was effective (for some) until the discovery of radioactivity at the turn of the twentieth century. Extracting historical arguments from their historical context obscures these disciplinary and theoretical interactions and, thus, distorts understanding of how science works. Contexts are local, so the focus in the classroom might likewise be limited in historical case studies. When the historical context is lost, so, too, is much of the nature of science.

THE VIRTUES OF TEACHING "WRONG" IDEAS FROM HISTORY

For some educators, history not only suggests what to teach, it also tells us what not to teach. Teachers should avoid scientific errors at all costs. The concepts are, after all, "wrong": how can one learn from them? Of course, most mistakes were once considered "right": why?

How could knowledge have been transformed from fact into error? History helps us understand, first, the evidential contexts in which wrong ideas were once considered right and, second, how (and why) such contexts changed. History thus shows how the process of science can sometimes lead to erroneous conclusions (the case of van Helmont, above; also see Chapters 10, 11, and 12). Addressing the problem of error historically, therefore, is central. It contributes to understanding the nature of scientific justification, as well as its limits.

Consider, for example, the case of teaching the fluid model of electricity. The use of water analogies in textbook presentations of electrical current has been strongly criticized.[24] Electricity is not a fluid, after all. One may justly question the authority of history in setting an agenda for teaching. Still, many outstanding scientists, such as Franklin, Ampère, Coulomb, and Cavendish, did (in their own time) consider electricity to be a fluid—in some cases, two fluids! The fluid concept must have had validity in at least some context. Of course, to recover that context, one must draw on history.

Fluid models of electricity are, in fact, generally appropriate for considering the "flow" of electricity in circuits. They are not appropriate when one also wants to consider other electrical phenomena, such as field effects or the action of individual electric particles. The fluid model has a specific scope or domain of application within which it is justified. Outside that particular domain, the justification fails. For many people, in fact, this domain is all they will encounter in their daily lives: the fluid model of electricity actually suffices for them. But the deeper lesson about science is how, historically, that domain was found to be limited and the fluid model declared false. The fluid model of electricity can be both right and wrong at the same time, depending on context. The tension between these two perspectives is why students can benefit from studying the fluid model, even as a wrong idea. The history helps to convey the role of models and theories in science in general (also see Chapter 8 on the limits of laws).

It may seem perverse to teach wrong ideas. Yet without a full understanding of error, students cannot learn how to distinguish between simple ("wrong") scientific models and more sophisticated ones. Consider, again, the exercise on phlogiston introduced above. What is the value in teaching a concept that was abandoned two centuries ago? Here, as in the case of the electrical fluid model, the historical perspective offers a simple, general framework for thinking about a group of causally related phenomena. Students can use phlogiston to map the relationships of burning, calcination, rusting, tarnishing, corrosion (all oxidations), reduction, and photosynthesis (or, as one group of students saw it, 'reverse combustion')—all on a macroscopic level. They can eventually contrast their understanding in these terms with knowledge they gain on electrons and emission spectra. They can see how the simple concept is limited—even misleading by today's standards—yet fully justified within a certain domain. They know what it means to call the concept of phlogiston "wrong" (Chapter 11).

The process through which scientists determine and deal with error may at first seem peripheral to science. Yet the characterization of fact versus error is clearly not, and the occasions in history when one is transformed into the other are central to deciphering how science works.

In other cases, it is the wrong idea itself, rather than the process, that is important. Students typically bring to science courses preconceptions that often linger after they leave the classroom, even when taught the correct concept. Teachers who know the historical context of such ideas can appreciate more vividly exactly how these perspectives offer a way of interpreting the world. They can validate such views, even while helping students work towards a deeper understanding. Teachers may find it appropriate, therefore, to acknowledge all that was "right" in Lamarck's ideas before turning to Darwin's criticisms. If Darwin's arguments are ineffective, then perhaps we need to reevaluate our own conclusions. Similarly, our culture supports a host of assumptions about brain size and intelligence that once dominated nineteenth-century anthropology. It may be worth recognizing how scientists once justified their views on this subject, as an entry into investigations that transformed those views historically. Ultimately, in order to fully understand why scientists now endorse certain concepts, students must also understand why alternative explanations (perhaps their own) are wrong. Teachers must start from the historical context.

The lessons from exploring error can sometimes be unexpected. Consider, for example, the study of primates in the eighteenth and nineteenth centuries. Europeans were introduced to orangutans in 1641, chimpanzees in 1738, and gorillas in 1847. One can imagine how the new awareness of these apes sparked discussion. They were human-like, yet not human. What, then, distinguished humans? The major questions were: Can apes walk erect? Can they think? Can they speak? Can they create culture? For the first, the answer seemed a tentative "yes"; for the next two, "no." Opinion on the last was divided. The answers, however, do not tell us as much as the questions themselves. That is, they reveal the values of European scientists' culture at the time. For males, at least. When it came to characterizing female apes, the questions changed. The human/ape differences were based instead on sexual anatomy. Scientists focused on menstruation and mammary glands, the clitoris and hymen. That difference in focus reflected the scientists' views that women did not represent the "ideal" form of a species. Indeed, to be a female (of either species) was apparently based chiefly on sexual anatomy, not behavior. What made male apes different from humans and what made female apes different from humans were based on wholly separate categories. In a similar way, male apes were portrayed as rude and lascivious, female apes as modest and demure: projections of the scientists' views onto nature.[25]

The errors of gender bias seem obvious to us today. But they were surely invisible to the male scientists at the time. For contrast, consider one of the few women during this period to comment on monkeys and apes, Priscilla Wakefield. In her view, the key feature concerned maternalism. That seemed to have escaped the notice of the men. Ultimately, the gender of the scientist in a gendered society can apparently matter. In a similar way, primatology changed significantly in the middle twentieth century when women entered the field. For example, Jane Goodall focused on female behavior rather than on male aggression and social hierarchy. Her work exposed gendered assumptions and transformed the understanding of chimpanzee social organization.[26] Gender bias was present, too, in Linnaeus's classification of flowers, in the naming and description of mammals, and in views on hysteria.[27] Reliable scientific conclusions, it seems, may depend on who participates in the science. That would be a hard lesson to learn without a striking historical example.

The exploration of wrong ideas is potentially far reaching. For example, some educators would banish astrology, alchemy, phrenology, craniology, mesmerism, etc., from the science classroom because they represent mistakes of science. Some contend that even mentioning such unscientific or pseudoscientific practices gives them unwarranted credence. Historically, of course, each of these practices was once considered science—in some cases, exemplary science. It is hard to imagine how we should expect students to know better than these scientists without teaching them why. What has changed? If testability or falsifiability are benchmarks of modern science, for example, then students should discover, as scientists did historically, how those philosophical principles are important. By tracing the historical context of wrong ideas, students learn what makes science science (elaborated further in Chapter 5).

SUMMARIZING HOW NOT TO TEACH HISTORY

In researching the history of science, historians frequently encounter ideas from the past that strike us today as strange, wrong, or even incomprehensible. For example: van Helmont's *blas*, *gas*, *archeus*, and *spiritus sylvestre*. Yet historians endeavor to make sense of apparent absurdities. Just like scientists, they try to find underlying order amid apparent chaos. They examine historical context—available theories, previous ideas, biographical events, personal perspectives, styles of reasoning, records of observations, and so on—and then apply their imagination in inferring what the historical character was likely thinking. How may the strange or outlandish seem obvious and natural given the right *context*? The thinking combines evidence with creativity. Historians call this interpretive skill *the historical imagination*. Yet this skill is not really limited to historians. It is about sympathetic understanding of another perspective or context. That is familiar to good teachers, who seek to work from a student's perspective in trying to *transform* their scientific or NOS understanding. Understanding scientists in their historical context can be a resource for working with students, too (Chapter 2, sections 2 and 3).

Too often, science teachers assume that students who learn about evolution have thereby unlearned creationism; that those who learn about the cause of the seasons have corrected misconceptions about Earth's proximity to the Sun; or that those who can solve problems about inertia have changed intuitions about releasing an object spun on the end of a string. In all these cases, negative knowledge is important: we must *know* that the error *is* an error. For complete learning, one must notice, articulate, and be able to regulate the potential mistake. This chapter acknowledges the same challenge in education itself: in knowing how *not* to use historical case studies to teach the nature of science. Some intuitions can unwittingly lead one astray. Learning to teach NOS involves recognizing them and developing new patterns of responses and judgments. Properly understood, then, this chapter is as much about knowing how to *teach* history in science as it is about how *not* to teach history in science. The aim is to master the sometimes hidden errors in order to foster NOS learning more effectively. Just as one might learn to modify tendencies towards myth-conceptions, as discussed in the last chapter.

Science teachers (and scientists, too) often enjoy their profession because our culture associates science with certain knowledge. They may envision themselves as guardians and dispensers of the truth. Such teachers typically make awful historians. They tend to view all science in the past as either triumphant discovery or pathological error. They often filter history through the lens of current knowledge and disregard the context in which scientists worked. They may *judge* the history or try to *correct* it. However, such views obscure the *process* of science and, with it, the *nature* of science. Reshaping these perspectives is the core aim. The challenges are deep: one more variant of misshaping history is addressed in the next chapter.

What, ultimately, are the take-home lessons? First and most fundamentally, respect historical context. Refrain from sorting scientists of the past into heroes and fools on the basis of whether they were right or wrong. Seek to understand how science works, without imposing modern expectations. Second, be as wary of cookbook history as of cookbook labs. Avoid rational reconstructions. Target instead historical simulations or narratives that illustrate the degrees of freedom in choices and decisions. Neither science nor history follows prescribed paths. Third, while one may adapt history for educational purposes, nevertheless respect the integrity of science-in-the-making. Finally, note the value of rendering scientific error, as well as how scientists remedy it. History can offer many insights, but only to those who listen carefully to what it tells us.

5 | Pseudohistory and Pseudoscience

A little learning is a dangerous thing;
drink deep, or taste not the Pierian spring:
there shallow draughts intoxicate the brain,
and drinking largely sobers us again.
  —Alexander Pope, *An Essay on Criticism*

Appropriate history vs. appropriating history • Lawson's shoehorn • bad history vs. pseudohistory • pseudohistory as pseudoscience • history of pseudoscience • strategies for teaching

Every science teacher, it seems, knows the dangers of pseudoscience: parapsychology, astrology, New Age healing, creationism, alien visitors, and the like. While defined variously, pseudoscience essentially tries to claim scientific authority where there is no science. Individuals may "conjure" science using only the emblems of its authority.[1] They may mislead their audience by using evidence selectively. Science teachers often endeavor to teach the nature of science, so that students will not succumb to the illegitimacy of pseudoscience.

In this chapter I discuss a variant of this challenge: *pseudohistory* of science. Pseudohistory, like pseudoscience, uses facts selectively and so fosters misleading images—but here about the *nature* of science. In this sense, pseudohistory of science *is* pseudoscience, on a deeper level. It fosters misleading ideas *about how science works*, often in impressive-sounding philosophical terms, while appealing to history to make the claims seem plausible. In particular, many apparent NOS lessons rewrite science of the past into a preconceived vision of how science "should" work. Rather than articulate how scientists achieved certain discoveries, they presume that the discoveries occurred according to an imagined universal method. Essentially, pseudohistory tries to claim NOS authority where there is no (or bogus) historical evidence.

Here I explore several cases of shoehorning history into a particular view of scientific methodology. In each case, preconceptions of the philosophy of science were allowed to trump historical facts. To clarify the teaching challenge, one may distinguish between pseudohistory and bad history, and understand how pseudohistory is a form of pseudoscience. Ironically perhaps, teaching about pseudosciences in their historical context may deepen understanding of the nature of science for students.

APPROPRIATE HISTORY | APPROPRIATING HISTORY

History is an excellent vehicle for conveying NOS lessons (Chapter 2). But the history may be distorted by storytelling tendencies (Chapter 3). Or it may be so tainted with retrospect and modern knowledge that one loses the fundamental uncertainty of science-in-the-making, along with lessons about how to reason from evidence to conclusion, rather than the other way around (Chapter 4). One may thus feel inclined to abandon history. Indeed, historian and science educator Leonard Nash admitted more than a half century ago:

> To be sure, it is easier to construct an arbitrary imaginary account of a scientific development, that will well illustrate the points that are to be stressed, than it is to dig into the raw material of scientific history, to disentangle the shifting threads of fact and fancy that made that history what it was.[2]

In other words, one may be tempted to freely adapt the history. We have already glimpsed some of the unwelcome consequences of this approach—in accounts of Priestley, Harvey, and Mendel in Chapter 3. Altering the history subverts the NOS lessons. To illustrate further the role of genuine versus imagined history, I consider here two other cases, in astronomy and earth science: the discovery of Jupiter's moons and the development of the meteorite-impact hypothesis of mass extinction.

First, consider Galileo's discovery of the moons of Jupiter. In 1610 the observation that a planet (other than Earth) could have satellites of its own upset assumptions about the structure of the solar system. It was dramatic, although not definitive, evidence against the Ptolemaic system and helped open more serious consideration of Copernicanism. In addition, the story of how Galileo parlayed the discovery into patronage from the wealthy Medici family holds a great lesson in the funding of science.[3] So one may wonder how Galileo made his famous discovery.

Most histories give a central role to Galileo's newly fashioned "perspective eyepiece," today's telescope. Galileo himself described it as follows:

> Finally, sparing neither labor nor expense, I succeeded in constructing for myself so excellent an instrument that objects seen by means of it appeared nearly one thousand times larger and over thirty times closer than when regarded with our natural vision.

Galileo perceptively turned the new instrument to the night sky. With it, he was able to see the irregular surface of the Moon (earlier assumed to be smooth), innumerable stars that had never been seen with the naked eye, and, of course, the moons of Jupiter. Equally important, the telescope allowed others to confirm his discoveries. The telescope was key to Galileo's discovery. Not long ago, philosophers of science regarded elements of experimental practice as peripheral to discovery proper. However, they now realize how profoundly such instruments—electron microscope, ultracentrifuge, spectroscope, particle accelerator, and others—contribute to the process.[4] Moreover, researchers invest substantial time in learning how to use new instruments and in validating observations made through their use.[5] As Galileo himself noted, creating his instrument involved much labor and expense. Galileo's case exemplifies nicely how technological developments factor into discovery.[6]

Imagine, then, a historical account for science educators that portrays Galileo's discovery as due foremost to the tired and oversimplified "scientific method" of textbooks. For educator Anton Lawson, the essence of Galileo's discovery was hypothetico-deductive thinking. No doubt this is because Lawson believes, as he details throughout his presentation (and elsewhere), that "many, if not all, scientific discoveries are hypothetico-deductive in nature." It should hardly surprise anyone, therefore, that his history bears witness to being fitted and trimmed just to match this overarching preconception. For example, in Lawson's account, Galileo's development of the critical scientific instrument is peripheral. It was merely "at his disposal." The reader may wonder how, then, without telescopic observations, the hypothesis of Jupiter's moons could emerge. It would have been dramatic, indeed, if Galileo had used a Copernican model to *predict* the occurrence of moons around other planets, then invented a telescope with the expressed purpose of looking for them. But he did not. Rather, Galileo observed the moons on successive nights—perhaps at first just to persuade himself that they were real, not some accidental features of the telescope or night sky. He saw their position change—quite unexpectedly: no prediction there! After many observations, he could discern the pattern of orbits seen on edge. Lawson, however, contends that Galileo had to blindly guess what he was seeing, then test his idea. Ironically, Lawson admits at one point that "we can not know if this is what Galileo was really thinking." Still, he blithely continues, presenting this as plain evidence of how scientific discoveries happen in general.[7] Biased speculation is presented as historical evidence. Imagination, as Nash cautioned, can easily substitute for historical understanding of how discoveries occur.

Lawson's analysis draws exclusively on Galileo's 1610 text, *Sidereal Messenger*. He casts it as a "report" that "chronologically reveals many of the steps in his discovery process." Referring to original historical texts directly can be fruitful, of course. But perhaps not without understanding the historical context: where the expertise of experienced historians matters. Galileo's rhetorical skills, in particular, are widely documented.[8] His published narrative is ultimately an artful construction designed to persuade. One can easily be misled (as Lawson was) by treating it as describing his experience or actual thinking. Here, Galileo used a tactic that he applied widely elsewhere. He led his reader through a series of hypotheses and rejected each in turn. That is, in order to dramatize his claims, he first considered other contrasting explanations. He showed how each was not only plausible, but also apparently supported by the evidence. Only then did Galileo introduce contradictory evidence. The strong sense of irony in reinterpreting the evidence made his own insight seem more impressive.[9] Galileo displayed clever rhetoric, not necessarily his own thinking process.

So, too, today. Nobel Prize winner Peter Medawar famously profiled how modern scientific papers misrepresent scientific thinking. They are persuasive texts, not narratives.[10] They follow numerous rhetorical conventions.[11] The format of a modern scientific paper does indeed seem to parallel the "scientific method" of science textbooks. But this is the structure of an argument formally arranging the available evidence in retrospect. It does not describe the process of discovery or of acquiring the evidence. Indeed, sociologist Karin Knorr-Cetina has shown that the order of presentation in a scientific paper typically unfolds in the *reverse* of the process that generated it.[12] One must read any scientific publication carefully, aware that it is an argument, not a narrative of research. Thus, they do not automatically "help students learn how to do science," as Lawson instructs his readers.[13]

Lawson approaches this historical episode already equating how to *do* science with just how to *think* scientifically. This can blinker one to many relevant dimensions of what Galileo's discovery of Jupiter's moons can tell us about the process of scientific discovery. For example, for Lawson the process of discovery apparently began only *after* Galileo had turned his telescope to Jupiter and only *after* he had seen the new "stars" in its vicinity. (For most of us, surely, the discovery has already happened by that time!) Consequently, calculating the geometrical optics in crafting lenses for his telescope does not seem to count as part of formalized science. Nor does *observing the night sky* with the new telescope for potentially new phenomena (without an explicit hypothesis). Or *noticing* the new celestial bodies as significant or worth further consideration. *Generating new hypotheses* or *revising* them on the basis of new results are also excluded.[14] Instead, Lawson focuses on successive testing, or a context of *justification*. There seems to be no context of *discovery*. The process of discovering Jupiter's moons (in Lawson's account) thus omits much of both historical and philosophical relevance.

Interpreting Galileo and his methodology is challenging, as illustrated in the history of such efforts. Nicholas Jardine, in reviewing several books by prominent scholars, noted how Galileo seemed adaptable to each. For pragmatist Joseph Pitt, Galileo relied essentially on common-sense experience. For the flamboyant Mario Biagioli, Galileo was a savvy politician and showman. For Christian scholar William Wallace, Galileo exhibited the Aristotelian method, notably of a Thomist flavor. For rhetorician Jean Dietz Moss, Galileo was an expert in persuasion. To these four, Jardine added "Mach's Machian phenomenalist, Koyré's Koyrean metaphysician, and

Feyerabend's Feyerabendian anarchist." All these interpretations seemed to support Alistair Crombie's observation that "philosophers looking for a historical precedent for some interpretation or reform of science which they are themselves advocating have all, however much they may have differed from one another, been able to find in Galileo their heart's desire."[15] Thus, for Lawson, Galileo appears in yet another guise: as a hypothetico-deductive thinker. Galileo, it seems, exemplified many methods. He is a protean scientific hero who can be cast to fit many molds. Ultimately, some histories seem to tell us more about the writer than about what happened in the past.[16] Lawson's account is, by his own admission, *entirely speculative* (although his initial tentativeness wanes conspicuously as he goes on). *Nothing substantive historically* is established from *imagining* that Galileo used the hypothetico-deductive method, *speculating* that he *could* have, or even reconstructing how he *might* have. Historical documentation matters. If we want to understand how science works, we need to heed Nash's advice and "disentangle the shifting threads of fact and fancy that made that history what it was."

In his Pulitzer Prize-winning book *Wonderful Life*, Stephen Jay Gould coined the expression "Walcott's shoehorn" to label a significant bias in the work of paleontologist Charles Walcott. Early last century, Walcott characterized and classified organisms from the Burgess Shale, a trove of fossils of some of the earliest life on Earth. The Burgess organisms (by today's reckoning) reflect a striking diversity of life forms at a very fundamental level. But Walcott was extraordinarily conservative: he classified them all within known groups (phyla), representing organisms today. The effect, Gould notes, was a misleading impression of progress, of an increasing diversity within established groups (the cone of diversity), rather than of the "diversification and decimation" of many types of early life forms. The apparently modest but pervasive flaw thus had substantial consequences. For Gould, shoehorning data into preconceived categories where they did not fit was a "cardinal error" in science. Preconceptions were allowed to speak louder than the facts.[17]

One can see a corresponding error in science education: philosophical preconceptions are allowed to trump historical facts. One may regard it as a cardinal error in NOS education. The error, as a general type, is fundamental— and, alas, all too common. Lawson's history vividly illustrates the error. It shows evidence everywhere of being altered. Relevant facts are omitted, false or imaginary details are added, emphases are misplaced, and qualifications are abandoned—all to accord with his model of hypothetico-deductive reasoning. One can hardly portray the nature of science authentically with such distorted history. By analogy with Gould's analysis of Walcott, we may call this *Lawson's shoehorn*. It is a potential trap for the unwary science teacher to recognize in reviewing prospective curricula.

The same pattern of shoehorning history into a particular view of scientific methodology appears all over again when Lawson reconstructs the development of the meteorite-impact hypothesis of mass extinction, which includes the famous demise of the dinosaurs.[18] Lawson draws from a popular book, *T. Rex and the Crater of Doom*, written by one of the scientists, Walter Alvarez. His strategy is the same: to *reconstruct* the arguments as fundamentally hypothetico-deductive.[19] That statement itself should be a critical warning signal to the science teacher. The aim is not to discern the nature of science from a historical episode, even with the firsthand testimony of the scientist readily available. Rather, it is to characterize it according to a predetermined philosophical model.[20] It is to *shoehorn* the history into an NOS preconception.

Alvarez's book leads the reader through the exciting series of unexpected findings that eventually led to the hypothesis. In his reinterpretation, Lawson takes each conclusion along the way and, using a hypothetico-deductive format, imagines a plausible hypothesis and a test that incorporate the results that were found. The research appears to be a logical succession of hypothesis, test, hypothesis, test, hypothesis, test, and so on. It *seems* quite plausible. Historically, however, the concepts frequently *followed* the unanticipated findings in the field or the laboratory. Lawson's history is consistently *backwards*. He fails to explain the initial development of hypotheses or the revision of models. Time and time again, Alvarez encountered anomalous results and was forced to modify his concepts or assumptions. Uncertainty, not clear logical direction, dominated the episode. Indeed, that is what made the ultimate discovery so striking and significant.

Those who have read Alvarez's book will find that Lawson omits several important NOS features exhibited in this case. These include

- encountering novel phenomena (chance, contingency, or luck)
- perceiving their relevance to other information (prior knowledge, noticing)
- generating new patterns or explanations (creativity)
- capitalizing on opportune observations (problem-shifts)
- applying experimental skills (technical know-how)
- interpreting evidence effectively and addressing criticism (social discourse)

Here are some examples. First, the role of contingency (chance or luck). While looking for fossils with a characteristic magnetic orientation (in a routine investigation), Alvarez found some that pointed in the opposite direction. Although unexpected, the fossils proved an important tool for dating the rocks the researchers were examining and led to new questions. As they continued their observations, they noticed that another group of conspicuous fossils seemed to go extinct at one layer. They wondered why and again reoriented their research. After that, while collecting data, they noticed a clay layer above the extinct fossils. There had been no hypothesis that one should find, or even look for, clay layers. It was just something they noticed. But it was consistent. And it reoriented their research again. These clay deposits later proved the vestige of a meteorite impact. A series of chance observations had led Alvarez and coworkers to something substantially new. In the context of the original research, the observations did not fit what they were looking for and, hence, were logically irrelevant. But this is precisely where new discoveries often begin. Science includes logic, of course, but its overall trajectory in episodes of discovery is not so strict as the "scientific method" implies. Happenstance that reorients research plays an important role.[21]

Second, consider the role of prior knowledge in interpreting unexpected observations. Connections between disparate facts are not automatic. In many instances, specific individuals with particular backgrounds were able to recognize how two apparently unrelated observations seemed, in fact, related. That critically opened possibilities for further exploration. For example, when Alvarez concluded that a catastrophic event must have led to the extinction, he could draw on his prior awareness of meteor impact craters as indicating a possible cause. That was not a deduction or prediction: who could have guessed in advance that reading a certain few papers might later prove relevant in this way? Far more dramatically, when Alvarez later began to consider whether the critical clay layer was deposited gradually or all at once, he consulted his father, who had won a Nobel Prize for his work in particle physics. The elder Alvarez knew that meteorite dust contained signature amounts of iridium. Here was a critical convergence of information from different fields—due to a personal relationship, not mere logical thinking. Later, the problem became how to link a meteorite to a global iridium layer. Again it was Walter's father who recalled reading about the 1883 explosion of Krakatoa and the enormous amounts of dust it produced: a meteorite would also generate substantial atmospheric dust, blocking vital sunlight. Again, the key event in advancing the research at this point was the memory of a prior fact, now suddenly made relevant in another context. The convergences seemed obvious once found, but they were hardly derived logically.

One may similarly highlight other factors that played a role in the unfolding discovery in this episode: the exceptional analytical skills of a chemist who helped measure the iridium in some of the clay layers; previous findings from oil explorers who had identified a submerged crater in the northern Yucatan peninsula; alternative theories about volcanic dust; criticism from a Princeton geologist, who ensured that the evidence was robust. We know that all these factors were important because if they had not occurred, the history would likely have followed another path. The history of Alvarez's meteorite-impact hypothesis is marked by contingency: unanticipated and unplanned convergences of people, places, and events.

Paleontologists had been puzzled by the extinction of dinosaurs for decades. They had posed causal hypotheses and predictions to test. Despite this, however, they did not find the answer. Hypothetico-deductive reasoning is limited. Paradoxically, Alvarez wasn't even looking at the problem of dinosaurs when he encountered the key clues to their demise. And that is a profound lesson in the nature of scientific discovery.

There are many other accounts of this same episode, besides the Alvarez book, and they all present the more complex (and more interesting!) view of the process of scientific discovery in this case: David Raup's *Extinction: Bad Genes or Bad Luck?*, William Glen's *The Mass Extinction Debates: How Science Works in a Controversy*, and James Powell's *Night Comes to the Cretaceous*.[22] Lawson's analysis for teachers, by contrast, omits important information about how science happens. Historical facts were selected and interpreted to fit a philosophical preconception: Lawson's shoehorn yet again.

Lawson's two "histories" exemplify many accounts offered to science educators under the banner of NOS education. They parade idealized concepts about science rather than delve into the realities of scientific practice. That is, they are selective and incomplete. Relevant facts are missing. Assumptions, and the limits they imply, are ultimately suppressed. A perspective seems to masquerade as fact. A little learning, as Alexander Pope cautioned, can be a dangerous thing.

In these cases, one might even contend that the historical facts are irrelevant to Lawson's conclusions. But the history is not idle. It functions rhetorically. "The intent is…," Lawson comments, "to reveal and model some of the key elements of scientific discovery in general."[23] The hidden purpose in using Galileo's or the Alvarezes' discoveries, then, is to borrow on their fame to promote a particular, narrow view of science (here, as *essentially* hypothetico-deductive). We have seen, however, how those interpretations depend on misrepresentation of the past. These are not just streamlined accounts that omit needless detail. By eclipsing certain historical details and context, they actively mislead. Neither was an *appropriate* use of history. Rather, each case *appropriated* history for ideological ends. When historical integrity suffers, so too do lessons about the nature of science.

BAD HISTORY | PSEUDOHISTORY

In the cases above, the historical errors may tend to stand out. But the lessons for NOS education are not about simply misinformed or bad history itself. Rather, the teacher should take note of how *certain types* of historical errors can corrupt NOS lessons. The primary concern is not bad history per se, but *pseudohistory*. Pseudohistory conveys false ideas *about the historical process of science and the nature of scientific knowledge*. Pseudohistory may even borrow from acknowledged facts. Fragmentary accounts of real historical events that omit context can mislead, even while purporting to show how science works. By eclipsing relevant information, "textbook histories" perpetuate a caricature of science.[24] Romanticized tales of discovery may overemphasize the contributions of one individual, minimize the role of accident or errors, simplify the investigative process, disguise less-than-noble motivations, and hide the effect of personal or cultural values. They turn real science into imaginary science. Such a misleading selective history masquerading as responsible history is justly called *pseudohistory*.

Not every science teacher can be a professional historian too, of course. While teachers should certainly care about historical accuracy, they cannot always sort bad history from good. For example, many popular anecdotes are apocryphal: the apple falling on Newton's head, Galileo dropping balls from the Tower of Pisa, Archimedes shouting "Eureka!" as he ran from the baths naked through the streets of Athens, among others. These are relatively harmless. Some such stories are still debated by historians, who cannot ascertain conclusively whether some historical testimony can be trusted. We need not worry about such errors. These are not pseudohistory.

Other false stories can be misleading. For example, Darwin did *not* deduce natural selection upon seeing the finches on the Galápagos Islands. The Church during the time of Galileo *supported* astronomical investigation, and many challenged Galileo's claims *scientifically* (Chapter 12). When Columbus set out on his voyage, educated persons did *not* believe that the world was flat. The photoelectric effect did *not* inspire Einstein's concept of the photon.[25] While many science teachers have retold such false stories, most abandon them once they learn otherwise. Science teachers tend to respect historical fact, I think. They seek to avoid historical error as much as scientific error—when they know the history. They typically learn about new historical discoveries through teacher magazines, TV shows, science blogs, and informal networking. These stories are not pseudohistory either.

Yet without historical expertise, how is the science teacher to proceed? The first strategy is simple: be aware of the potential for pseudohistory. The cases profiled above are examples that can help alert the teacher to the possible dangers. Second, of course, one should assess the credibility of the source, as one does with scientific claims (Chapter 3). Check sources and the scholarly credentials of the author (remembering that philosophers and journalists are not always good *historians*). Next, one can learn to recognize the typical warning signs of pseudohistory (Figure 5.1). Most notably, a conspicuous ideological or persuasive context might signal the reader that the author is trying to use history to naturalize a particular set of values. One can then look more closely for qualifications and view the evidence critically. An analytical posture is just as important in reading history as it is in assessing scientific claims. Stylistically, pseudohistories tend to disguise their persuasive intent. They rely on our intuitions that historical facts, like facts of nature, should appear transparently. But romanticism and one-sidedness are strong clues. Finally, because the nature of science is revealed most honestly through science-in-the-making, the account should convey a deep sense of historical context. Equipped with the proper analytical perspective, a reader may begin to recognize what is reliable history and

FIGURE 5.1. Warning signs of pseudohistory of science.

- Intent to justify values using history
- Author with a narrow agenda
- Romanticism
  - Flawless personalities
  - Monumental, single-handed discoveries
  - "Eureka"-type insight
  - Absence of any error
  - Unproblematic interpretation of evidence
  - General oversimplification or idealization
- "Crucial" experiments only
- Sense of the inevitable (plot trajectory)
  - Context is missing
  - No cultural or social setting
  - No human contingency
  - No antecedent ideas
  - No alternative ideas
  - Uncritical acceptance of new concept

what is pseudohistory. One need not be an expert on the history itself.

PSEUDOHISTORY AS PSEUDOSCIENCE

Few will dispute that everyone, including any teacher, has a minimal responsibility to respect historical fact. By contrast, pseudohistory may seem like a subtlety relevant only to historians. But as portrayed above, pseudohistory is more than just a historian's parallel to pseudoscience. Pseudohistory of science *is* pseudoscience. Like pseudoscience, it conveys false ideas about science. But unlike pseudoscience, which typically deals with facts or concepts, it concerns misleading ideas about *the nature of science.* Pseudohistory misportrays *the process of science,* rather than its content. The science teacher should thus be as concerned about pseudohistory as about well-worn pseudoscience topics. The dangers of hagiography, Whiggism, myth-conceptions, and rational reconstructions in science (Chapters 3 and 4) should rank alongside cautionary tales about the Loch Ness monster, dowsing, and cataclysmic planetary alignments.

Many educators may use historical stories strictly to teach science content, in a narrow disciplinary mode, apart from any intended overtones or lessons about NOS.[26] They may feel, therefore, that they can escape or minimize the dangers of pseudohistory.[27] But every narrative of science is *implicitly* explanatory. *Every* history of science teaches the nature of science. The pseudoscience of pseudohistory is thus always a concern.

History of Pseudoscience

Educators, then, should purge science classrooms of pseudohistory of science. But this may not be true for *history of* pseudoscience. Indeed, history of pseudoscience offers an ideal opportunity for teaching the nature of science. That may seem paradoxical. If teaching pseudoscience is reprehensible, how can teaching its history be any better?

Consider, first, that many of today's pseudosciences were yesteryear's sciences. Astrology, alchemy, craniology, and others were once pursued by notable scientists. For example, Johannes Kepler, renowned for finding that planets trace elliptical orbits (not circles), was committed to astrology, which indeed fostered many of his discoveries. Robert Boyle, of "Boyle's law" fame and cofounder of the Royal Society, pursued alchemy. So, too, did Isaac Newton, otherwise known for his three laws of motion and his monumental achievements in gravitation, calculus, and optics. Paul Broca, who identified a language-processing region of the brain now known as Broca's area, advocated craniology. He engaged members of the Anthropological Society of Paris in a sustained debate over the size of Georges Cuvier's brain, defending its importance in measuring intelligence. Given the pursuits of such distinguished scientists from history, teachers certainly might pause before disparaging students who themselves entertain such topics seriously.[28]

When teaching the nature of science, therefore, one might appropriately begin by acknowledging how pseudoscientific claims *seem* entirely plausible or reasonable, at least at first. Eliciting such views is preliminary to engaging them and transforming them into something more sophisticated. Here, the educational exercise is also historical: to recover the (good) reasons once advanced for what is now pseudoscience. For example, consider Robert Boyle's views on the healing power of gems. Boyle believed that the salubrity, or healthfulness, of the air was due to "subterraneal Effluvia," which carried the distinctive mineral properties of a region.[29] The effluvia would explain local disease epidemics, for example. Further, when gems crystallized, he said, they were "imbued with Virtues by subterranean Exhalations and other steams." That is, in solidifying, a gem could trap particles of, say, healthy vapors. Later, rubbing the gem would release them, just as rubbing amber excited its electrical attractions. The gem's "virtues" would "exert their power by the copious Effluxions of their more agile and subtle parts."[30] It was no different than deer leaving behind subtle effluvia that hunting dogs sensed, or perfume causing people to faint, or odors causing headaches.[31] Here, Boyle explained a commonly accepted fact of the time: that certain gems have specific types of healing powers. To our ears this all sounds dangerously New Age. But for Boyle it expressed his belief that matter is composed of distinctive minute particles. Indeed, what had once been attributed to some transcendental "sympathy," he reconceived as mediated mechanically by these particles. Now we tend to regard that as a scientific achievement, not pseudoscience. If Boyle was wrong, it was not because the idea was intrinsically unscientific. One can only say that further study of gems did not bear out their healing power. The historical transformation holds the critical lesson: the difference between science and pseudoscience.

Consider, too, the notorious history of craniology. For several decades in the nineteenth century, anthropologists such as Paul Broca tried to use skull measurements to prove sexual and racial differences in intelligence. Science would interpret and validate why social power and privilege had developed as it did.[32] Now, the whole enterprise is a blemish on science. At the time, however, craniology seemed like a straightforward application of the principle of structure and function: if mental functions take place in the brain, then shouldn't the brain's size reflect mental capacity or ability? Likewise, the brain's shape should reveal significant features of specific, localized mental functions, such as personality, rational faculties, and emotions. This would naturally affect the size and shape of the skull as well. Phrenology, the study of cranial shapes and proportions, thus also seemed eminently plausible—as it does to many even today. Moreover, craniology was *quantitative*, following one oft-cited hallmark of science.[33] Craniologists used more than six hundred instruments and five thousand measurements. For historian Elizabeth Fee, it was a "Baconian orgy of quantification."[34] Of course, the prospects of craniology and phrenology went unfulfilled. When women eventually entered the field, they challenged claims earlier deemed acceptable by men. Standards of evidence rose. The whole field soon dissolved. In retrospect one can see that the community of (white) European male researchers was culturally biased (not that any practitioner recognized his own bias). Now, the episode is a persuasive example of how diversity in a scientific discipline can contribute to its objectivity. A historical perspective makes the pseudoscience seem a little less "pseudo," although no less wrong. Subsequent data showed that the plausible approaches were, ultimately, unfounded. Craniology was wrong, not inherently misguided. History thus offers complementary lessons in science and pseudoscience. It helps reveal vividly how science works and why, sometimes, it errs.

A historical perspective highlights that many topics typically branded as pseudoscience are not *self-evidently* pseudoscience. That status of understanding is the outcome of historical investigation. Identifying error in various claims or assumptions involved work, *scientific* work. Claims characterized as pseudoscientific reflect *negative* scientific knowledge. Such negative knowledge continues to be important (for example, in ending the searches for a perpetual motion machine or for parapsychological abilities). Negative discoveries generally raise the standards of evidence or interpretation. Indeed, scientists disregard past error at their peril, lest they repeat it. Just because something "is known" does not mean that each individual automatically knows it. Therein lies the very rationale for education. So: every generation must relearn what is scientific, what pseudoscientific.

The evidence that shaped knowledge of pseudoscience historically must be reinstantiated in each student. The history itself, of course, may be a valuable tool (Chapter 2).

By advocating a historical approach to pseudoscience, one should not, of course, succumb to pseudohistory. Turning to history merely to bash pseudoscience with today's apparently obvious conclusions achieves nothing.[35] History is valuable, rather, for showing students how they might *challenge* what seems obvious. Educators can help them probe evidential claims and show them how historically, with further evidence, later scientists found them ultimately to be without merit. Indeed, the very understanding that something may appear reasonable until it is considered more deeply is a powerful lesson worth offering to anyone. "A little learning is a dangerous thing"—whether it is about the history of science or current scientific knowledge.

STRATEGIES FOR TEACHING

Concerns about pseudohistory or "quasi-history" in the science classroom are not new.[36] Yet most such concerns address preserving the integrity of *history*. The focus here, by contrast, is preserving the integrity of *the nature of scientific practices*. That is, while many accounts may be historically misleading, they can also be *scientifically* misleading—where science is construed as its methods and practices, not just its conceptual content. When past science is reconstructed to fit someone's idealized model, it becomes *pseudoscience*.

As noted at the end of Chapter 3, we do not simply need more history in the science classroom. We need *different* history. Much classroom history, as exemplified in Lawson's accounts of Galileo's discovery of the moons of Jupiter and of Alvarez on the meteorite-impact hypothesis, is *pseudohistory*. As such, it is really *pseudoscience* trying to borrow illicitly the authority of history through a false narrative. One problem is that Disneyfied, G-rated pseudohistories of science promote undesirable or negative role models.[37] The idealized scientific method of pseudohistory provides an unrealistic model for what citizens can expect of scientists in our society. It distorts the nature of scientific knowledge by concealing its limits and oversimplifying the interpretation of evidence. How can students schooled in pseudohistory of science ever make informed decisions where complex science is involved—for example, in cases of global climate change, genomics, cloning, alternative energies, or biological and chemical weapons (Chapters 1 and 9)?

But without historical expertise, what can a science teacher do? As noted above, teachers should first understand pseudohistory, perhaps just by a few clear examples. They may then recognize and be alert to their warning signs (Figure 5.1). Ideally, they learn more about analyzing texts rhetorically (Chapter 3). Perhaps most important, teachers should master at least one historical case study in depth (see Part II and Figure 14.2). A single well-developed case study can be far more valuable in profiling the nature of science than, for example, a comprehensive "greatest hits" survey course in the history of science. Teachers will then adapt this case for their students. Familiarity with its complexity will allow discussion and probing. Ideally, a single complex case, well rendered, can illustrate for students how simple stories can be misleading. Teachers and students then expand their repertoire to other cases. Remedying pseudohistory in science education is not that difficult. It begins with simply recognizing it for what it is. Echoing the conclusions of Chapter 4, one thereby sets appropriate standards for what history is welcome in the classroom for informing NOS teaching.

6 | Sociology, Too

Contested nature of science • sociology of science as an NOS resource • normative versus descriptive NOS • reconciling philosophy and sociology of science • strategies for teaching

In 1974, historian of science Stephen Brush reflected on the virtues and risks of using history in science education. Not all scientists of the past, he observed, seem good role models. Might exposing students to their sometimes ambitious and unsavory behavior disillusion emerging science enthusiasts? Might recruitment of future scientists suffer from glimpses into unvarnished science? Brush playfully adopted the new film rating system and asked (now famously), "Should the history of science be rated X?"[1]

Brush did not wholly discount the value of his own discipline, of course. He also highlighted the importance of seeing scientists as human and of situating science in its cultural context. Such lessons offered valuable insight, especially for non-scientists interpreting science in society today. History of science could help bridge what C. P. Snow once called the "two cultures," the apparent gulf between the sciences and the humanities. The use of historical perspectives thus had multiple outcomes, with conflicting values, and posed a dilemma for science educators. Should students learn only an idealized version of science or science as it is actually practiced, with all its blemishes?

Decades later, the consensus has fallen on the significance and values of using history (Chapter 2). Educators now seem to agree that history is an essential vehicle for conveying the nature of science (Chapter 1). Yet the dilemma is hardly resolved. Disagreement persists now, instead, about what the nature of science *is*, or *which features* of NOS belong in the science classroom. The tensions Brush noted are now captured in the ambiguity of the very phrase "nature of science." Does it describe science as it *ought* to be or science as it *is*? Does one interpret NOS *normatively* or *descriptively*? Does it refer to scientific *ideals* or to *real* science, as found in practice? The same ambiguity equally pervades other popular phrases, such as "scientific practices" or "how science works." Each perspective indicates, of course, a different view of scientific authority. The ideal version does not include error, chance, or bias: scientific claims, and the scientists who make them, seem beyond question. On the other hand, the concretely grounded version acknowledges that achievements are more human in character and open to critical analysis. The debate is thus ultimately political, often deeply emotional, and, at times, quite testy.

The contrasting views have disciplinary overtones. As Brush observed, historians seek to document and understand episodes of scientific practice and discovery as they occurred with all their complex context. Philosophers, by contrast, are primarily concerned with methodological ideals, norms, and simplified abstract rules. They strive to articulate good practice. Hence, one finds much interdisciplinary wrangling. Here, one may also encounter an important third field of study: the sociology of science. Sociologists focus on the social, political, and cultural contexts of science, not always finding pristine perfection. Indeed, many of the most striking and insightful studies of science in the past few decades have emerged from sociologists. Their work has further inflamed debates about the nature of science.

For example, sociologists of science have profiled how politics within science can shape consensus.[2] Some have stressed the "interpretive flexibility" of data and suggested that struggles for credibility dominate science, quite apart from objective standards of the quality of evidence.[3] Others have underscored how cultural values permeate scientific knowledge. For example, one finds that a dispute about phrenology in Edinburgh in the early 1800s was divided along economic class lines—with each side arguing for its own social power perhaps more than some objective "truth."[4] Robert Boyle and Thomas Hobbes, in debating the nature of a vacuum and the possibility of an air pump in creating one, mirrored their political views about governance and the monarchy in seventeenth-century England. Hobbes was convinced that admitting to a vacuum opened the existence of an immaterial soul and, with it, religious mischief. Social order could only be maintained, he claimed, through a sovereign power, not the collective action of atom-like individuals, as implied in Boyle's mechanical philosophy.[5] The idealized and impersonal scientific method found in textbooks seems not to describe how science actually works in practice. Many see sociology of science as thereby threatening the legitimacy of science. The fear seems to be that science will be reduced to nothing but politics. Social interests will be viewed as eclipsing rationality. If one cannot establish that science follows objective, universal standards for knowledge, then it cannot escape the awful fate of utter relativism. And if there are no scientific laws, how can one justify laws of any kind? All will be chaos. Anarchy will reign (also see Chapter 8). For some, then, sociology has no place in the science classroom.[6] Brush's concern about history has been transformed. The question today instead seems to be "Should the *sociology* of science be rated X?" And for some, the answer is decidedly "yes."

Others, by contrast, feel that sociological studies of science offer important cautionary tales about the limits of scientific knowledge and the inappropriate use of scientific authority. History exhibits numerous cases of science giving unwarranted authority to cultural prejudices. For example, early in the twentieth century, science seemed to legitimize IQ testing as an objective measure of intelligence.[7] These tests, in turn, were used in the United States to deny immigration to Eastern Europeans (who could not speak English) and, later, to set immigration quotas that discriminated against these predominantly Jewish immigrants. Henry H. Goddard, Charles B. Davenport, and Karl Pearson, the leaders in the field, were well aware of the political overtones and saw them as supported by their science. Science may have corrected itself in the long run, but the discrimination supported by the science for decades affected perhaps six million individuals who tried to escape Nazi Europe. As Stephen Jay Gould has poignantly observed, we know what happened to many of them. For some, then, sociology of science—as vividly illustrated in such cases—helps illustrate the dangers of unchecked deference to science. Scientific authority can be and has been misused. Prudential caution is warranted. Sociological case studies thus seem critical in the science classroom for developing informed citizens (the benchmark goal of scientific literacy outlined in Chapter 1).[8] In this view, sociology of science should be rated "E": Essential for Everyone.

In the 1990s, an informal consensus on central NOS features emerged (Figure 1.1). One item highlighted understanding how science affects, and is affected by, a larger social and cultural milieu. This consensus seemed to open a wide role for sociology. Yet views on the mutual effects of science and society can differ markedly. For some, the effect of science on society is illustrated by the positive contributions of electrodynamics to the emergence of radio communication or of the understanding of germ theory to rescuing the wine industry and preserving livestock through vaccinations. For others, however, it is illustrated by the adverse role of knowledge about fission and nuclear weapons in stoking the Cold War or about nitrogen fixation and fertilizers in creating oceanic "dead zones" around the world. For some, the effect of society on science may be epitomized by the role of the steam engine in contributing to the development of thermodynamic concepts or by the threats of World War II that spurred the development of antibiotics and DDT for controlling disease. For others, it is the knowledge of genetics that led to hybrid corn and the subsequent dependence of farmers on seed companies, or of special high-yielding crops that shifted power to those with the requisite farm machinery and money for fertilizers. The corresponding fractures in the NOS consensus were evident in one project that monitored how participants reasoned about each NOS item.[9] Consensus seemed to break down precisely on questions of the reliability and authority of science. Even experts seemed to disagree on "the status of scientific knowledge" (whether it can be relied on as a basis for action); "the empirical basis of scientific knowledge" (as a basis for its reliability); and "the cumulative and revisionary nature of scientific knowledge." Just how the sociology of science should inform an understanding of NOS can itself be contentious.

The different perspectives on sociology of science parallel, of course, larger cultural debates. The science classroom remains contested turf in the lingering Science Wars of the 1990s. For some, sociology of science draws us down into an inescapable whirlpool of relativism, like the Charybdis of ancient Greek mythology. For others, disregarding sociological cases engenders a monstrous scientism that devours human values in the name of science: like the Scylla of myth, treacherously positioned just opposite Charybdis. Practicing science teachers, like the ancient Greek mariners, must chart a course between the joint threats. To succeed, one must first acknowledge that each perspective has some warrant (next section below). The ensuing challenge, then, is to map how to accommodate the central concerns of both (following sections). Aware of both positions in the debate, teachers must reject the either–or choice so often presented as inevitable. Nature of science is *both* normative and descriptive, *both* idealized and real: a synthesis of the history, philosophy, and sociology of science.

SOCIOLOGY AND THE NATURE OF SCIENCE

How, then, might the science teacher interpret the sociology of science productively, not just as critical or negative? How might sociology be important in characterizing the nature of science in the classroom and contributing positively to scientific literacy?

To begin, one may sort sociology of science into three types, each posing different challenges.[10] First, there is macrosociology of science, which focuses on the institutions of science and the social framework of peer review, rewards, funding, and so on.[11] As long as sociologists talk, like Robert Merton, about institutional norms or, like Derek de Solla Price, about "big science," all may be well. Here, the student may be healthily engaged in contemporary issues in science policy, from genomics to large particle accelerators. (Although many might argue that such topics belong in a *social studies* class.) One may even discuss how scientists guard their professional boundaries from those seeking science's cultural authority.[12]

Here, one also encounters some notorious cases of fraud, where the system of trust and credibility fails.[13] Is fraud part of science? Many consider it external to science *proper*, even though it is perpetrated by scientists, is found in the scientific literature, and can affect day-to-day research activities. Its significance depends, in part, on its frequency. It would be foolish to suggest that the whole of science is a lie because it is nothing but fraud. Yet it would be equally blind to contend that cases of fraud are so rare as to be hard to find. The gossip runs thick and heavy in the pages of *Science* magazine, epitomized in the cases of erstwhile wunderkind Jan Hendrik Schön of Bell Labs and ecologist Anders Pape Møller.[14] So, do students get the real story? It would be dishonest to deny or discount fraud. It would be equally inappropriate to condone it or legitimize it. Here, as noted above, one encounters some of the fundamental tension in whether one construes NOS normatively or descriptively, or both. The U.S. National Academy of Sciences, at least, considers fraud significant enough to address substantively in its public brochure *On Being a Scientist*.[15]

Second, there is microsociology of science, which addresses the interpersonal dynamics in the lab. How does one interpret the behavior of scientists and their cryptic "inscription devices," such as amino acid analyzers? How do scientists' activities yield knowledge? These studies often irritate persons steeped in science because the anthropologists or ethnographers generally adopt an agnostic perspective. They aim to dissect facts, not to read them transparently. Accordingly, they tend to focus on behavior and language over empirical content. Arch-realists cringe. But this is precisely the view of many students, who are not privileged to the jargon or the meaning of all the black-boxed instruments. Nothing in these studies ultimately excludes a realist stance, although it may be absent. Like the naive student, one needs to pause to consider how the day-to-day practice of science yields knowledge. Science-in-the-making looks dramatically different than the ready-made science found neatly edited and arranged in textbooks (Chapter 2). As Bruno Latour, in particular, notes, our understanding of nature is the *outcome* of a deliberative process. Data are rarely unambiguous, and empirical "facts" are not established without considerable "negotiation" among alternative interpretations. Determining "the simple fact of the matter" is not always as simple as suggested in widespread images of science. Considerable interpretive resources are deployed in ascertaining the meaning of experimental results.[16] Sociologists show how much *work* is involved in science. From an epistemic perspective, they tend to profile the many factors, viewed as potential sources of error, that must be addressed before scientists feel secure about the facts (Figure 1.3). Many apparently trivial events may be significant. Sociologists also underscore that one cannot justifiably read the outcome of the debate anachronistically onto the process that led to its resolution. Rational reconstructions hide the uncertain horizons and contingencies in how real science proceeds (Chapter 4). If one is interested in teaching process of science, not merely science content, microsociological studies are important resources.

The language of microsociologists can be inflammatory because they claim to explain scientific knowledge causally. In their jargon, scientific knowledge is "constructed," or "socially constructed." But, again, the prudent observer does not succumb to mere rhetoric. Ultimately, *of course* knowledge is constructed. Scientists articulate facts where before there was uncertainty or ignorance. Justification, in particular, must be cobbled together from disparate shards of evidence. And *of course* science is socially constructed. Scientists work *collectively* as a *community*. Indeed, as underscored further below, diversity of perspectives often serves as a vital system of checks and balances in detecting bias or error. Hence, we need to dismantle the mythic impression that to be social or constructed is to be irrational or relativist.[17] The rational/social dichotomy is false. Science cannot be divorced from its human context. Science is a human activity. And if humans err, then science will as well. We need to understand how and when social factors foster reliability, not disavow the social character of science.

The third (and perhaps most challenging) type of sociology is cultural studies of science, which probe the relations between social ideology and scientific ideas and methods. Scientists inevitably draw ideas and values from their culture. Bias can result, carrying the imprimatur of science, as illustrated in the case of IQ and Jewish immigration (above). Even something as apparently innocuous as the taxonomic name of mammals can betray bias. Carl Linnaeus introduced the term *Mammalia* in 1758, in the tenth edition of his *Systema Naturae*. But unlike other names he used there, it is remarkably charged sexually, referring to women's breasts. Indeed, it ultimately reflects certain cultural values. Mammary glands and mammae (as external structures) are hardly the only unique or most distinctive structure of this group of animals. They also exhibit hair, the basis for the earlier names *Pilosa*, *Pillifera*, and *Trichozoa*. They also give live birth, hence another designation: *Vivipara*. In addition, they have two ventricles in the heart: hence, *Tetracoilia*, a later suggestion. One might also have focused on nourishment of offspring by milk, rather than a structure that typified only one sex, and only at certain ages. There were certainly viable alternatives. Why, then, did Linnaeus choose to name them 'mammals'? The name appeared at a time when there were lively cultural debates about the value of wet nursing and the domestic role of women. Only six years earlier, Linnaeus had penned a short tract critical of the widespread custom of wet-nursing and advocating breastfeeding by mothers. His taxonomic name thus helped render a mother's milk as essential to the organism's "natural" identity. That is, the biological name had political overtones. Numerous others during the period also portrayed a mother's nursing as "consonant to the laws of nature," even to the extent that it was responsible for a society's moral order. Such a role simultaneously helped exclude women from public discourse. One politician expressed it in the 1790s:

> Since when is one permitted to abandon one's sex? Since when is it decent for women to forsake the pious cares of their households and the cribs of their children, coming instead to public places, to hear speeches in the galleries and senate? Is it to men that nature confided domestic cares? Has she given us breasts to feed our children?

Linnaeus's name for mammals thus seems to have "infused nature with middle-class European notions of gender."[18] As exemplified in this case, cultural ideas—including values based on gender, race, class, and religion—can sometimes shape scientific concepts and research.

Cultural influences are certainly not all negative, however. The same sources (or biases) also lead to discovery. So, for example, Michael Faraday was inspired to look for the relationship between electricity and magnetism because his religion espoused the unity of nature. (He was not so lucky investigating the relationship between gravity and electricity.) William Buckland found fossils half buried in a cave and reported that the relics were proof of the Biblical flood. His research earned the Royal Society's highest honor, the Copley Medal, and inspired similar research across the continent. Charles Darwin was steeped in Victorian ideas about competition in society that helped shape his thinking about natural selection. William Harvey believed in the Renaissance notion of the microcosm, which guided his reasoning about circulation of the blood (Chapter 3). Ideas from culture can sometimes be productive. One cannot universally equate cultural influence with error.

Cultural studies of science have amply demonstrated that scientists reflect their culture in their thinking. This should surprise no one, really. Scientists work with the cognitive resources at hand. They will draw ideas and perspectives from the social context in which they live. Cultural context may thereby shape what topics are pursued, what questions are asked, what observations are noticed, how evidence is interpreted, and what theoretical virtues are preferred. The scope and content of science emerge from culture—for better or worse. The image of science typically promoted in science education—as pure and isolated from culture—thus needs substantial transformation. Accordingly, teachers might profile, for example, how the fervor for genomics fits into an economic and political milieu that supports discredited notions of genetic determinism.[19] Likewise, many claims in evolutionary psychology, a trendy but immature science, can be shown (like craniology) to express sexist and ethnic stances.[20] Critics of global warming and anthropogenic climate change—or of ozone depletion, or of acid rain, or of the dangers of secondhand smoke, or of regulating occupational health—may have political affiliations that signal intentionally misleading conclusions.[21] By consensus, skepticism is part of good science. Cultural studies of science help students learn precisely when, and perhaps how, to exercise such skepticism. The easily hollow slogan, "limits to science," becomes concrete.

NOS: NORMATIVE OR DESCRIPTIVE?

One common response to sociology of science, especially cultural studies, is to say "Yes But": "Yes But…that's not *science*. That's *pathological* science."[22] In this approach, all error is deftly carved away from science and attributed to interfering, supposedly extra-scientific factors: social and psychological. Science is thereby preserved pristine. Such a response is rightly dismissed as mere rhetoric, no more than a thin political stratagem. No mechanism ensures against pathological science, nor does any reliable diagnostic test distinguish it from pure science in context. But the appeal to pathological science is itself informative. It reveals an important guiding conception of science. That is, science is—or should be—perfect and error free. It can thus guarantee its claims in every instance. Here, science is idealized: an abstracted essence. *Authentic* science, in this view, does not err *by definition*. The contrasting view, of course, acknowledges science as actually practiced. Scientists do err, sometimes with significant social consequences. Each view, of course, leads to a different assessment of the authority of science. In one case, science—appropriately defined—is insulated against, or immune to, error. Science is thus the ultimate authority, especially where values enter discourse. In the other case, error is a part of science. Biases inherently limit its reliability. Here, the authority of scientists must be kept in check, especially where values enter discourse. The dilemma is clear. The challenges of sociology are intimately related to the very conception (or caricature) of science itself.

So: is the fundamental nature of science the ideal or the real? Is science characterized by its epistemic norms or by its historical practice? As suggested earlier, one may reject the either–or choice. The nature of science embraces *both* the ideal and the real, as well as the tension between them. The nature of science is compound. In the classroom, idealized science must be distinguished from—and not substitute for—science as actually practiced. One needs to be informed by *both* the philosophy and the sociology of science.

Coupling the two views requires a framework for reconciling them. One approach is to adopt a descriptive philosophy of science. Here, the philosopher documents epistemic strategies and perhaps links them with various outcomes historically.[23] A complementary approach is for sociology to adopt a normative stance. This was certainly the case for Robert Merton. He postulated four benchmark norms for "the growth of certified knowledge" in the 1940s: universalism, communalism, disinterestedness, and organized skepticism.[24] They are not always followed. For example, scientists do not always work selflessly for the benefit of all. But such instances do not threaten the status of Merton's social principles as *goals*. The challenge may be, instead, to imagine concrete institutional structures or practices that can help ensure that such sociological norms can be met.

Below I profile recent work that endeavors to bridge the concerns of philosophy and sociology. All address how to establish reliable knowledge. But they simultaneously acknowledge and address the findings of sociological studies about the factors that shape how scientists think and act. As a result, they are already significantly expanding the scope of philosophical conceptions of scientific methodology. In these approaches, descriptive and normative perspectives complement each other.

RECONCILING PHILOSOPHY AND SOCIOLOGY OF SCIENCE

Feminism might at first seem irrelevant to understanding how science works. But it has led to many insights and inspired much fruitful philosophical work. Initially, feminist critiques of science documented underrepresentation of women in science. Then they revealed gender bias in the interpretation of results (as in the case of Linnaeus's naming of the mammals). Surprisingly, the two features turned out to be closely related. On the basis of sociological studies, philosopher Sandra Harding has clarified two types of errors. In some cases, evidence was grossly misinterpreted. This was just recognizably bad science—although male researchers typically failed to *recognize* it as bad. More importantly, in other cases, claims seemed consistent with the evidence collected. All seemed well by standard methodological norms. But the evidence was essentially incomplete and, hence, misleading. Harding thus distinguished between *weak objectivity* and *strong objectivity*. Harding also underscored how women—indirectly disadvantaged by the findings—noticed the gendered errors. Their *standpoint* was relevant. Thus Harding linked epistemics with who participates in science. An abstract philosophical goal involves a concrete social component. In this view, to achieve strong objectivity, a scientific community must have the appropriate diversity of perspectives, or standpoints—especially by those potentially disempowered by the conclusions. Harding has generalized the argument, noting in particular the role of race/ethnicity and non-industrialized cultures as relevant to assessing scientific claims. Gender equity in the science classroom is not just a matter of fairness. It is also about fostering good science through the recruitment of multiple perspectives into the scientific profession.[25]

Helen Longino was also inspired by feminist studies. While maintaining a largely conventional orientation to justifying knowledge, she shifted focus to the social level. Namely, how do scientific *communities* reach trustworthy conclusions? In particular, she asked how one accommodates criticism, especially from a dissenting minority. Several factors seem important. Do opportunities exist for voicing criticism? Are cogent criticisms addressed responsibly and weighed equitably? Can a community fairly judge the arguments? These mark conditions for productive critical discourse *at the social level*. To ensure objectivity, then, the methods of science must go beyond empirical evidence and individual belief. They must also include *appropriate interactions among scientists*. Thus, a field that has reached consensus may have failed to fully validate its claims, if it has not addressed certain critical perspectives. For example, the community that endorsed IQ testing failed, even though it seemed to respect the empirical evidence, because critical perspectives were dismissed or not addressed. One consequence of Longino's analysis is that philosophers need not focus so strictly on what constitutes good evidence. Rather, such norms will inevitably arise through mutual criticism and be validated by the scientific community itself (again, when it functions fully). Norms may even change as research develops. Standards for evidence have certainly risen over the past several centuries. Like Harding, Longino has shifted the rules for good science beyond the individual, to include the social. Many educators, similarly, have promoted a social component in learning (Joan Solomon, for example, was an early advocate). In Longino's system, the cultural values and background assumptions that different participants bring to science are corrected through social checks and balances. The widespread dichotomy between the rational and the social thus seems unwarranted.[26]

Miriam Solomon also addresses epistemology at the social level, as indicated in her ideas about *social empiricism*. Unlike Longino, however, Solomon insists that rationality can emerge *only* at the social level. Individuals each have particular perspectives, or biases. None has great enough scope. Solomon borrows substantially from cognitive science, especially studies of decision making. That is, success in science must be possible with the limited cognitive abilities that humans bring to their work, not according to some utopian reasoning structure. Solomon analyzes a wide array of factors that shape scientific decisions. These "decision vectors" range from empirical consistency and salience of data to theoretical simplicity and peer pressure. They echo the "agonistic forces" described earlier by microsociologists. Decision vectors embrace motivational factors, social ideology, theoretical virtues, non-logical heuristics, personality, gender, and appreciation of the data. All can influence a scientist's thinking. One cannot stipulate, however—as in the cases of religion promoting the discoveries of Faraday, Buckland, and others—whether any vector is always beneficial or detrimental. Any outcome is contingent on the circumstances. Still, one can assess the aggregate of multiple decision vectors. Consensus is rational, Solomon says, when a theory demonstrates empirical success *and* the decision vectors show a balanced distribution. So, for example, if Boyle and Hobbes had agreed about the vacuum, despite their contrasting political beliefs, one could express increased confidence in the rationality of their joint conclusion. Solomon, Harding, and Longino all see the social level as providing a system of checks and balances against particular cultural and other biases only when the scientific community is appropriately diverse. Hence, objectivity (or strong objectivity) may be a property of scientific communities, not achievable by individuals. Effective science may be inherently social.[27]

One may also consider the case of Andrew Pickering. His 1984 book *Constructing Quarks* was an early sociological study adopting an agnostic posture about scientific realism. But as Pickering's research has deepened, his position has shifted. Like many sociologists, Pickering appreciates that representations of nature are not cast preformed upon a shore merely to be collected. Human agency is involved. No science occurs without scientists. No knowledge emerges that is not performative. So a story of discovery is ultimately about human behavior. Yet Pickering, once a particle physicist, now also sees that researchers respond to how their *apparatus* behaves. He now refers to "the dance of agency." Experimentalist and experiment act reciprocally, in tandem. The experimentalist, as primary agent, crafts an experiment. Then he "tunes in" to its material performance. He responds differently according to the outcome. In that differential response, agency is shared with the material world. For Pickering, the world is necessarily filtered through an intentional, human dialectic. The human and nonhuman are melded in a machinelike "mangle." Fellow sociologist Bruno Latour, too, has adopted realist tendencies. His landmark 1979 study of *Laboratory Life* epitomized arch-relativist ethnographic studies. But Latour soon started talking about "resistances" in the laboratory or field as harbingers of reality, at least as scientists interpreted them. Originally, he allowed only humans to serve as "spokespersons" for experimental events. Science was a network of human actors. But the politics of those human conversations and negotiations were influenced by the resistances in the lab. Soon science became an *actant* network, where objects or processes of nature seemed to have a role, even if still mediated by human interpretation. Pickering and Latour reflect transformations of many sociologists who now accept how human actions and interactions are shaped by what others would call plain observations or evidence. They differ from conventional philosophers, however, in refusing to erase the ineliminable role of human agency and interpretation. The microsociological jargon of "construction" does not exclude a role for some form of materialism. But it also refuses to eclipse the mediating role of humans.[28]

In conventional philosophical interpretations, career ambitions and political maneuvering among scientists, while inevitable perhaps, weaken the reliability of scientific knowledge. Not so for philosopher David Hull. He revels in an evolutionary view of science red in tooth and claw. Hull leaves judgments of methods and quality of evidence to the scientists themselves. He addresses instead the system by which they will be motivated to demand rigor from each other and even to escalate standards. Hull thus endorses old-fashioned competition. As long as there are professional rewards for making discoveries and reliable accounts, scientists will aim to produce them. Other scientists, desirous of the same rewards, will try to do better. Thus, the Edinburgh phrenology dispute was not a blemish on science, but a mechanism for indirectly improving the antagonists' ideas. As in the economic system envisioned by Adam Smith, reliability in science is ensured by the "visible hand" of competition. Hull's vision depends on context, of course. Do scientists uniformly act with the competitive motives he ascribes to them? Does a system of competition for reward truly function as effectively as he portrays it? Hull demonstrates, nevertheless, how (contrary to Brush's assumptions, perhaps) even unsavory motives might serve science through appropriate social organization. Even if his own system is wanting, Hull raises the important question of how social organization may guide native human motivations towards productive ends.[29]

More problems about the reliability in science arise when considering the social transfer of knowledge. Alvin Goldman, like Solomon, acknowledges the limits of humans as cognitive agents. He thus advocates the role of strategies for regulating reasoning—that is, for working towards counteracting our inherent cognitive flaws. Goldman does not pretend that some idealized method will prevent all error. Yet this does not preclude us from adopting methods to minimize them or to find and fix them. More importantly, Goldman has considered the problem of distributed expertise and knowledge. To say that something "is known" is not to say who knows it. Knowledge has *distribution patterns*. Problems can arise in a system where intellectual labor is divided—a relatively recent problem for science. How can non-experts judge the quality of knowledge if they cannot evaluate the evidence themselves? In particular, can someone who cannot justify the knowledge directly be said to know something themselves (the conventional philosophical stance)?[30] The challenge in its more general form is clearly familiar to teachers who aim to develop a citizenry of non-experts able to assess and apply expert scientific knowledge in social decision making.[31] Here, the problem is *social epistemology*, how we can know things through the knowledge of others. For example, we learn from others only by exercising trust in others and through a complex system of communication (media).[32] But trust can be misplaced. Claims may be fraudulent or incomplete and misleading. In a social setting, whose testimony can be trusted and when? Justifying someone's credibility may be as important as justifying the evidence itself, for scientists and non-scientists alike. Goldman's strategy—the regulation of potential error—underscores the importance of how experts are trained, certified, monitored, and sanctioned. The same applies to communication systems and to the dissemination of information through the media. As Goldman notes, a free market of ideas does not always foster trustworthy public knowledge. Educators may recognize that the critical-reading skills they have long taught may be constitutive of science, not merely a feature of consuming science. Hence, the problems addressed by Longino and Solomon are only the beginning of understanding the social dimension of science, interpreted in a cultural context. They extend to problems of testimony, expertise, credibility, and how facts travel.[33]

In all these examples, sociology helps generate a more robust *philosophical* view of scientific method. One more example may help conclude the discussion here. Philosophers and sociologists have argued at length about how scientists make judgements: whether they assess evidence or each other's credibility. Perhaps the debate reflects again their divergent emphases on ideal methods versus social practice. Using an accommodating posture, however, one finds that scientists do both. Anyone who listens to informal lab talk is aware that researchers frequently comment on the authors of journal articles and whether the work in their lab is any good. This is not mere gossip. Evaluating the skills of a researcher (his or her track record) proves to be a good strategy for assessing the quality of their work. Namely, should one accept the results uncritically or scrutinize them more thoroughly? No one can afford the time or resources to assess every detail or repeat every experiment, however ideal that may seem. Repeatability is an abstract principle, not a description of what scientists typically do. Researchers frequently economize by using such heuristics.[34] Sometimes, such heuristics fail. Then—and only then—do researchers adopt the bulkier but surer methods. They plod through the lab protocols, perhaps trying to isolate any experimental problem themselves. Credibility is thus a *proximal* criterion, used heuristically as a surrogate for an *ultimate* criterion, the evidence itself. Evidence may be more fundamental, but credibility is faster and generally applied first.[35] Both criteria are used, each in specific contexts. Notably, philosophical and sociological views may be reconciled. Together, they provide a more realistic and nuanced view of how science works, in practice, towards the ideal of reliable knowledge.

Perspectives are still under development. For example, reflecting on Harding's views on standpoint, philosophers wonder whether diverse standpoints can be achieved cognitively, by individual creative thinking, or only through social diversity. Reflecting on Longino's and Solomon's model of interaction, how can we know *which* critical perspectives are relevant in a particular case or *when* diverse perspectives are appropriately balanced? Specifying complete criteria for gauging expertise or the credibility or testimony of others (Goldman's aim) may be equally problematic. Puzzles remain. Positions are evolving. However, the examples here illustrate the fruitful intellectual work in resolving once apparently conflicting perspectives. Normative goals may be coupled with realistic expectations and pragmatic management skills.

STRATEGIES FOR TEACHING

Well, where does all this work on the intersection of philosophy and sociology of science leave teachers? Foremost, *teachers should clearly differentiate between normative and descriptive perspectives of the nature of science—and address both in teaching NOS fully*. An idealized version of science—especially of romanticized Science Triumphant—is an irresponsible lie, even if it embodies important guiding principles.[36] On the other hand, an image of science as bumbling through endless error, without articulating how scientists manage errors, is equally misleading.[37] To be complete, NOS must encompass both the normative *and* the descriptive.

Thus, while sociologists have documented some less-than-inspirational episodes of science, their findings about how culture shapes scientific conclusions are important to understanding fully "how science works." Sociological studies yield a much richer and more complete description of science. This is the broader focus encompassed in the concept of Whole Science (Chapter 1). The scientifically literate individual needs to learn about scientific error and the boundaries of scientific authority. At the same time, awareness of failures and flaws needs to be coupled with an understanding of the methods that might help make knowledge more reliable or trustworthy. How do such methods work? When do they work? Why do they sometimes break down? What may be the consequences when they fail? How do scientists, collectively, recognize the lapses? How do they remedy errors when they are exposed? All these questions may emerge from studying cases of science-in-the-making and reflecting on their NOS dimensions.

The Scylla of unbridled scientism that disturbs sociologists is indeed monstrous. It must be tamed and cultural expectations of science deflated. The ideals of science are misleading if they are not qualified. On the other hand, the Charybdis of cynical relativism that riles defenders of rationality is equally unacceptable. One must not conflate instances of actual practice with all of science or its aims. Nor must flaws in the past eclipse the prospect that methods may help regulate bias, even cultural bias. To chart a course between the Scylla of scientism and the Charybdis of relativism, teachers must adopt a dual nature of science.

Should the sociology of science be rated "X"? Decidedly not. But neither should one therefore abandon the philosophy of science. Understanding of the actual practice of science, with its occasions of error, through sociology must be coupled with the philosophical principles by which we endeavor to build reliable knowledge. As mapped by recent work reconciling these two fields, educators need to engage the sociological material far more deeply than indicated in simple profiles of NOS or of "scientific practices" for the classroom.[38] Concepts in science studies deepen (as they do in all fields), and responsible educators will rightly consider them as one further dimension in their ongoing professional development.

Thirty years after initially asking "Should the history of science be rated X?," Stephen Brush revisited his hallmark question. He recanted somewhat. Some scientists, he noted (offering several examples), did act with propriety and could indeed be considered good role models.[39] Sometimes, it seems, we need not choose between the ideal and the real.

7 | Kettlewell's Missing Evidence: A Study in Black and White

The peppered moth as a case of "black and white" • reducing nature to black and white • reducing scientific practice to black and white • reducing school science to black and white

Even non-biologists know the peppered moth, *Biston betularia*. The widely reproduced images of the moths against different backgrounds—black against mottled white, and mottled white against black (Figure 7.1)—vividly visualize natural selection.[1] What half-witted predatory bird would not notice the difference? The lesson for "survival of the fittest" is clear, even without words. The images are a virtual "before" and "after" view of Britain's forests, polluted during the Industrial Revolution. They explain evolution in an instant. No wonder these icons "pepper" biology textbooks. One text even includes the same paired images twice![2] Students readily recall these images long after leaving the classroom, a tribute to their potency—and to their importance for understanding science education more generally.

FIGURE. 7.1. "Peppered" and melanistic forms of the moth *Biston betularia* on contrasting backgrounds.

FIGURE 7.2. Various forms of the peppered moth: *insularia* (left, 1–5), *carbonaria* (right, 1), and *typica* (right, 2). Most textbook accounts omit *insularia*. (From Kettlewell, 1973, plate 9.1; reproduced by permission, Oxford University Press.)

The image of the peppered moths gained renown through H. B. D. Kettlewell. He investigated the survival rates of the moths in the contrasting forests of Birmingham and Dorset (ostensibly portrayed in Figure 7.1). Kettlewell presented his studies, including the now familiar images, to a lay audience in *Scientific American*. Reprints of that article became a stock feature of biology classrooms. In this 1959 article, a centennial tribute to *The Origin of Species*, Kettlewell underscored that Darwin's evidence for natural selection, while persuasive, had been only circumstantial. The peppered moths and other cases of industrial melanism, on the other hand, dramatically demonstrated evolution in action. They were, Kettlewell argued in his title, "Darwin's Missing Evidence."[3]

Consider, then, the image (Figure 7.2) that appears in Kettlewell's monograph, *The Evolution of Melanism*.[4] On the top right are the two familiar forms of *B. betularia*: *typica* (no. 2), the once common "peppered" form, and *carbonaria* (no. 1), the nearly black form that proliferated later. Arrayed on the left, however, are five other specimens of the same species, all intermediate in darkness: a third form, known as *insularia*. Here, one witnesses the whole series of light and dark between the two extremes. How would they fare in different environments compared with the others? How does this change our "image" of natural selection? Why does the popular story, promoted in part by Kettlewell, exclude them? Ironically, might we call the *insularia* moths "*Kettlewell's* missing evidence"?

REDUCING NATURE TO BLACK AND WHITE

Kettlewell was well aware of the *insularia* forms. Indeed, in his now famous field studies, he tallied the survival rates of all three forms.[5] In addition, he catalogued the frequency of the three forms in various locations around Britain for two decades (Figure 7.3). The incidence of *insularia* was some-

times as high as 40 percent or more, especially in southern Wales and the Isle of Man.[6] *Insularia* is no minor player. Although Kettlewell documented *insularia* in his scientific publications, the form disappeared from subsequent renditions of his work. For example, Figure 7.3 is redrawn in the *Scientific American* article, but the main text fails to refer to *insularia*.[7] Textbooks, too, sometimes reproduce this diagram, but usually without any gray section in the pie charts (treating *insularia* as a wholly melanic form).

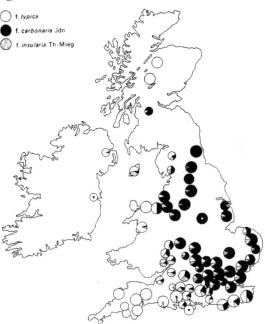

FIGURE 7.3. Frequency map of the forms of *Biston betularia*. (From Kettlewell, 1973, p. 135; reproduced by permission, Oxford University Press.)

The dust jacket of Kettlewell's monograph, similarly, sports a simplified image (akin to Figure 7.1).[8] Like the textbook versions today, it omits *insularia*. By hiding the real complexity, the now canonical presentation of the peppered moth reduces nature to black and white.

If accounts of Kettlewell's evidence are incomplete, of course, one should not conclude that the results no longer support natural selection in the wild.[9] However, in the classroom the simplified image can still be significantly misleading. It shapes how non-biologists think about evolution. It affects metaphors about nature and, hence, perceptions about what is deemed "natural" in human culture. Unschooled individuals already tend to conceive natural selection in stark terms. Survival is life-or-death. The struggle for existence has just two categories: the fit and the unfit. Competition has only winners and losers. The simplified peppered-moth case, especially as a benchmark example, merely reinforces this harsh stereotype. There seem to be, after all, only two types of moths. And only two types of environments. (Worse, each is also portrayed as homogeneous.) The choice, posed visually to the student as a vicarious predator, is pretty "black and white." A more textured view of selection, with *differential* survival and *differential* reproduction in sometimes *heterogeneous* environments, never even emerges as a possibility. Nature is cast in black-and-white simplicity.

Culturally, this either–or framework has powerful overtones. Images of competition and "survival of the fittest" pervade our society, from the Super Bowl, "American Idol," and school athletics to the job market, political elections, and national economies. Construed as a natural scenario, the peppered-moth case implicitly guides our thinking. A win–lose model of competition is subtly inscribed into the moth images. They indirectly legitimize that win–lose model. They convey visually (not logically!) that natural selection functions through clear, dualistic choices. Humans cannot escape the fundamental laws of nature, students think. The iconography embodies ideology. It strongly affects what we consider natural or normal and, thus, deem acceptable or unchangeable. (Perhaps this sounds unjustly cynical. If so, ask students yourself. Is competition "the way of the world"? Is that how nature works? Is that therefore how human society works?) The simplified version of the peppered moth is not idle. It helps reinforce notions of either–or, win–lose competition on the basketball court, in business, in Congress, and elsewhere in our culture.

The misleading imagery is only compounded when teachers treat the peppered moths genetically. With just two forms—dark and light—one assumes that one allele is dominant, the other recessive. That is the traditional Mendelian model, as described too by Kettlewell.[10] This assumption also fits conveniently into basic models of population genetics, which are frequently taught using the peppered-moth case as an example. Unfortunately, the diversity of moth forms (Figure 7.2) immediately implies greater complexity. Kettlewell imagined multiple alleles.[11] Other experts now suspect polygenic inheritance. We cannot really say that the expression of one allele suffices. In this case, one simple, false interpretation leads directly to another simple, false interpretation. The genetics of peppered moths is not "black and white," either. Indeed, the case is only one of many that help perpetuate the mistaken notion that Mendelian dominance itself is fundamental (Chapters 3 and 8).

Of course the peppered-moth case is hardly unique. Science teachers frequently teach not only simple concepts but simplified views of nature. Frictionless planes? Ideal gases? These are rarely presented as useful heuristics in a complex world, but rather as the essence of nature. If students learned thoroughly the nature of modeling in science, one might feel more confident that they would interpret the simplifications *as* simplifications.[12] But more often than not, nature is reduced to laws and simple lawlike behavior (for more on laws, see Chapter 8).

REDUCING SCIENTIFIC PRACTICE TO BLACK AND WHITE

Nature is not the only feature simplified in the Kettlewell case. The process of science is, too. Many biology texts honor the now classic study by going beyond the concepts. First, they name the scientist(!), giving a human dimension to science. Second, they celebrate the elegant design of his experiments. Many textbooks detail Kettlewell's key comparison of complementary environments. Some describe his experiments, which measured the differential survival of the moths that were marked, released into the wild, and then recaptured. Some even include tables or graphs of the original published data. However, these "textbook histories" are greatly streamlined. Like the image of the moths themselves, they leave out important information—with profound effect (Chapter 3).

For example, many textbooks highlight Kettlewell's exemplary scientific practice in using two contrasting environments to show the effects of natural selection. In the dark, polluted woods near Birmingham, the melanistic forms (*carbonaria*) were recaptured twice as frequently. On the other hand, in the lichen-covered woods of rural Dorset, the speckled forms (*typica*) were twice as likely to survive. The coupled investigations exemplify a colossal controlled experiment, with the selective environment as the single variable. That's the familiar textbook story. As historian Joel Hagen has observed, however, originally (in 1955) Kettlewell presented only data from Birmingham.[13] At first, he made no reference whatsoever to Dorset. Nor did he give any hint that his study was incomplete or preliminary, or that readers could expect forthcoming complementary data. Why? If the Dorset data were so crucial, was the first study flawed? What does Kettlewell's "missing" evidence mean in this instance?

Hagen considers several possible scenarios for why Kettlewell published the Birmingham results alone. Originally, perhaps, Kettlewell did not see the control as important. This seems likely, given subsequent criticism of his work, his personal correspondence, and the apparent timing of his plans to add Dorset the following year. In this case, the need to address criticism would have motivated Kettlewell's extended study, not an initial perception of the need for clarity. That is, he did not conceive the entire set of experiments in a single flash of insight. Rather, he patched together two separate studies. This scenario transforms the stereotypical image of great scientists working by way of "eureka!" moments into one involving less extraordinary modes of thinking and working.

In another scenario, Kettlewell could not afford to run both full-scale experiments simultaneously. The release–recapture method is labor intensive and he was working alone. (He had no funds to hire field assistants). Travel between the two sites would have been problematic. Later, Kettlewell enlisted his wife and son, for instance, to help conduct the research. Or perhaps Kettlewell began with a pilot study, which yielded unusually favorable results. Might he instead have rushed to publish simply to establish his priority? One of the hidden tasks in a recapture study is raising the hundreds of organisms for release—and having them all ready at the appropriate time of year. This would have meant breeding moths in cages and sorting each form—apparently not as simple a task as one might imagine. Was Kettlewell limited by sheer logistics? All these possible alternatives help reveal the complexities of doing science—labor, cost, ambition, developing reliable technique, maintaining lab organisms, and responding to peer criticism. They show that the process of science is not so "black and white" as the textbook stories typically lead us to believe.

Textbooks also tutor students to see Kettlewell's studies as well designed, definitive, and (hence) beyond all doubt. This earns them their classic status. At the same time, teachers often identify skepticism as a hallmark of science. Are Kettlewell's conclusions open to analysis or criticism? David Rudge offers an example. He focuses on Kettlewell's central claims about bird predation. At the time, many doubted whether birds preyed on peppered moths at all. Polluted versus unpolluted woods, Rudge contends, was not the key variable. Rather, Kettlewell made various ancillary observations to ensure that his recapture rates reflected predation, not some other environmental factor. Kettlewell checked possible bias in the traps, for example, and monitored migration from the study area. He enlisted ethologist Niko Tinbergen to film birds eating the moths in the wild. Texts rarely discuss these tests, Rudge notes. To further isolate differential predation as the chief causal factor, Kettlewell would need to have controlled for the presence (or absence) of birds, even if it required an unimaginably large exclusion enclosure. Yet alternative explanations seem not to have been ruled out. For instance, were the two sites parallel in all relevant respects? Nearly twice as many moths recaptured in Birmingham as in Dorset: why? Did the release of a large number of moths alter predation rates? Can we safely generalize from the limited studies to real nature? When viewed more closely, the path to secure conclusions is more complex.[14] The textbook narrative reduces the experimental reasoning, too, to black and white. Ultimately, a simple account distorts the process of science and perpetuates misleading stereotypes about scientific genius and experimental evidence.

Finally, the popularization of Kettlewell's work raises interesting questions about the ethics of reducing science to black and white as it moves from professional to lay contexts. Currently, omitting or hiding significant data violates norms of research conduct in scientific circles. But what are the relevant norms for interpreting science for the public or for other scientists outside the field? Did Kettlewell have an ethical responsibility to discuss *insularia* in his own popular (and now widely read) *Scientific American* article? Or is this precisely where professional judgment alone is responsible for simplifying scientific lessons for non-experts?

Parallel ethical questions arise in education: to what degree should teachers simplify the process of science for students?

The Kettlewell case, of course, is not isolated, nor even atypical. As an additional example, consider what John Cairns once described as "the most beautiful experiment in biology": Meselson and Stahl's work on DNA replication. It, too, is celebrated for its elegant, simple design. Yet the *process of science* en route to that ultimate simplicity was itself anything but simple.

Meselson and Stahl gained renown for demonstrating in 1958 how DNA replicates. Their experiment (described now in many introductory college textbooks and websites) explored Watson and Crick's model of DNA: if the double helix splits to replicate, does each strand indeed serve as a template for its own new complementary strand? Meselson and Stahl's achievement was twofold. First, they conceived how to label and identify the new versus parent strands. Second, they developed a method to separate the different forms of DNA resulting from successive replications. Labeling was done with isotopes: in this case, their different weights were just as important as their radioactivity. Separation occurred in a density gradient, established with a heavy salt solution in a high-speed centrifuge. Macromolecules of modestly different molecular weights float at distinct levels in the gradient. The resulting bands at each generation were visually definitive: "perfect Watson-Crickery," as described by one researcher.[15]

The experiment was remarkable in several ways. First, it captured a central theoretical question in a single experiment. The problem of DNA replication was certainly not new. Watson and Crick's model had puzzled researchers for several years. Imagining possible events at the molecular level is relatively easy. Manifesting them in the lab is quite another thing. Sometimes, the molecular realm is revealed piecemeal, in clues and partial glimpses. Here, one well-oriented probe sufficed. Second, the experimental design addressed all the alternative theoretical models of replication simultaneously. Failure to confirm one model would not lead to further tests exploring another. Third, the experiment was marked by laboratory expertise. Material skills matter as much as thinking. The results were clean and unambiguous. Finally, the team also introduced a new method of wide scope. The technique of using heavy isotopes to differentiate macromolecules, once demonstrated, could be applied elsewhere. Meselson and Stahl's experiment thus exhibited creative arrangement of laboratory conditions, theoretical import, clarity, and craft skills, all while pioneering an important new method. Rarely do all such elements come together in one work. When they do, scientists (and science teachers) justly celebrate.

Stahl later commented on the perceived beauty in his experiment: "It's very rare in biology that anything comes out like that. It's all so self-contained. All so internally self-supporting. Usually, if you are lucky to get a result in biology, you then spend the next year doing all those plausible controls to rule out other explanations; but this one was just a self-contained statement."[16] However, the final structure of the experiment, as a product, hides the process that led to it. It took historian Larry Holmes a whole book to document this one experiment fully.[17]

How did it begin? Was there a clear vision of the experiment? Matt Meselson and Frank Stahl met as graduate students in the summer of 1954 while at Woods Hole Biological Laboratory. Matt was a course assistant for James Watson himself. Frank was taking another course not available at his home institution. Stahl was drinking a gin and tonic under a tree. Watching from the main building, Watson remarked on his reputed fine lab skills. Matt, curious, went to introduce himself. Frank had been considering a statistical problem requiring calculus. Several days later, Matt offered a solution, impressing Frank in turn. An enduring friendship developed. Before the summer was out, Meselson had mentioned a prospective study on DNA replication and Stahl had structured it experimentally. It all started, then, with a chance encounter.

Where had Meselson's idea come from? Early in 1954, he had been working on problems on deuterium for a course on chemical bonds. He wondered whether organisms would live in heavy water (made with the hydrogen isotope). Later that spring, he attended a lecture by Jacques Monod on the synthesis of inducible enzymes. Meselson applied his new idea, wondering whether one could label the new proteins with the heavy isotope. If so, one could then separate new and old proteins by density in an appropriate solution. One would float to the top, the other sink to the bottom. The innovative design for the later experiment on DNA thus first emerged in an entirely different context! A few months later, when Max Delbrück introduced him to his recent models of DNA replication, Meselson saw another application of his scheme—the one he shared with Stahl at Woods Hole in the summer of 1954.

But still the route to the last run of the experiment three and a half years later was hardly direct. Before long, 5-bromouracil replaced the heavy deuterium in their design. It would substitute directly for thymine in the

DNA and provide a more dramatic weight difference. This, in turn, led the team to a second line of investigation on mutagenesis, which soon became primary. Finding a solution with an appropriate density involved trial and error. Would KBr work? No. $MgSO_4$? No. $Ba(ClO_4)_2$? No. CsCl? Perhaps. But at what concentration? In trying the new technique, they discovered to their dismay that centrifugation destroyed the homogeneous density, creating a gradient instead. They had planned to separate the DNA in *discrete* layers! Fortunately, the gradient was gradual enough. The key molecules would still separate. Regardless, they explored electrophoresis as a possible alternative. In August 1957, Meselson saw an advertisement for the nitrogen isotope, ^{15}N. They had rejected it earlier, assuming that the weight difference with DNA using ^{14}N would be too small to measure. The new density-gradient method, with its potential for separating molecules with small differences, now made that possible once again. Suddenly, 5-bromouracil was abandoned. The mutagenesis inquiry was set aside. The experiment now so celebrated by history finally emerged. But there were many practical challenges as well: for example, competing for time on the centrifuge, finding the right spinning speed and chemicals to break open the bacteria, getting theses written (on other topics), going on job interviews. If the final experiment was simple, the process was anything but.

Real science hardly resembles the cookbook labs one frequently encounters in school classrooms. Nor is it the formulaic "scientific method" enforced on many science-fair projects. Science is a creative enterprise, filled with metaphoric thinking, chance encounters, false starts, tinkering, and plain hard work. The final idealized "textbook experiment" tends to disguise how it developed from a convergence of contingencies. Appreciating Meselson and Stahl's experiment—as much as appreciating science in general—includes understanding its convoluted history as well as the beauty of the product.

The history of Meselson and Stahl's "simple" experiment thus turns out to be quite complex. Indeed, it is likely too advanced to share in all its details with introductory science students. Yet the NOS lesson for teachers remains: even *"the most beautiful experiment in biology," when viewed in the context of the process of science, exhibited complexity.* The same complexity of process hides behind the ultimate elegance of Mendeleev's periodic table, Newton's "crucial" experiment on the nature of white light, Heezen and Tharp's mapping of earthquake epicenters, and so on.[18] The elegance of experiments so often paraded in textbooks may well be worth celebrating. But the aesthetic goal need not—and should not—eclipse an honest rendering of how science actually works.

REDUCING SCHOOL SCIENCE TO BLACK AND WHITE

Much as the images of the peppered moths typically serve as a visual epitome of evolution, the Kettlewell case can serve as an epitome of common approaches to science education. Most important, it highlights the difference between real science, as performed by research scientists, and *School Science*, the reconstructed version that appears in teaching. The current goal of most School Science, exemplified in the peppered moths, seems to be to reduce real science to black and white. But why? Is the goal of simplicity warranted?

The motivations for simplicity are obvious. School Science is easier to learn. It is easier to teach. It is easier to test. One might even contend that it is a professional responsibility. Reducing science concepts or the process of science to its simpler forms may seem a necessary form of respect for students. School Science certainly conveys an impression of learning.

However, the consequent danger is that we convey a false image of the world—of nature and of science both. In a sense, we condition students to expect simplicity. When they later encounter complexity, as they must, they can feel betrayed. They will "simply" lack the skills to interpret the circumstances. Recently, several outraged individuals have sued scientists for making mistakes.[19] What black-and-white frame of mind did our educational culture produce that expects science never to err? Did cookbook school labs and textbook celebrations of famous experiments help shape their thinking?

Many teachers assume that one must begin simply and introduce the complex only gradually, layer by layer. "Basics first," so the maxim goes. This approach appears to be justified. Yet how often do teachers address "Complexities later"? The implicit promise is rarely fulfilled.

The goals of science education and scientific literacy consistently refer to the role of science in social and personal decision making (Chapter 1). Most such issues are quite complex. The safety of Vioxx as a pain reliever, the prediction of earthquakes, the reasons for the hole in the ozone layer: these are not resolved with simple experiments like Kettlewell's or Meselson and Stahl's—or any of the other classic experiments typically profiled in textbooks. By focusing only on such examples, School Science prepares students only for simple problems. That is, a diet of School Science does not yield scientific literacy.

No one should be surprised, therefore, when the science in public issues is reduced to black and white. A prime example is the history of fluoridation of public water supplies in the 1950s and 60s and beyond. Originally, research indicated that modest fluoridation could help prevent tooth decay. Yet some people found the introduction of a chemical into their water, without their individual consent, to be a bureaucratic imposition reminiscent of totalitarian states. So anti-fluoridationists arose. What might have unfolded as a discussion about the role of government in public health, however, was transformed into a debate about the science. That is, each political position appealed to the scientific data as the objective arbiter. The evidence therefore became dichotomized into the simple categories of "pro" and "con." For example, critics pointed to the risks of excess fluoride (fluorosis), possibly including cancer. The voice of science was polarized. At this point, the scientific evidence might have informed a discussion of how to manage the "pros" without experiencing the "cons." But each side presented evidence as speaking *exclusively* for them. That is, *both* sides adopted an NOS posture that science could provide a unequivocal, black-and-white answer. And so debate festered.[20] With no easy resolution at this level, arguments shifted from the credibility of claims to the credibility of the spokespersons who made those claims. The NOS assumption was that you were either a good scientist, who defended the evidence, or a bad scientist, who was biased by political factors (also see Chapter 4). Again, the science was reduced to stark either–or, black-or-white terms.

The fluoridation controversy bears witness to common NOS views. Science is regarded as always certain and unambiguous. Yet, as in this case, science may reach multiple conclusions that can conflict in policy decisions. Here, fairly reliable findings indicated both benefits and risks. The values of dental health and risk of cancer were not commensurable, however. Science cannot assess conclusively whether fluoridation is acceptable. One must first resolve the various relevant *values*. The science does not yield black-and-white answers, despite the textbook images. Again, one should not see the case of fluoridation as unusual. One may find science cast in similar extremes in cases involving nuclear power, a 1986 referendum in California on mandatory HIV testing, the cold-fusion fiasco of 1989, or the teaching of evolution.[21] These are the types of occasions that should inspire science educators to convey the subtleties and complexities of science.

It seems that we have conditioned our citizenry to regard all scientific evidence as black and white. But to be well informed means in part to understand that science can involve uncertainty, ambiguous results, and mixed evidence. That is, studies may be incomplete. Data may be subject to contrasting interpretations. Different studies may support contradictory conclusions.

Perhaps we serve our students better by exposing them, on a few requisite occasions, to a messy, unordered, complex world. How do we negotiate a way through it? Sometimes, the primary challenge in science is not even to solve the problem, but to tease out a clear question from a tangled network of processes. Students may need lessons in how to address unruly complexity.[22]

Others no doubt teach simple concepts intending to equip students with the fundamental tools for interpreting a world that is far too complex to master in its entirety. Unfortunately, the curriculum rarely includes how to apply the simple concepts in complex scenarios. Students are thus underprepared.

The solution is simple: teach complexity.[23] At least on occasions. That means, of course, setting the textbook aside periodically and going beyond it. Students need to see that nature does not always fit the simple models in the textbook. They need to encounter ambiguity, qualified judgement, and the limits of reasoning in science. They need to understand science as *work*. That is, sometimes they need the *real* peppered-moth case, not the diluted black-and-white one of School Science. Even one example, articulated well, can dispel entrenched views by showing that the norm of simplicity is artificially imposed.

Ideally, teachers will also explicitly contrast the simple and complex pictures. That is, simple models can mislead us. To understand that, one must analyze the differences. What complexities can be hidden, and how do they matter? Through experience with samples of actual complexity, students can learn when to be wary of simple claims and how to pose the right questions that probe deeper (see Chapter 8 on Boyle's law).

Another strategy is to teach questions instead of answers. Replace prepackaged science-made with science-in-the-making. When students confront genuine questions without obvious answers, they begin to understand the scientist's challenge of making sense of the world. They will soon find the need to target, organize, and filter observations. Guided through an investigation, they begin to exercise experimental reasoning with all its subtleties. Here, historical case studies prove effective vehicles for situating students in rich problem scenarios (Chapter 2; for the Kettlewell case, see http://doingbiology.net/kettlewell.htm). The aim need not be to teach *about* Kettlewell. Rather, one poses Kettlewell's problem of industrial melanism. What alternative explanations are possible? What observations would help us determine which explanation is most reliable? How do we interpret various findings and draw conclusions, with what degree of confidence? Again, even one experience through a textured problem sets a context for interpreting all scientific claims.

For the experienced teacher, the details of the peppered moths and Kettlewell's research can be entertaining. They can provide fascinating anecdotes to add depth while teaching this particular case in a biology class. But they should be more than that. They are an occasion to reflect on the difference between real science and School Science—and on teaching the nature of science, so often reduced to black and white.

8 | Teaching Lawless Science

Laws and Boyle's law • what's not in Boyle's law • rampant lawlessness • the law metaphor • lawless causality, material models, and case-based reasoning • laws, maps, models, and theories • teaching lawless science

Imagine teaching lawless science. Perhaps you might envision science red in tooth and claw, with scientists, motivated only by competition, at each other's throats. Or nature itself in chaos, careening out of control. Or gangs of science-teacher vigilantes roaming through school hallways wielding huge 20-gallon Leyden jars, ready to unleash electrical havoc on any unruly student. Each image, more laughable than the next, seems worth a chuckle. But the whimsy should not obscure the substantive lessons encrypted here. Each wild image betrays a particular view about nature, science, and science teaching, all based on how we conceive laws of nature. And each view adversely affects how we understand the nature of science. We need to reassess this core concept. As provocative as it may sound, we may indeed want to teach lawless science.

The concept of scientific laws has a venerable tradition. Laws are empirically substantiated regularities. They are regarded as universal. And invariant. For many, they are the basic units of scientific knowledge. Laws reflect a conception of nature as lawlike or machinelike, as expressed in the mechanical philosophy, famously advocated by Robert Boyle in the late 1600s.[1] Nature may thus be described by reducing phenomena to their parts and the laws that govern them. Elucidating these laws is widely portrayed as the goal of science. And a foundation of its authority. Accordingly, laws earn a central place in science education. Familiar examples include Snell's law of refraction, Galileo's law of the pendulum, Newton's laws of motion, Ohm's law of electrical resistance, and Mendel's law of independent assortment. The list goes on and on, of course. These laws are so integral to teaching that it may seem nigh impossible to conceive of science education without them. Yet we may want to rethink this stance.

The specific case to be considered here is Boyle's law. It is among the most standard concepts in introductory chemistry and physical science. Nearly every science curriculum gives it central place. For example, in the past three decades of the *Journal of Chemical Education* one can find the law mentioned in 142 articles: that's an average of twice every five issues. One may indeed wonder at the law's prominence. It can hardly be because students need to understand how gases behave. What consumer choice or citizen decision requires knowing the relationship of gas pressure and volume? No, the importance of Boyle's law is as a cultural icon. It epitomizes the concept of a law of nature, especially as a hallmark of science. For someone promoting healthy NOS understanding, Boyle's law provides a prime occasion for reflection.

Many science teachers, even with no formal history background, learn about Boyle and the origin of his law.[2] Namely, Boyle was interested in the "spring of air." Air exerts pressure, felt when it is compressed. Borrowing from a device of Otto von Guericke, Boyle enlisted Robert Hooke to build an "air pump" (image opening the chapter, in the right hand). They did not generate a true vacuum, but the vessel did exhibit extremely low pressure. In 1660, Boyle published his findings on its many effects—on magnets, sound transmission, sealed bladders, burning candles, the life of small animals, and more. In response to criticism, Boyle further demonstrated the strength of the spring of air. He and Hooke compressed a small amount of air trapped in the end of a glass J-tube with increasing amounts of mercury. In a similar second set of experiments, they used the new air pump to draw up the column of mercury to dilate the trapped air. In both cases, they recorded the volume of the trapped air and the corresponding height of the column of mercury (an indirect measure of its weight, or pressure). Even today, students can graph his figures. This helps visualize the law, which describes the relationship of pressure and volume in gases.[3]

Boyle's law indicates that as the volume of a fixed amount of gas increases, the pressure decreases, and vice versa. The relationship can be expressed in many ways. For example: $PV = k$. Or: $P_1V_1 = P_2V_2$. In the 1662 original, Boyle referred only modestly to "the Hypothesis, that supposes the pressures and expansions to be in reciprocal proportion."[4] Hence, the version used in the analysis here:

$$P \propto 1/V$$

The pressure is inversely proportional to the volume.

The formula or equation, in whatever form, has become an icon of natural order and the aims of science—and, thus, of science education. Yet this paradigmatic law is misleading, if not false—a lie that contains important clues for teaching the nature of science.[5]

WHAT'S *NOT* IN BOYLE'S LAW

Boyle's law is an indispensable tool for calculating pressure or volume within certain systems when one of those two variables changes. Yet, as noted, in the context of science education it becomes much more: it is an epitome of scientific laws. As typically presented, such laws are universal and invariant generalizations.[6] That is, they apply everywhere and without exception. Indeed, their universality and invariance seem integral to their value and authority. However, Boyle's law, so frequently paraded as an exemplary law, ironically does not meet these very standards!

As the original 1661 data indicate, Boyle's law certainly holds true in Boyle's J-tube. But it does not hold in all cases. It depends on context. At high pressures, the direct inverse relationship of pressure and volume breaks down. Henri Regnault and, later, Louis Cailletet noted this variation two centuries after Boyle's work.[7] In modern terms, we might say that there is a limit to compression. Under very high pressures, a gas begins to behave more like a liquid than a gas. The scope of Boyle's law is limited. It only holds in the restricted domain of pressures up to approximately 10 atmospheres. Never mind that the high pressures are rarely encountered in daily human life: the law is plainly not universal.

Perhaps the behavior of gases at high pressures is a rare, minor exception. Suppose one stipulates explicitly, then, that only certain pressures apply:

$$\forall \, P \not\ggg 0 \;\Rightarrow\; P \propto 1/V$$

For all pressures not substantially greater than zero,
the pressure is inversely proportional to the volume.

Is Boyle's law universal now? Well, no. The inverse relationship breaks down at low temperatures as well. Historically, this was noted by Thomas Andrews in the 1860s (Figure 8.1).[8]

Here, another context qualifies scope. But this case also introduces something unexpected: a new variable. The equation or formula does not even refer to temperature. An experimenter, for example, would need to be aware of temperature, if only to ensure that the temperature was not extremely low. The simple expression of Boyle's law hides this relevant variable entirely.

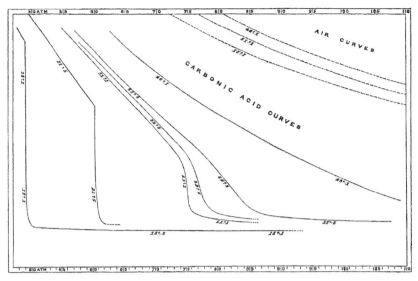

FIGURE 8.1. Pressure–volume curves for carbon dioxide as temperature decreases. Note the distorted curves at lower left (Andrews, 1869, p. 44).

Well, then, let us add both these conditions, or provisos:

$$\forall\, T \gg 0 \;\&\; \forall\, P \ggg 0 \;\Rightarrow\; P \propto {}^1\!/_V$$

For all temperatures well above zero,
and for all pressures not too high,
the pressure is inversely proportional to the volume.

Is Boyle's law fixed now? For example, with the provisos of scope, is the law now invariant? Well, no. Temperature is indeed important. Not just the range of the temperature, but also its holding constant. As Boyle himself (and others of his era) noted, gas pressure is sensitive to changes in temperature, as exemplified in several apparatus for measuring temperature. A few decades later, of course, Jacques Charles (in 1787) and Joseph Louis Gay-Lussac (1802) formalized this in yet another law, which now bears (alternately) their names.[9] Hence, while one may address the relationship of pressure and volume independently of temperature, temperature is nonetheless relevant. Constant temperature is a boundary condition. Sometimes it is expressed as a *ceteris paribus* clause: "all else being equal." But not all other things need to be equal. The amount of illumination, the relative motion of the system, or the gravity exerted on it have no effect on gas behavior, so far as we know. No one need stipulate these as boundary conditions or imagine them in a *ceteris paribus* assumption. So specifying what must remain constant is important indeed if one expects the law to hold. The invariance of Boyle's law, ironically, depends on context.

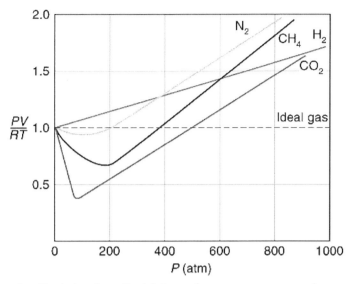

FIGURE 8.2. Deviation from Boyle's laws of some common gases due to intermolecular forces. (Image courtesy of Prentice Hall.)

Well, let us now add our additional boundary condition:

$$\forall\, T \gg 0 \ \&\ \forall\, P \not\gg 0 \ \&\ \forall\, \Delta T = 0 \ \Rightarrow\ P \propto 1/V$$

For all temperatures well above zero,
and for all pressures not too high,
and at constant temperatures,
the pressure is inversely proportional to the volume.

Is Boyle's law expressed fully now? No, still not yet(!). If the volume is very low (or the density very high), the volume of the gas molecules in relation to the volume of the space between them becomes significant. Intermolecular interactions (London forces) become relevant. The behavior of the gas changes noticeably. Moreover, because the size of the gas molecule matters, the variation is specific for each gas. Yet another factor qualifies Boyle's law. Even the qualifications are not regular. Moreover, any gas with strong polarities may also exhibit intermolecular forces, although of another sort. Such gases—carbon dioxide, for one—also vary from the simplified law (Figure 8.2). Ultimately, then, we must also take into account the nature or identity of the gas (or gases). Johannes van der Waals investigated these various dimensions of gas behavior in the 1870s and 80s, as well as the nature of some of the intermolecular forces, work that was recognized by a Nobel Prize in 1910.[10] One can "correct" for the subtleties of molecular size and interactions, but the corrections (known as van der Waals constants) differ for each gas (Figure 8.3). Thus, even his now well-known generalized

FIGURE 8.3. Van der Waals constants for various gases.

Compound	a (L^2-atm/mol^2)	b (L/mol)
He	0.03412	0.02370
Ne	0.2107	0.01709
H_2	0.2444	0.02661
Ar	1.345	0.03219
O	1.360	0.03803
N	1.390	0.03913
CO	1.485	0.03985
CH_4	2.253	0.04278
CO_2	3.592	0.04267
NH_3	4.170	0.03707

form of the gas law included variables specific for each gas:

$$\left(P + \frac{a}{V^2}\right)(V - b) = nRT$$

To be accurate, or realistic, Boyle's law must sacrifice universality. Boyle's law, like so many others, is lawlike only when qualified:

$$\forall \text{ elastically compressible gases,}$$

$$\forall\, T \gg 0 \ \& \ \forall\, P \not\gg 0 \ \& \ \forall\, V \gg 0 \ \& \ \forall\, \Delta T = 0 \ \Rightarrow \ P \propto 1/V$$

For all elastically compressible gases,
for all temperatures well above zero
and for all pressures not too high,
and for all volumes well above zero,
and at constant temperatures,
the pressure is inversely proportional to the volume.

A critical analysis of Boyle's law exposes and highlights this double irony: universality comes at the cost of limited scope; invariance, only with conditions. The lawlike world of "for all..." is inseparably coupled with a set of contingent "if-and-only-ifs." This is the lesson typically missing when teaching Boyle's law: what's *not* in Boyle's law—a world full of context and contingency, not simple rules.

RAMPANT LAWLESSNESS

One may well be tempted to imagine that Boyle's law is an exception and that most scientific laws are indeed universal and invariant. Not so. Consider Galileo's "law" of the period of the pendulum, another standard favorite in introductory physics classes[11]:

$$t = 2\pi \sqrt{\frac{l}{g}}$$

(where t = the time of the period, g = 9.8 m/sec², and l is the length). The mathematics are simple enough that teachers and textbooks can, and often do, lead students through the derivation. There is a critical assumption in that reasoning process, however. At one step, a value for the sine of the angle of the swing is required. By assuming that the angle is very small, one can approximate the value of $\sin(x)$ with x, which makes everything much(!) easier and, equally important, much cleaner. Students are cautioned (sometimes) that the formula works only for very small angles. Strictly speaking, though, the simplified formula is false. The precision implied by the mathematics (and with it, the image of scientific authority) can easily mislead. The formula delivers, at best, an approximation. Understanding fully the nature of this approximation is important, but this involves a level of subtlety not usually found in the classroom.

The familiar equation is based on the broader characterization:

$$\frac{d^2x}{dt^2} + \left(\frac{g}{l}\right) \sin(x) = 0$$

(where x is the angle of swing). This may be universal, but it is rarely used. It cannot be solved analytically. One typically approaches a solution by resorting to some rather advanced mathematics—expanding the Taylor series approximation for $\sin(x)$ to the desired precision:

$$\sin(x) = x - x^3/3! + x^5/5! - x^7/7! + \ldots$$

One then tests values iteratively by substitution—a form of brute trial and error. No wonder that one encounters this only in far more advanced classes (for college physics majors). In this context, one can see more clearly the small-angle formula for what it is: a lie. A convenient and useful lie, perhaps. But a lie nonetheless.

Even the "correct" expression sells reality short. It assumes that the mass is concentrated on a single point. It assumes that no friction affects either the fulcrum or the interaction of the pendulum and its medium. These are not just boundary conditions, as advanced physics students learn, but unrealizable idealizations. The law of the pendulum, like Boyle's law, is neither universal nor invariant. Most physics students who measure the motion of a pendulum in the lab encounter damping. They recognize the deficits. They will also likely never need to entrust their lives to knowing the limits of the formula for the period of a pendulum at small angles. Few will ever try to build a timepiece requiring a frictionless pendulum. The value of the classroom law is not about the pendulum itself. Rather (again, as in the case of Boyle's law), it seems to be about framing nature in a simple mathematical expression. When the law of the pendulum replaces the motion of real pendulums, however, as it inevitably does in the interest of simplicity, the School Science *lies* about the simplicity and order of nature (also see Chapter 7).

Ohm's law of electrical resistance, too, has numerous exceptions: for example, at high current densities. Many common materials "violate" Ohm's law: temperature-sensitive resistors (such as filaments in incandescent light bulbs, or sensors in digital thermostats), air (whose threshold resistance results in bolts of lightning), diodes (common electronics components), light-sensitive resistors, piezoelectrics (used in touch-sensitive switches), weak electrolyte solutions, varistors, and high-vacuum electron tubes, as well as other more technical variants. Newton's laws of motion do not apply at relativistic velocities (approaching the speed of light) and (like the pendulum law) exclude friction. Mendel's law of independent assortment breaks down, as most students learn, for genes linked on the same chromosome (see Chapter 3). Snell's law of refraction does not apply to Icelandic spar, or calcite. Nor does it accommodate angles of total internal reflection. It, too, comes with numerous *ceteris paribus* clauses.[12] In all these cases, laws seem very unlawlike. Scope circumscribes universality. Boundary conditions and exceptions limit invariance. That is, in case after case of laws presented in School Science, context matters.[13] The implicit message of pervasive lawlike order, echoed again and again, is highly misleading.

Thus, one may be impressed by an apparatus under special conditions in a laboratory, where Boyle's law is indeed observable, such as Boyle's J-tube. A change in volume is observed only when the system is closed. In an open system, such as the atmosphere, a gas under pressure does not change volume. It moves. *Wind* happens. Indeed, the cases where Boyle's law can usefully apply are few: pistons (say, in internal combustion engines), industrial boilers, some ventilation systems. Boyle's law is not basic at all. It is an esoteric scientific fact. Few students will ever need to know or apply Boyle's law in either their personal or professional lives. Yet it is *de rigueur* for science education. The ultimate lesson of Boyle's law is not about how gases behave. Like virtually all the laws in School Science, it is about nature being simple and lawlike. Ironically, all those "laws of nature" grossly misrepresent nature. We live in a world where weather happens. We live in a world of friction. We live in a world of lightning, chromosomes, large-angle pendulums, and unruly complexity.[14] School Science, centered on laws, fundamentally misleads students. Accordingly, we need to teach about lawless science.

THE LAW METAPHOR

The mere mention of lawless science bristles with images of anarchy in nature. It reeks of chaos and disorder.[15] The implication can be so unsettling that it eclipses thoughtful reflection about the nature of laws. However, the effect, even when deeply emotional, is based largely on word connotations and metaphor. The *language* of "laws of nature" is a metaphor that can powerfully shape thinking.[16] That is, all the familiar meanings of laws *in the human political realm* are transferred *to nature*. Thus, laws in science, like laws in society, seem to dictate "expected" behavior and how nature "should" act.

They are not merely descriptive. They are cryptically *prescriptive*. Note the use of specific words and their overtones. For example, one hears that a gas "behaves according to" Boyle's law. A system is "governed by" Newton's laws of motion. A resistor "obeys" or "follows" Ohm's law (or does not). Nature is subtly anthropomorphized. As Boyle himself stressed, however, air does not have free will nor act intentionally. Still, the language conveys a standard of sanctioned behavior. Laws seem to ensure order in nature, because that's what laws in society are supposed to do.

The metaphor is pervasive. And largely hidden from consciousness. Using the normative connotations, "laws of nature" implicitly sort natural phenomena into two categories: the sanctioned lawlike, and the unsanctioned unlawlike. Again, the words and phrases are telling. Thus, Icelandic spar *violates* Snell's law. A large-angle pendulum, or one dampened by friction, *deviates* from Galileo's law. Mendel's law *breaks down* for linked genes. Diodes, air, light bulbs, etc., are *exceptions* to Ohm's law. At high pressures and low temperatures, carbon dioxide *fails to adhere to* Boyle's law. It refers to *ideal gases*. The other gases are implicitly less than ideal. In these ways, the language of violation, deviation, failure, and breakdown expresses and subtly reinforces a view that something is "wrong" if nature is not lawlike. In such cases, nature is, ironically, deemed *un*natural.

The tropes of lawlike language were well appreciated by Bugs Bunny. When the high-dive plank he was standing on was sawed off its support, his pursuer and the support fell, not him. "I know this defies the law of gravity," he commented, smiling. Then he added wryly, "but you see, I never studied Law." Indeed, the very phrase "law of nature" should strike one as peculiar. How did legal terminology become applied to science? How did non-intentional nature get mired with normative rules? The perhaps surprising culprit is natural theology. Originally, in the late Middle Ages, laws referred to basic principles in formal systems. They functioned in mathematics and astronomy, as in legal contexts, to ground logical derivations and to justify explanations. God's world was rational. With the emergence of windmills, watermills, and other self-actuating mechanical technologies, it became possible to envision God exercising His will indirectly through causal principles, rather than directly in each separate motion. Faith in a rational, divine order led René Descartes in the mid-1600s to call these basic principles "laws of nature." The search for laws of nature, for Descartes and many others in the seventeenth century, was thus motivated by a desire to understand God's ordained order in the world. Science was religiously motivated.[17] Boyle adopted this view in his mechanical philosophy, a belief that God operated through basic rules of motion that could be discovered through experimental investigation:

> For, for aught I can clearly discern, whatsoever is performed in the
> merely material world, is really done by particular bodies acting ac-

cording to the laws of motion, rest, etc., that are settled and maintained by God among things corporeal.[18]

Boyle's notion of the mechanical "spring of air" (its pressure) reflected belief
in an autonomous natural order instilled by God.

Western science, of course, has since mostly shed this early theological
framework. Still, potent vestiges of these perspectives persist. For example,
we still have faith in a natural order, although we no longer associate this
order with an omnipotent creator. Scientists generally assume that the
world *is* fundamentally orderly and lawlike. But as physicist and philosopher
Nancy Cartwright observes, that is an empirical question. She entertains
the possibility of a "dappled world," where laws are only part of the story.
Simplicity, for example, is widely endorsed not just as a pragmatic preference,
but as a conviction that nature is inherently simple (Occam's razor).[19] In the
case of understanding how bodies cool, however (as just one example), this
assumption of simplicity misled scientists for many decades.[20] Laws, such as
Boyle's law, imply that nature is simple and orderly. Yet, as noted earlier, the
pressure–volume relationship is quite complex. Thus, when School Science
focuses primarily on laws, it inherently endorses and inculcates the ideology
of the law metaphor.

A fascinating irony lurks in Boyle's law. That is, although Boyle believed
in laws, he never presented "his" law itself as a law. He certainly produced
data commensurate with the law, yet he did not write an *equation* for the
relationship. He surely understood the mathematics of "what the pressure
should be according to the hypothesis," yet he was generally disinclined
to characterize nature mathematically.[21] Also, while Boyle embraced the
concept of laws in science in a theological framework, he did not regard
his findings about the condensation and rarefaction of air as a universal
law in the sense now accepted. Rather, he referred to the spring of air
more modestly, as a "habit of nature" or "custom of nature": local, perhaps
contingent, in nature.[22] He was well aware, for example, that temperature
and "atmospheric tides" (today's weather systems) affected the Torricelli tube
(today's mercury barometer).[23] Boyle's law, as a law, is not strictly Boyle's.
Teachers may thus want to reflect on Boyle's own modest posture whenever
they refer to "Boyle's" law.

Given the effects of the law metaphor and the vestiges of history, it is all
too easy to overstate the role of laws in science. For example, in presenting
the pendulum as a case for the classroom, former schoolteacher Michael
Matthews rhapsodizes about laws as the foundation of modern science.
Idealizations, such as Galileo's law of pendular motion, he contends, were
necessary for mathematical approaches and, thus, the emergence of science
in the 1600s—what he calls "the *raison d'être* of the Scientific Revolution."
Anything else, he claims, leads one to abandon science and embrace
relativism. For Matthews, as for others perhaps, you cannot have science

without laws.[24] Ironically, this view discounts the vast achievements and the methods of prehistoric, indigenous, and non-Western science.[25]

Matthews's history, however, is shaped by his veneration for laws, another case of Lawson's shoehorn (Chapter 5). Expert scholarship on the emergence of modern science offers a much different image. For example, Boyle's experiments with the air pump were certainly hailed by his contemporaries. When the Royal Society was founded as an institution "for the improvement of naturall knowledge by Experiment" in 1662, Boyle's work was paraded as a paradigm. However, what members valued, what they considered new and significant, was the style of extended experimentation and detailed reporting of methods, not his law. As vividly documented by Lisa Jardine in her aptly titled volume, *Ingenious Pursuits*, the Scientific Revolution of the 1600s was not built on laws or mathematical idealizations, so much as on commerce, new instruments, voyages to new lands, a spirit of experimentalism, and a new social technology, including communication among investigators.[26] The mathematization of nature, although adopted by some during this era, was ultimately not central to the transformation that yielded what we know today as science.[27] Indeed, many rejected mathematics as an approach that privileged artificial logic over empirical data and experience. The presumed centrality of idealizations, mathematics, and laws is another ideologically guided myth-conception (Chapter 3). Science can—and did—thrive without laws.

LAWLESS CAUSALITY, MATERIAL MODELS, AND CASE-BASED REASONING

To interpret lawless science, one needs to go beyond familiar patterns of causal thinking. But how? Can one even think causally without a law? What alternative is possible?! While the language of lawless nature certainly seems to imply indeterminism, science without laws need not be any less deterministic. What differs is how one characterizes, or expresses, causality. Lawless science is no less causal. It just requires broadening one's view of causality.

So, strip the lawlike status from Boyle's law and it may seem as if nothing remains. What remains, of course, is Boyle's J-tube and its mathematical description. It is a concrete model. Thomas Kuhn called such experimental exemplars *paradigms*, in the sense of established patterns that could be followed.[28] It fits into a larger category of model systems, cases, and exemplary narratives, all used in science to develop knowledge.[29] Such a focus underscores the importance of material culture and experimental systems in investigation. In this way of doing science, one does not appeal to general laws. Rather, one compares other cases to Boyle's original. To the extent that the conditions are the same—including pressure, temperature, etc.—one expects to find similar behavior. The reasoning is primarily analogical, not deductive. It is also direct, not indirect. One reasons from case to case. One does not detour through a general law. That is, one does not first abstract Boyle's J-tube to a law, then judge whether a new case is a legitimate example of that law. Reasoning from case to case does require, however, clear attention to the variables and details of context. One quickly learns the basis and limits of similarity, an integral part of the exemplar, or paradigm. The lawless alternative to Boyle's law is Boyle's J-tube. It is eminently teachable.

Although lawless science challenges lawlike thinking, the legal system (on a cultural level) ironically provides an instructive metaphor. In legal practice, especially in juridical contexts, one distinguishes between statute (or code) law and common law. Statute law is based on rules, or codes: like the laws of nature. One assesses actions in reference to the law's general and explicit statements. One reasons deductively and often abstractly. Common law, by contrast, is case based. One assesses actions on the basis of precedent, or similar past cases. Interpretation emphasizes the basis for similarity. One reasons chiefly analogically and concretely. Of course, numerous variables, or bases for similarity, may be possible. The effectiveness of an analogy is highly contingent. Context plays a major role. Yet the multitude of benchmarks can be beneficial, especially in interpreting complex cases. Under statute law, statutes may overlap and indicate conflicting interpretations. Case-based reasoning can often resolve this. Both frameworks provide viable systems of law and of interpreting justice. Here, lawlike science resonates with statute law, lawless science with common law.

Historically, Boyle's J-tube seems more significant than Boyle's law. The scientists who subsequently articulated the limits of Boyle's law often did so by using mercury in J-tubes. For example, with the construction of the Eiffel Tower in 1889, Louis-Paul Cailletet was able to build and support a column of mercury of unprecedented height: an impressive 300 meters. With it, he could examine the same system, but now with pressures up to 400 atmospheres. He studied the limits to Boyle's law by probing variations in the original setup. Even today, to apply Boyle's law one depends on being able to match the particular conditions of Boyle's J-tube. While Boyle's J-tube may be local and contextual, it is no less valuable on that account. Indeed, its particularity functions to guide reasoning through analogy—and to help keep it in check. Historically, Boyle's J-tube—not the law—may be the more fundamental benchmark, and that may serve as the critical clue for what one teaches about the nature of science.

The tradition of working with material model systems and reasoning from them continues today. Indeed, scientists often focus primarily on experimental systems, rather than on specific concepts.[30] The lineage of investigation is often shaped by the materials and opportunities at hand, not by a theoretical master plan. Accordingly, biologists often work with model organisms, such as fruit flies, the nematode *C. elegans*, or baboons. The strategy applies from the small flowering plant *Arabadopsis* to the vertebrate zebrafish, *Danio rerio*; from maize and mice to yeast, the bacterium *E. coli*, and the tobacco mosaic virus.[31] Geologists and climate scientists reason from simulations; psychologists and physicians from clinical cases; political scientists from historical cases (such as ancient Athens); and anthropologists from cultural rituals. Ecologists, too, will use location-specific models in contrast to general law approaches.[32] Analogical reasoning, sometimes from simulations or thought experiments as well as material models, all prove extraordinarily fruitful and pervade modern scientific research.[33] Laws may emerge on occasion. But they are special cases. Again, science can thrive without laws.[34]

A corresponding strategy for the classroom is to engage students more actively in experimental systems and other examples of model- or case-based reasoning. An excellent exemplar of this style of teaching was developed by Falk Reiß and colleagues at Carl von Ossietzky University of Oldenburg in Germany. The program there taught physics teachers through historical perspectives and replicas of original apparatus. Students thereby learned, first, about the skills that are required in ensuring that an experimental apparatus works as intended. They also learned how to "tinker" with the system to investigate phenomena. This work formed the foundation for taking measurements and thinking about the observations conceptually. Students learned the concrete basis for the now familiar textbook laws, including their contexts and limits.[35]

The replication efforts demonstrated their value in part by deepening an understanding of the original historical experiments and, thus, of how science works. Some of their discoveries were surprising. In one case, the group tried to recreate the work of James Prescott Joule on the mechanical equivalent of heat. In that experiment, a paddle rotates inside a water calorimeter, releasing heat. The change in temperature can then be measured and correlated with the amount of mechanical work required to rotate the paddle (image opening the chapter, in the left hand). Experience showed that several variables were unexpectedly important: the temperature of the water and the surrounding air needed to be almost equal; vibratory oscillations of the equipment due to imbalances needed to be controlled; and the observer's own body heat needed to be shielded by a wooden screen. The most significant variable—completely unanticipated—proved to be the dimensions between portions of the rotating paddle and the calorimeter's stationary framework: a matter of millimeters. Joule's published account did not provide enough information, and the group needed to track down his original apparatus in the Science Museum in London as a model. That's how exacting experiment can be. Joule's findings were context dependent indeed. In other work, the Oldenburg group tried to replicate Coulomb's work on electrostatic attraction. Work with the apparatus was so delicate that they ultimately questioned whether Coulomb could have produced the results he reported.[36] Both cases underscore the lesson that the materiality of experiment is more complex than the simple equations and laws that educators often present as the foundation of science. Here is one form of inspiration for how to teach science without laws: the experimental realm of Boyle's J-tube.

LAWS, MAPS, MODELS, AND THEORIES

One should not discount the significance of laws. Neither should one forget how they lie. In that conundrum one may find a central lesson for teaching the nature of science. One can at least be honest about lying.

Indeed, the reason for delving into the nature of laws in science here is not to ignite irresolvable metaphysical debate, nor to develop some abstruse philosophical position on anti-realism. Rather, the nature of laws can help inform how the nature of scientific theories is taught. Laws are perhaps the most basic, or primitive, form of scientific theory: an arithmetic map relating measurable variables. It is a drastically reduced view of the world. It highlights certain elements. At the same time, it backgrounds, or completely eclipses, others. That twofold nature is inherent in any informative representation: all are inevitably selective. That is both their strength and their weakness. That is, laws both reveal truths *and* lie, simultaneously.[37]

Laws share this ambivalent feature with other forms of abstract representation: maps, models, and theories.[38] Maps can be especially informative in understanding the relationship between the representation and what it represents. The map and the territory can be easily and concretely visualized and compared. How do maps work, then? Crudely, they depict a certain geographic territory, but they highlight only certain features of the landscape. For clarity, they actively suppress other elements that would tend to confuse comprehension. Some maps are topographical. Others are political. Some show soil types, some biotic zones. Some show population densities, some highways, some watersheds, and so on. All may be important (for example, in resolving an environmental issue). But none individually is complete. This does not make maps useless. Nor does it mean they have no correspondence with reality. Only that the correspondence of each is limited and mediated. Maps are highly context dependent. Moreover, interpreting a map beyond its intended form of representation can lead to error. The map can be misleading—even grossly wrong—when used to reconstruct the complete landscape. At the same time, the map itself cannot report all its own limitations. One needs to read maps in context. Just like Boyle's law. Maps reveal truths *and* lie, simultaneously.[39]

These many properties apply to scientific models as well. Such models might include a "rain room" as a model for the effect of the monsoon on soil; a cockpit flight simulator as a model for operating a jet; a set of tubes as a model for blood flow in studying arteriosclerosis; a computer simulation as a model for the world's wind patterns; a monkey's response as a model for the effectiveness of a malaria vaccine; a set of precisely sized balls as a model for protein structure; a scaled-down version of Japan's Inland Sea or the Chesapeake Bay as a model for water currents; an aquarium as a model for a coral reef; and so on. Models only capture certain variables. At the same time, they ignore other dimensions of the same reality. Thus, they are selective. Like maps and laws, models reveal truths *and* lie, simultaneously.[40]

As noted by evolutionary biologist Richard Levins, models may aim to achieve different virtues of representation: generality, realism, or precision. Unfortunately, they cannot achieve all at the same time. There is a threefold trade-off. As a result, for example, the simpler and more general—the more elegant and more encompassing—a concept, the less real it becomes. That is, the more it lies. The challenge is to recognize where or when and how the model lies.[41]

By correspondence, one can find all these features in scientific theories, as well. For example, like laws, theories are selective. They address only certain features of the world. Like laws, theories are also contextual. They have limited scope, or domain. Thus, the popular image that a theory is either absolutely true or patently false seems strange. One cannot even approach such judgments without a list of conditions about where and when and how the theory is supposed to apply. Like laws (and maps and models), scientific theories also reveal truths *and* lie, simultaneously.[42]

The features of scientific theories and models apply in school settings as much as anywhere. Consider, for example, the Bohr model of the atom, another benchmark in most chemistry courses. Bohr envisioned electrons orbiting around a nucleus at various energy levels. Historically, the model was especially effective in explaining, or mapping, spectral lines (the Balmer series) and chemical bonding. But according to currently accepted quantum theory, the Bohr model is wrong. There are no discrete electrons moving in discrete orbits. They are a fiction. Should teachers thus abandon the Bohr atom as a lie? Most do not. They accept it as a model, for the truths it does reveal. We tolerate the inevitable lies. But teachers often do not sufficiently elucidate the nature of models.

The fluid model of electricity serves similar functions. As a model or mechanical metaphor for understanding the "flow" of electrical "current" in circuits, it is effective. It makes sense of voltage, resistors, capacitors, switches, and circuits arranged in parallel versus in series. Circuits do really behave in fluidlike ways. But electricity is not a fluid. It does not *flow* in a *current* like a river, despite the connotations of even the scientific language. Electricity in circuits is about displaced electrons. Many texts acknowledge the lie but exploit the fluid analogy anyway. The solution is not to teach electricity differently.[43] Rather, it is to complete the lessons about models. The tensions between fluid and electron-based models of electricity are central to understanding how science works.

Teachers need to convey that lies and truths can coexist. For example, although Copernicus seeded a revolution in astronomy and, with it, a dramatic change in world view, he did not wholly eliminate a role for geocentric thinking. Even today, celestial navigators and solar time-reckoners are Earth-centered astronomers.[44] They use a geocentric model appropriate to their limited domain. Within that model, there are right ways and wrong ways to navigate or tell time. One need not claim that heliocentrism or geocentrism is absolutely true—or false. Each perspective captures enough reality appropriate to different occasions. In the same way, relativistic physics does not wholly eclipse Newtonian physics, although adopting the former makes the latter, strictly speaking, false.[45] No one includes a factor of $\sqrt{1-(v^2/c^2)}$ when calculating the velocity of automobiles in crash tests. The allowance for imprecision gives contextual justification to classical mechanics in nonrelativistic domains, where objects travel well below the speed of light.

Finally, consider the common caricature of phlogiston as the quintessential embarrassment of science. In its original role during the eighteenth century, phlogiston was the matter of fire. It explained combustion and other oxidation and reduction reactions at a macroscopic level. No one needed to refer to electrons. Indeed, like many chemists after the discovery of oxygen, we can still talk meaningfully in terms of phlogiston, regarding it, like energy, as outside the domain of weight. The phlogiston model helps unify combustion, calcination, tarnishing, rusting, explosions, reduction of ores, animal respiration, and photosynthesis (and even electrochemistry and acids) in a common conceptual framework. Accordingly, the model can prove valuable in an introductory class. Among its values, it can provide an occasion to discuss how apparently incompatible frameworks, such as the phlogiston model and the oxygen theory of combustion, can be reconciled by differentiating their contexts (see sample classroom case study in Chapter 11).

Laws of nature lie. Scientific theories lie. School Science lies. But as exemplified in these four familiar cases—the Bohr model of the atom, the fluid model of electricity, earth-centered astronomy, and phlogiston—we can at least be honest about the lies. The ultimate lesson of lawless science for education is not to abandon laws or the teaching of laws and School Science. Rather, it is to embed them fully and appropriately in the nature of science. One needs to characterize the role, meaning, and limits of scientific knowledge, whether in the form of theories, models, mechanisms, or laws. Only thus will students learn what one may properly expect of science when it becomes relevant in personal and social decision making. One needs to teach the theory and its corresponding lie *in tandem*. Accordingly, the analysis of Boyle's law above focused primarily on the importance of what is now *missing* from School Science: what's *not* in Boyle's law. Both the truth and the lie are needed in science education.

TEACHING LAWLESS SCIENCE

One could, of course, just bluster through, teaching the laws in a conventional style, as Matthews contends. That is, one could brush aside the *ceteris paribus* clauses as irrelevant and embellish the law with philosophical rhetoric about idealizations upstaging reality.[46] Such an objectivist approach tends to make science remote, isolate scientific discovery to a privileged few, and render a scientist's authority as impregnable. It mitigates against the core NOS lessons by promoting myth-conceptions and pseudohistory (Chapters 3 and 5) and erasing the role of cultural context in theoretical knowledge (Chapter 6). This is why attending to the way laws are taught is so important to teaching NOS. One needs the unlawlike to understand the full scope of science and the limits and context of its theories.

An equally ineffective approach is merely to mention the exceptions and limits without exploring them. Conant's case study on Boyle's air pump spans sixty pages, ten devoted to "the discovery of Boyle's law." In two short paragraphs—almost an aside in closing—it acknowledges a role for temperature, and the "deviations" at high pressures and low temperatures (near the condensation point).[47] This is akin to a teacher commenting, "of course, things are more complicated than this, but we'll end here." This is hardly adequate for understanding what the complexities mean—or just how they transform the meaning of the simple laws or theories.

The concept of models is sometimes introduced in the science classroom through various "black box" activities. That is, students construct models of a mechanism hidden inside some sealed device. Such exercises are not uncommon adjuncts when teaching about unseen entities, such as atoms or genes. But these lessons function primarily to profile the role of hypotheses or how to reason about phenomena inaccessible to direct observation. The lesson about the limitations and contexts of models requires more: the comparison of different or successive models. For this purpose, one might benefit from mock forensic activities that introduce evidence in a stepwise series. One such activity invites students to reconstruct historical events from a random sample of canceled checks. Students revise their interpretations as more checks become available at successive stages of the investigation. By reflecting explicitly on the revisions—and especially on early errors—one can begin to learn the lessons of simple versus more complex models.

One can also introduce the map metaphor explicitly. Students may learn from examining, comparing, and discussing various types of maps. David Turnbull's *Maps Are Territories: Science Is an Atlas* provides an excellent series of visual exhibits designed to instruct largely through the inquiry mode. For example, using the map of the London Underground, one can contrast its clear distortions and unrealities with the critical information it provides (and how it does so). Analysis can highlight the roles of selectivity, context, and conventions of representation. Likewise, comparisons of maps of the

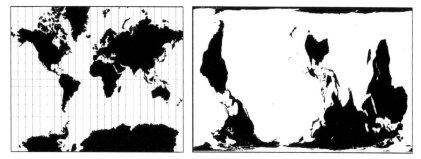

FIGURE 8.4. An "ironic diptych" of world maps: two contrasting representations, each indicating the particular perspective of the other. (Mercator projection image courtesy of Bruce Jones Design.)

same area at different scales—as simple as engaging the zoom function of maps online—may be especially valuable in highlighting how maps can differ and depend on context.[48] Donna Haraway introduced the concept of an *ironic diptych*: paired images that implicitly expose each other's biases.[49] This can be an illuminating strategy in the science classroom. For example, placing a Mercator projection map next to a recentered and inverted Peters projection map (Figure 8.4) can open reflection and fruitful discussion on the reality and context of each. The map metaphor can thus be a stepping stone. It helps frame the themes that ultimately must be addressed with scientific laws, theories, models, and mechanisms.

But as with all NOS lessons, the learning is deeper with actual cases from the history of science (Chapter 2). Here, the strategy is not just to describe an episode of conceptual change or reveal the process of discovery. Rather, it is to compare the "before" and "after" versions of a theory or model. It is to contrast the law with the unlawlike context often articulated only later. How is the original theory or law supported by the evidence? How does the revised theory or later discoveries alter that apparent truth? Haraway's ironic diptychs are powerful tools for teaching the nature of scientific laws, theories, and models.

Lawless science ultimately contributes to understanding NOS in many ways (in contrast to this chapter's opening images). First, it illustrates that scientists' aims and motivations extend well beyond the search for laws. They explore experimental systems for their potentialities. They search for interesting patterns and regularities in diverse ways. Second, nature is complex. Scientific concepts will not all reduce to simple equations like Boyle's law. Nor do laws exhaust the ways to think about causes in nature. There are material models, cases, simulations, and other forms of causal reasoning. Both of these themes are significant for scientific literacy: what exactly does science provide when informing public policy or personal choices? Finally, lawless science can open reflection on the aims of science teaching. How many science teachers envision themselves as Guardians of the World Order or as Stewards of Lawlike Thinking?[50] There is ideology at work in disciplining students to memorize laws and then use them to solve problem sets. Justifications for teaching mainly content falter if the content lies. Addressing lawless science is, in part, an occasion to liberate science education from the self-contradictions and NOS misconceptions that govern the common preoccupation with content. There is so much more to discover, so much more to learn. We want to humanize scientists, appreciate their work in all its dimensions, and broaden awareness of the many wonders in the world, included in lawless science.

9 | Nature of Science in an Age of Accountability

Tests as explicit goals • a well-informed analysis • prototypes • other forms of assessment

We live in an Age of Accountability. Testing seems utmost—for school administrators, district supervisors, state officials and legislators, and national politicos and pundits. Pragmatically, this seems to set a firm boundary on contemporary educational practice.

At the same time, teaching to the test is widely viewed as a problem. Why? If the test truly reflects what we want students to learn, it should merely provide guidance and a framework for motivation—and, ultimately, reward as well. If there is a problem in education (and I concur that there is), it is with the tests themselves, not the inevitability of teaching to them. Our tests should be *meaningful, transparent,* and *authentic.*

To be meaningful, tests should address knowledge relevant to the ordinary citizen and the kinds of personal and public decision-making cases that involve science. To be transparent, they should be explicit, and the standards clear and relatively concrete. To be authentic, they should reflect the kinds of complex, real-life situations where scientific and nature-of-science knowledge are used or applied. That is, they should reflect the ultimate purpose of science education. Current tests, unfortunately, are typically oriented to piecemeal factual knowledge. They are largely irrelevant. They hardly contribute to evaluating the preparation of informed citizens or consumers of science.

The problem of appropriate tests is especially acute for teaching the nature of science. There are few—if any—ways widely acknowledged as acceptable for evaluating understanding of how science works and what contributes to the reliability of its claims. The prospect of NOS testing is greatly impoverished when compared to the prominence of NOS as a goal in national and state curriculum standards. How, indeed, do we effectively test beyond content?

So, consider: emissions from a proposed local waste incinerator; reported links between the measles vaccine and autism; revised vitamin D recommendations; underground seepage from a chemical waste site into the groundwater; or a new Earth-like planet with the potential for life. In these cases (as noted at the opening of Chapter 1), an informed judgment requires understanding the nature of science, or scientific practice, not just scientific concepts. But how do we assess this type of understanding? In particular, how do we get past the pervasive multiple-choice format, with its inherent tendency to reduce knowledge to trivial factoids? If a primary goal is to prepare scientifically literate citizens—able to participate in public and personal decision making that involves science—more sophisticated assessment formats are needed for assessing NOS knowledge.

A WELL-INFORMED ANALYSIS

Efforts to probe student understanding of the nature of science now extend back more than half a century. Over two dozen instruments have appeared. Educators would seem to have plenty of options. Choose your favorite acronym, perhaps: VNOS, NOSS, NOST, COST, VOSTS, VASS, FAST, TOUS, TOES, TOSRA, WISP, at least.[1]

Most such instruments are inappropriate for classroom use, however. Most were designed for educational research—for assessing the effectiveness of particular educational strategies or interventions—not for evaluating student achievement in a classroom setting.

In addition, many of the tests are based on a benchmark set of beliefs, values, or "views." That is, the student's beliefs are expected to align with those that guided the construction of the test. Yet even experts disagree about many of the claims. In addition, the benchmark standards seem to have shifted with time. These tests thus seem arbitrary. That is, they do not assess NOS knowledge or understanding objectively.

Other tests reflect an effort to be more neutral. They adopt a diagnostic posture. That is, they aim merely to characterize student belief, opinion, or view. They do not seek to assess the responses—say, on a scale of depth of understanding. They do not *evaluate*. In particular, the currently most popular instrument, VNOS, suffers from this blind approach. These alternatives, by contrast, imply that we cannot hope to establish an objective standard for measuring knowledge of NOS.

Both these conventional approaches seem problematic in that they focus on NOS as a set of beliefs or personal perspectives. As profiled throughout this book, however, understanding how science works is a form of knowledge. Knowledge, not personal belief, is the goal. In evolution education, at least, biology teachers have long recognized that the goal is not to shape beliefs or attitudes. Rather, it is to inform. The aim is to develop understanding of divergent lineages and natural selection, not beliefs in human nature. Just so for NOS. One aims to help students understand scientific practice, not to indoctrinate them into a set of scientific values. Nor instill beliefs about the authority of Science writ large. Any NOS test that asks a student whether they agree or disagree with a particular view about science is probing personal belief. That is irrelevant. Rather, we want students to understand how science works—and to recognize the limits of science and cases where it fails. Students should ideally be able to conduct *a well-informed analysis* of

NOS features—say, where they affect the reliability of some scientific claim.

Recall, too, from Chapter 1 the overarching goal of scientific literacy, often lost in the various existing NOS surveys:

> *Students should develop a broad understanding of how science works to interpret the reliability of scientific claims in personal and public decision making.*

That is, as noted above, the test should be *meaningful.* Abstract generalities, such as "science is tentative" or "scientists are creative"—the focus of available tests—mean little without context.[2] We want to know if the students' understanding is functional, not academic. An assessment will thus ideally focus on analytical skills, not decontextualized concepts. The aim of teaching NOS principles, after all, is to inform cases from day-to-day life. Students should thus be evaluated for how well they can apply and articulate them in such concrete, cultural contexts.

Accordingly, an NOS assessment will not focus exclusively on some idealized model of science, or the way science "should" work. What matters is understanding science in practice: how it *does* work (or, sometimes, doesn't work). If scientists make mistakes or are biased by their cultural context, for instance, then such facts must help inform students' interpretations of cases in their lives. Normative and descriptive dimensions of NOS exist in tandem (as introduced in Chapter 1 and echoed in Chapters 3–6). One must acknowledge the tension between them, while not conflating them—especially in an assessment.

Moreover, where the ultimate aim is interpreting science in society, NOS assessment should be *concrete*, or case based.[3] A student should be able to apply any relevant principle to inform analyses of *particular* cases. For example, consider the revised mammogram recommendations issued by a U.S. task force in late 2009: can a student compare the status of these specific claims to that of earlier ones? Are all the claims equally tentative? Are there good reasons for a change in consensus? Or has science been eclipsed by efforts to cut costs at the expense of women's health? Such an assessment strategy complements an approach to teaching NOS through historical and contemporary case studies (Chapter 2).

Cases in the news or advertising seem appropriate vehicles for assessment.[4] For example, test questions might address a new Climate Change Vulnerability Index or new studies of MRI scans of smokers or of causes of teen suicide. General questions about "science affected by its social milieu" or the "nature of experiment" will not suffice. Rather, an appropriate target is the nature of investigation into, say, what caused the Gulf oil spill in 2010. Or scientific claims about long-term psychological effects of violent video games. Namely, can a student interpret a case in current events, not merely echo abstract NOS principles rendered in other cases? Students must demonstrate thoughtful analysis, not merely choose among preformed answers. The cases are real. That is, the test will be *authentic*.

Working with cases in the news is not without potential pitfalls, however. At first, it seems simple enough: tell each student to adopt and defend a position on a particular issue. Yet assessing such views (students will be the first to note) is subject to teacher bias. Many teachers try to avoid this problem, I think, by focusing just on the student's argument: How well has a student justified their judgment? Is the reasoning logical? Are the assumptions valid? Is the evidence sound? And so on. The teacher can thereby sidestep contentious judgment in grading. Yet this approach (one I once tried, too) has a hidden cost. One assesses only an ability to assemble an argument. A well-*formed* argument is not necessarily a well-*informed* argument. It need not demonstrate depth or breadth of NOS understanding. Worse, perhaps, one cannot differentiate between healthy reasoning and unfruitful rationalization. Do students reach judgments that are based on the evidence, or do they compile the evidence on the basis of preformed judgments? Cognitive science reminds us that our minds typically follow the second pattern. We tend to adopt a position first, often relying on emotion, then cherry-pick evidence to "justify" it.[5] Effective education, however, should not reinforce the inherent tendency to short-circuit the process of thinking through evidence. Rather, lessons should foster informed analysis and reflection.

The benchmark, as noted earlier, should be *a well-informed analysis*. That is, the student should interpret the case fully and analytically for a friend or public official, who is to decide independently. The responsibility demands thoroughness and clarity. Can the student recognize and articulate the NOS features and explain their relevance (for example, as listed on the inventory in Figure 1.3)? Here, there is no position, no rationalizing. Instead, there is analysis. Notably, *all* the relevant information and perspectives are essential. All must be fully explained. Omissions matter. Evidence in such a context must be *complete*. One can thereby measure *breadth* of NOS knowledge.

Another key feature of analysis is its degree of detail, or specificity. Is it coarse-grained or fine-grained? Are subtle but significant distinctions noticed? Are comparisons to similar cases introduced and explained fruitfully? This measures *depth* of knowledge.

Together, breadth and depth of NOS understanding become a basis for *evaluation*, not merely a diagnostic assessment. One can rate the quality of the analysis. One thereby dissolves the dilemma of arbitrariness or subjectivity on the one hand and lack of evaluative fulcrum on the other.

A well-informed analysis, then, is marked by both breadth of knowledge of NOS factors (completeness) and depth of NOS knowledge (specificity). Students must be able to recognize and identify such factors without explicit prompting or cueing. They must extract the relevant information from the profile of a particular case. Merely agreeing or disagreeing with certain statements is insufficient.

PROTOTYPES

Consider now a prototype question, based on the case (mentioned briefly above) about the revised age recommendations for mammograms in 2009 (Figure 9.1). This case is striking because of the contrasting interpretations presented in the major news media. In one view, the revision was primarily political. Raising the standard age to begin screening was seen as a measure to cut health-care costs, with women bearing a burden of increased risk of undetected breast cancer. Good science was apparently victim to economics and gendered politics. *Women's Health* magazine advised its readers to respect the older recommendations. For others, the claims in the new task force report were a standard example of science benefiting from better evidence, even if the conclusions changed. Longer-term data, larger sample size, and meta-analysis indicated that one could now lower the risks of exposure to X-ray radiation due to screenings, without also sacrificing timely detection of cancer. In an editorial, the journal *Annals of Internal Medicine* underscored the importance of the deeper scientific knowledge. How does NOS understanding help one interpret this disparity in perspectives?

In the sample assessment, the student is asked to provide a well-informed analysis. For background and reference, there are several documents, reflecting the information and views that are readily available via the Internet. In this case, the student is not asked to evaluate the clinical studies or evidence themselves. (Indeed, few can. That was the reason for convening an expert panel.) Rather, the crux of the case is NOS. How does one interpret the reliability of the claims in context? Nor are students instructed to give their own opinion, as in "What is your view?" or "What would you do?" Again, this tends to foster premature conclusions and rationalization. Rather, the

---

FIGURE 9.1. Prototype question for assessing NOS knowledge.

Revised Mammogram Recommendations [Feb. 2010]

A female acquaintance of yours is just turning 40. Concerned about the possibility of breast cancer, she had planned to get a mammogram in the next few months, despite her fears about excessive radiation. She has heard that a major national task force now advises waiting until 50, yet finds reassurance in *Women's Health* magazine about still following the old guidelines. You both knew a woman who was diagnosed unexpectedly with breast cancer at age 43 and died last year. Your acquaintance is unsure how to interpret the apparently conflicting information and asks your view. Provide an analysis of this reported change in scientific consensus that would help inform her decision.

Resource Documents:
- article from *Women's Health magazine*
- news item from *The New York Times*
- U.S. Preventative Services Task Force original report
- editorial from *Annals of Internal Medicine*

student must defer his or her own judgment in informing an acquaintance. Hype or spin must yield to substantive and balanced analysis. No biased "Well, *I* think that..." or "*I* recommend...."

Optionally, one might also ask the student to identify important information that is missing—and where one would likely find such information. One is aiming, after all, to expose the student's thinking fully and to make it visible for assessment.[6]

In scoring a student's written response, one is thus freed from addressing the student's own view. Rather, the aim is to assess: how informed and complete is the analysis of the available information? Relevant (scorable) factors here include (at least)

- the role of systematic study versus anecdote
- the role of probability in inference
- sample size
- the nature of conceptual change
- the role of prior beliefs
- emotional bias in interpreting evidence and risk
- gender bias
- sources of funding
- credibility of sources of information

In each case, the relevance must be articulated. Mere statements of general principles, without concrete examples or context (for example, "all scientific knowledge is provisional"), do not sufficiently demonstrate fully developed NOS skills. For each NOS factor, then, the analysis may be further ranked as simple or detailed, short or extended, minimal or well elaborated, rough hewn or nuanced. Perhaps one might add further credit for informative comparison to other cases of scientific change or gendered science (possibly encountered in class lessons). Here, one has a sample test of scientific literacy for NOS knowledge. In format and scoring, it is not unlike the AP exam essays familiar to many high school teachers. A sample question such as this, with its structured scoring rubric, can be presented to students and others in advance, allowing the form of evaluation to be *transparent*, as noted above.

Evaluation, of course, need not always involve formal testing. Assessing student performance in general is the aim, not performance specifically under testing conditions. A teacher could easily adapt this question for project work. Indeed, such an approach might allow a student to demonstrate skills in posing appropriate questions, identifying reliable sources of information, or collecting such information. Not all teachers will have such opportunities, however, and envisioning alternatives using familiar institutional constraints seems appropriate. For this sample question, one can easily imagine historical case studies that would contribute to developing the relevant NOS knowledge through classroom activities. The case of the cause of beriberi (Chapter 10) is a dramatic example of profound scientific change. Christiaan Eijkman concluded that beriberi was caused by a neurotoxin produced by bacteria present in rice. There was also an antitoxin in the rice coating, apparently removed in processing the white rice. Eijkman drew on extensive evidence from controlled experiments, and his results contributed historically to lowering the incidence of beriberi. Yet Gerrit Grijns subsequently concluded that beriberi was instead a nutrient deficiency. His studies were able to find differences not apparent in Eijkman's studies. The rice coating contained an essential nutrient, also present in other foods. More detailed data led to rejecting the former claim and adopting a quite different alternative. One could tell a similar story about the dangers of DDT, thalidomide, or Vioxx, exposed only after extending original studies. The mammogram case similarly relies on more data to make a better conclusion.

Many historical cases might also demonstrate the potential for gender bias. These range from studies of the female skeleton and physiology, rendered as indicative of women's "natural" role in society, to studies of female cranial capacity and the "natural" level of women's intelligence.[7] All were flawed, notably by excluding criticism from women's perspectives. One could equally well point to cases where sources of funding biased scientific results. It is not unreasonable to posit that in the contemporary mammogram case, the task-force study might plausibly have embodied bias. However, an analysis of the information on the context and conduct of the recent work, guided by such historical understanding, indicates otherwise. For example, many women participated in the task force. Women's views were not eclipsed outright, as in many historical cases. Indeed, the task force's concerns included minimizing a woman's health risk from X-ray radiation, not just considering the dangers of breast cancer alone. In addition, the group was convened as part of a regular periodic review. It was not prompted by the high-profile public debates on the growing cost of health care. Moreover, the task force's work was through an independent body, not subject to overt political pressure. That is, there are no telltale signs to damage their credibility. Also, they transparently presented their evidence for public inspection, even if the average person is not equipped to review it in detail. If one harbors doubts, one can look further for other experts who have reviewed the findings. Historical case studies can thus inform a student's interpretation.

In this prototype, one finds a way to evaluate a student's NOS understanding on the basis of their ability to make a well-informed analysis based on a real-life decision involving scientific claims. It demonstrates how the assessment can be based on knowledge developed through classroom study of historical case studies. In short, it illustrates how one might develop an NOS evaluation that is *meaningful*, *transparent*, and *authentic*.

OTHER FORMS OF ASSESSMENT

Although summative, or final, assessment seems the utmost concern in the high-profile political context, in recent years educators have shifted their primary attention to formative assessment.

That is, how is the student doing along the way? A teacher ideally monitors each student's learning as it progresses, in order to help guide it further.

One may begin, of course, with questions like the one above. When used *before* a series of lessons, such assessments can help probe students' baseline abilities. Responses, individually and collectively, may help to expose NOS preconceptions and alert a teacher to particular instructional challenges. Pretesting also provides a valuable benchmark for demonstrating and measuring improvement at a later date.

The first occasion to find out what a student has learned after instruction is at the close of the NOS lesson itself. Simple, direct questions seem sufficient: see Figure 9.2. Here, the NOS features are conveniently recalled, but their role in shaping the process of science and its conclusions is left open. (One wants to avoid cueing any target response.) These questions form a critical occasion for explicit reflection, so important to any learning (Chapter 1). These questions also function simultaneously as a form of review. Repetition and active synthesis help consolidate the lessons. Discussion can surely involve a whole class or small groups. For individual assessment, however, one will need independent written responses from each student. Can they articulate what they have learned explicitly in the context of their own concepts and prior learning? With this information, the teacher discovers how next to proceed, possibly to return to some feature of the case that was not clear, or possibly to plan more sophistication into an intended sequel lesson.

Another opportunity to probe students' learning occurs when discussing subsequent cases, where an NOS feature appears again. These might include cases that appear fortuitously in the news. Do the students remember the earlier lessons? Can they relate them to the new case? The comparison can

---

FIGURE 9.2. Sample NOS review question for a historical case study.

How does the case of Alfred Russel Wallace & the Origin of Species illustrate the following dimensions of the nature of science, or how science works?

- the influence of early encounters and life experiences
- personal motivation
- funding
- scientific communication
- diversity in scientific thinking (by different persons)
- priority and credit

enhance learning in both cases, by highlighting similarities and differences. The NOS concept is refined and further contextualized.

It is equally important to assess NOS learning after an extended series of interrelated cases or lessons. Perhaps the teacher uses a mixed series of reflective lab activities, historical episodes, problem-solving cases, scientist interviews, and contemporary news stories, distributed throughout the school year. The ensemble will surely be unique to the particular teacher or classroom. An optimal test will capitalize on those specifics rather than probe vague generalities. Here, one shifts the primary focus from the cases to the various NOS features, now exhibited in multiple examples. Again, those instances may be recalled, providing the student an opportunity to synthesize the lessons from the individual cases in a capstone exercise or assessment. Sample questions are presented in Figure 9.3. (These come directly from final exams I gave many years ago in a high school biology course, for students age fifteen.) The questions target relatively basic concepts: controlled experiment, making connections, and the process of inquiry generally. Students are given some choice in cases, allowing them to highlight their strongest knowledge. Once again, the essays are free response.

FIGURE 9.3. Sample NOS questions synthesizing multiple cases.

1. Discuss the role of controlled experiments using 4 examples:
    (a) You and enzyme catalysis
    (b) You and the effectiveness of exercise
    (c) You and spontaneous generation
    (d) You and nutrient indicators
    (e) Mendel and inheritance
    (f) Gause and the outcome of competition

2. Discuss how making connections is important in scientific discovery using 4 examples:
    (a) Darwin: domestic breeding and adaptation
    (b) Darwin: Malthus's essay on population and nature
    (c) Morgan: inheritance and chromosomes
    (d) You: circulation in humans and squids
    (e) Eijkman: beriberi and chicken diets
    (f) Ingenhousz: wind or sun and transport in plants
    (g) Paul Broca: speech impediments
    (h) Enzyme deficiencies and inheritance
    (i) Associative learning (Pavlov, Skinner)

3. Discuss how 3 of the following are important to the process of inquiry:
    (a) Microscope/electron microscope
    (b) Experimental model systems
    (c) Pea plants (for Mendel) OR fruit flies (for Morgan)
    (d) The social position of Darwin, Lamarck, or Mendel OR
         the budget of the National Science Foundation

The challenge for many may be in interpreting or assessing the quality of free responses. Here, three criteria were identified for the students:

- Organization. Unity. See that everything supports one central idea.
- Breadth. Completeness. Relate as many different ideas as possible together.
- Depth. Proficiency. Convey your ideas effectively by using specifics.

Such criteria are widely applicable for assessing how well informed someone is in any particular domain. The teacher is deeply familiar with the examples that a student uses in making comparisons or supporting general statements. The awareness of information introduced into class discussions can be a basis, in part, for interpreting the depth and breadth of responses. Scoring, where it is important, relies here on the individual teacher's judgment. But the ranking need only be crude to be effective.

Of course, evaluation need not be limited to conventional written tests. As noted above, project work might indicate student performance in the authentic contexts of applying one's knowledge in real-life scenarios. This could potentially involve questions for interviewing a scientist with relevant expertise. The creative teacher—or educational researcher, as the case may be—can develop alternative assessments to suit the occasion. Many educators have adopted student journals as vehicles for NOS reflection. These may also help a teacher monitor developing understanding. Interviews, while time-consuming and potentially bulky in large numbers, may also be informative. Observations of behavior can be telling, while also not relying on self-reporting. One strategy adopted by some educational researchers is to present students with problematic events and document their responses or tactics used to resolve the problem. For example, when an experiment does not yield an expected outcome, what is the next course of action? Teachers might demonstrate their own understanding of NOS by analyzing a historical or contemporary case, highlighting the relevant NOS teaching opportunities, and transforming the case into an NOS lesson.[8]

The ultimate educational goal, of course, is to apply NOS knowledge in interpreting science in a personal or public settings. One is aiming to prepare students to address new, unfamiliar cases. Such novel encounters may surely be modeled in class activities. The prototype profile above is a sample activity for developing such analytical skills, as well as a prospective assessment of those skills as an endpoint of the educational process.

ENVOI

National science education standards—and many state curricula, borrowing from them—now include nature of science or understanding of scientific practices as a major learning objective or thematic thread. Clear explanations of what that means, however, are far less common. Ironically, standardized methods for evaluating student understanding of NOS, despite a half century of numerous surveys and research instruments, have yet to be developed to fully address the widespread standards. Such tests are needed especially to concretely render the educational goal. One needs to clarify as plainly and as transparently as possible the ultimate aim of scientific literacy and what it means to understand how science works. Tests will identify and articulate just what about NOS is important for every citizen and consumer to learn.

Until such tests appear, teachers will need to plumb their own creativity, as they do so often throughout their practice. The examples here may offer a guide, while providing a model for further discussion. One may be encouraged, at least, that science education standards almost uniformly advocate the central importance of NOS. Teachers who can demonstrate their effectiveness through any kind of relevant student performance are surely on track. They will also be able to see the success of their own efforts in supporting student learning about the nature of science, one of the most important aims in science education.

Teaching the Nature of Science

Part II: Resources

10 | Christiaan Eijkman and the Cause of Beriberi

What does Whole Science (Chapter 1) look like in a classroom? Here I present a sample, ready for use. It is a historical case study that integrates science content, process-of-science skills, and NOS lessons. It highlights a small set of features from the expansive NOS inventory (Figure 1.3) and shows how they are related through a particular concrete case.

The case study also illustrates the use of history, rendered as science-in-the-making (Chapter 2). The student is situated alongside the historical scientist and invited to interpret and think through some of the problems in their original context (see the THINK exercises). The case study adopts an active learning strategy or guided inquiry approach for NOS themes (Chapters 2 and 14). Namely, it engages the student in a series of reflections (more THINK exercises).

Myth-conceptions (Chapter 3) are addressed indirectly. In particular, students learn that a Nobel Prize winner was wrong on a main idea related to the honor he later received. One must reconcile justified acclaim with error. Indeed, the case study avoids an artificial rational reconstruction (Chapter 4) in part by leading students through Eijkman's mistakes, based on his assumptions, demonstrating how easy it is to succumb to such errors. In addition, the triumph of science in reducing the incidence of beriberi is contrasted with the context of exercising and reinforcing colonial power and questions about research ethics (Chapter 6). Many of the complexities of conflicting claims, multiple alternative hypotheses, and policy implications are acknowledged and addressed, rather than hidden or simplified "for the sake of the student" (Chapter 7).

IMPLICIT TEACHING METHOD AND STYLE

Ideally—for making the case study most effective—the teacher will adopt a highly interactive and open teaching style, in a student-centered mode. One may conceive the primary objective as engaging the students in the various questions that appear throughout the narrative. Ideally, the narrative should function merely to motivate, inform, and connect student reflections and discussions. Although there are inquiry activities, it is a structured, guided inquiry. The teacher may adopt a format of class presentation, interrupted by questions and/or discussion (visuals are available online at http://ships.umn.

edu/modules/biol/beriberi.htm). Alternatively, the student may read the text, providing written responses for all or selected questions or preparing responses in advance for later class discussion. The final set of reflections on the nature of science form a critically important capstone, helping consolidate and close the lessons.

In a classroom scenario, students may address the think exercises as individuals, in small groups of two to five, or as a whole class. (I have led TV-talk-show-type discussion in a class of up to 40 or 200, after first providing an opportunity for think–pair–share). Different questions may well be treated in different ways. Writing (perhaps in a dedicated NOS journal) may be used to clarify thinking either before or after discussions.

The teacher's role in case-study discussions includes

- monitoring student engagement in group interactions,
- facilitating discussion,
- ensuring that comments are responsible to (a) the evidence and (b) other student comments, and
- providing additional information or perspective when appropriate.

Another role is

- assisting students in articulating their views for others—possibly inviting students to share their reasoning and relevant examples or to elaborate through more details (all these may also help foster better communication skills).

The teacher may foster deeper discussion by

- priming discussion with observations on the text that students have not introduced,
- entertaining alternative perspectives to enrich reflection, or perhaps
- linking to earlier cases that have been discussed in the same class.

That is, the teacher may combine the roles of a talk show host and a journalist in interviewing students—aiming to elicit and clarify views, perhaps echoing or rephrasing significant contributions. The teacher may also help to consolidate lessons by reviewing and summarizing comments.

One of the most important—and exciting—elements in guiding students through a historical scenario or case study is adopting a posture of not knowing the right answer—that is, sharing a position of ignorance with both students and the historical characters. How does a scientist reason, using only the evidence and knowledge at hand? Further, how does one ensure the reliability or certainty of a prospective answer? This is the crux of science-in-the-making versus ready-made science (Chapter 2). The challenge for the teacher in inquiry mode is to strategically adopt this posture of uncertainty.

OVERVIEW OF CASE

Christiaan Eijkman shared a Nobel Prize for the discovery of vitamins. His research on beriberi in the Dutch East Indies in the 1890s highlighted the role of the rice diet, leading to the understanding of a nutrient deficiency involving a previously unknown type of nutrient. Yet originally Eijkman was convinced that beriberi was caused by a microorganism, and this assumption shaped his work. This case study delves into the fruitfulness of his work, even as his preconceptions guided him into error. The context of Dutch colonialism also helps frame reflection on the cultural contexts of funding and ethics in science.

Major NOS features include

- the role of chance or accident
- theoretical perspectives in interpreting data
- the distinction between causation and correlation
- growth of knowledge through small cumulative additions versus through major conceptual reinterpretations
- the role of the individual versus groups in making a discovery
- scientific communication and communities of researchers
- the cultural and economic contexts of science

Christiaan Eijkman and the Cause of Beriberi

A MEDICAL RESEARCH ADVENTURE

FIGURE 10.1. Christiaan Eijkman

Meet Christiaan Eijkman (Figure 10.1), who shared the Nobel Prize in Medicine in 1929. Let's journey with him on the mission of medical research that led to his award.

The year is 1886. It's October. Eijkman embarks with two other doctors from the Netherlands. Their destination is the island of Java, almost halfway around the globe. They pass through the Suez Canal—opened only a few years ago— and arrive a few weeks later. Java is part of the Dutch East Indies, one of many important trading colonies around the world. Java and the surrounding islands, with their exotic forests and exceptionally tall trees, typically fascinate Europeans. There are dense thickets of fibrous rattan vines, harvested by the Javanese and exported to Japan to make tatami mats. One finds the crops that make the East Indies valuable to the Dutch, including sugar cane, coffee, cacao, and indigo. Many trees have been cleared to grow the crops, which have been imported from other tropical regions.

Life on Java is very different from home for the three doctors, even in the Dutch community of Batavia. The tropical heat is everywhere, and they have to develop a taste for rice, a staple in this region of Asia. Eijkman, age twenty-eight, has seen the sights of Java before—while serving as an officer for the Dutch Army. However, after two years he had contracted malaria and returned to the Netherlands. Malaria is one of many diseases common in the tropics. Cholera, influenza, dysentery, and plague are also widespread. So, too, is beriberi.

Beriberi is, in fact, the reason why the medical commission has been sent to Java. It is a debilitating disease, indicated by the name itself. In Sinhalese, the word *beri* means "weak," and doubling it intensifies its meaning. Beriberi involves weight loss and muscle weakness. Patients lose their sense of feeling and control of limbs, often leading to paralysis. Fatigue can give way to confusion, depression, and irritability. In some cases fluid collects in the legs, taxing the circulatory system, enlarging the heart, and causing heart failure. The disease can be fatal. Anywhere from 1 to 80 percent of beriberi patients have died in various epidemics.

Epidemics of beriberi in Asia have become more frequent. In Japan in 1880–1881, one doctor was swamped with so many beriberi patients that the hospital could not accommodate them all and they overflowed into nearby temples. The Dutch government is now particularly concerned because the local soldiers and even Dutch Navy sailors are suffering—which recently crippled an effort to quell a native uprising in a remote province. They want to find a cure for the disease or—better—prevent it. They have sent the medical commission to find the cause of beriberi. Eijkman will eventually share a Nobel Prize for his discoveries on Java.

DISEASE, GERM THEORY, AND EIJKMAN

Eijkman and his colleagues are not the first to study the cause of beriberi. Beriberi has been known in southern and eastern Asia for centuries. A Chinese physician described it four thousand years ago. In the East Indies it had been reported as early as the 1630s. But no one knew a cure.

> **THINK [1]:** What might have caused the epidemics? What are the possible causes of any disease? How would you confirm one cause versus another?

The Dutch medical commission has arrived with new ideas about disease from Europe. Indeed, Eijkman's career nicely reflects the discoveries. When Eijkman first visited the Indies in 1885, he was fulfilling a contract with the military that had helped pay for his medical education. After his return to the Netherlands, however, Eijkman became fascinated by the exciting new studies of Louis Pasteur and others on the role of bacteria in disease. He turned from practicing medicine to pursuing medical research. Eijkman went to Berlin to study with the world leader in the field, Robert Koch. According to Koch's germ theory of disease, disease results from microscopic organisms that infect the body.

In 1880, Koch developed an important method for culturing bacteria on a solid medium instead of in a liquid nutrient broth. By spreading out the bacteria on a plate, he could separate the different strains or species of a mixed culture, isolate each one, and then breed a pure culture. With this method it became much easier to isolate and identify specific disease-causing agents. In 1882 and 1883, Koch identified the bacteria that caused tuberculosis, cholera, and diphtheria.

Outbreaks of beriberi have been common in armies and navies and in prisons, all relatively closed communities. Is the disease therefore infectious, transmitted by some "germ"? That was the hope of the Dutch government, who sent two doctors to Germany to learn the latest techniques directly from Koch. There they met Eijkman, who joined the team. Also in that year, a prominent French researcher, using a method that he had pioneered a few years earlier, created a vaccine for rabies. The Dutch commission has now brought all these new methods with them. They are thus prepared to find the bacterium that causes beriberi, isolate it, and make a vaccine. The scientists are transferring germ theory from Europe to Java.

Not quite a year later, the group completes its work in Java. They characterize beriberi more precisely in terms of both its clinical symptoms and the nerve degeneration visible microscopically in the tissues. They confirm in their report that a bacterium causes beriberi. But they also discover a new infection pattern. They have not been able to infect one organism directly by injections of blood, unless repeated many times. While most diseases seem to be transmitted through a single exposure to the germ, a person has to reside in an area of beriberi for several weeks to contract the disease. For beriberi, the bacterial agent apparently must be transferred many times. The commission returns home. Eijkman remains, however, to continue the studies and work at the local medical school. He has yet to establish a pure culture of the bacterium and, from that, develop a vaccine.

Chicken-Feed?

Several years pass as Eijkman continues investigations in his small laboratory. His work has been frustrated because, even using Koch's techniques, he has been unable to isolate the beriberi bacterium in a pure culture. He continues injections of diseased blood, but the results are inconsistent. Are the transfers responsible, or is it just chance? He realizes that he needs many more organisms: some injected, some not. He switches from rabbits to chickens, which are cheaper and easier to raise. Before long, the chickens exhibit signs of the disease. They walk unsteadily and have difficulty perching. Later they do something chickens rarely do: lie down on their sides! They also have trouble breathing. Yet now the disease occurs among all the chickens, even those not injected. Eijkman decides that there must be contagion. He separates the chickens, at first in different cages, and then, when that fails, in different parts of the hospital grounds. No sooner are the chickens isolated than they all recover. Why?

> **THINK [2]**: Given this unexpected turn of events, what would be an appropriate next step? Where would you look next for clues?

Eijkman learns quite unexpectedly from his assistant that the chickens have received different food. When the experiments began, the assistant had obtained leftover cooked rice from the military hospital—a way to save costs. But later a new cook had arrived. He had refused to give "military" rice to "civilian" chickens! The assistant had returned to raw, feed-grade rice.

> **THINK [3]**: With this information, what would you do next?

Eijkman must now isolate the difference in the rice. Which factor is responsible? Does the cooked rice spoil overnight? No. Is it the cooking of the rice? No. Is it contaminated water used to cook the rice? Is it the hospital rice itself? There are different varieties of rice. Normally, the local rice, known as *beras merah*, has a reddish cuticle (or pericarp, in botanical terms). You can remove the cuticle, though, by milling or "polishing" the rice. Polished rice has a fancier white appearance and a taste that many prefer and was used at the hospital. Here, finally, is the relevant difference.

Soon, Eijkman is able to make chickens sick almost at will, simply by controlling their diet. When fed the polished, white rice, healthy chickens soon show symptoms similar to human beriberi. In addition, when fed red rice, they become well again. They recover as well when just the husks or cuticles of the rice—the rice polishings—are added to a diet of polished rice. In some cases, the sick chickens regain a normal gait and the ability to fly within a few hours of eating the rice polishings!

Eijkman now has an important clue for finding the bacterium. It must be in the polished rice. This would certainly explain why beriberi is so prevalent in nations where rice is a staple food. Eijkman clearly did not plan to change the chickens' diet, but the chance event has revealed valuable information that he and his colleagues had missed during many years of deliberate study.

Yet healthy chickens eating red rice remain healthy, even when living in the presence of other diseased birds. Cross-injections have also been ineffective. Eijkman reasons further that the bacterium must never enter the body. It must create a neurotoxin that is absorbed in the body. The cuticle of the rice must be a neutralizing agent or antidote.

Not everyone who hears of Eijkman's conclusions accepts them. Others agree that the rice Eijkman used was responsible, but perhaps not for the reasons he specified.

> **THINK [4]**: Imagine that you are among the skeptics of Eijkman's new discovery. How else might you interpret these findings? How might Eijkman design a test to respond to your criticism?

OF RICE AND MEN

Eijkman continues with his various administrative and teaching duties, while also finding time for his research on beriberi and the toxins it produced. Meanwhile, controversy over the new germ theory of disease continues worldwide. Two researchers (one Japanese, one French) independently seem to have isolated the bacterium that caused bubonic plague. In India, over 45,000 people receive a new cholera vaccine. Compared to those not inoculated, 70 percent fewer die. In 1892, a skeptic of germ theory in Germany swallows a vial of live cholera bacteria to demonstrate his belief that the bacterium does not cause the deadly disease. Indeed, he does not get sick.

Eijkman has still not demonstrated conclusively how polished rice is part of the process in which bacteria cause beriberi in humans. He needs a properly controlled experiment.

> **THINK [5]**: How would you construct such an experiment on human diets, while also following basic ethical principles about respect for persons?

Eijkman turns to institutions. There, diets are already determined. The large number of cases will also help ensure that the results are not due to chance or mere coincidence in a small group. He persuades the prison at Tolong, where 5.8 percent of the population suffers from beriberi, to substitute undermilled, or half-polished, rice for white rice. All cases of beriberi are cured. But as Eijkman notes later, this merely confirms the potential effectiveness of the cure. It does not demonstrate that a bacterium in the polished rice initially causes the disease. This requires comparing individuals who consume the different types of rice.

Eijkman thus enlists A. G. Vorderman, supervisor of the Civil Health Department of Java, to help survey the incidence of beriberi on a wide scale. In each prison on Java, prisoners eat either polished or half-polished rice, according to local customs. In some cases prisons serve a mixture. Here is a natural experiment, a case where the desired experimental conditions exist on their own. For Eijkman and Vorderman's purpose, the experiment is fortuitously already in progress. Between May and September of 1896, Vorderman leads an exhaustive study of beriberi in one hundred prisons of Java and the small neighboring island of Madura—a survey that embraces nearly 280,000 prisoners. He reports the distribution of beriberi in the hundred prisons and its frequency among prisoners as follows:

	Number of prisons	Number with beriberi	Percentage with beriberi	Frequency among prisoners
Half-polished rice	35	1	2.7%	1 in 10,000
Mixture	13	6	46.1%	1 in 416
Polished rice	51	36	70.6%	1 in 39

Vorderman also reported on other factors:

	Number of prisons	Number with beriberi	Percentage with beriberi
Age of buildings			
40-100 years	26	13	50.0%
21-40 years	32	11	34.4%
2-10 years	42	19	45.2%
Floors			
Impermeable	58	24	41.4%
Partly permeable	13	7	53.9%
Permeable	29	12	41.4%
Ventilation			
Good	68	28	41.2%
Medium	11	8	72.7%
Faulty	21	7	33.3%
Population density			
Sparsely populated	73	32	44.6%
Medium	1	1	
Overcrowded	26	9	34.6%

THINK [6]: If Vorderman is able to show a correlation between diet and beriberi, why are these additional statistics necessary? What purpose does each serve?

Vorderman's data further indicate that beriberi does not correlate with lower altitude (many other diseases were more prevalent among those on lower ground). Nor does the incidence of other diseases match the distribution of beriberi. In four prisons, Vorderman notes further, the number of cases of beriberi has increased with the arrival of a prisoner who already had beriberi.

THINK [7]: What conclusions can be drawn from Vorderman's study beyond what Eijkman could conclude from his study with chickens? (Reconsider especially your earlier assessments.) How do Vorderman's results support Eijkman's and/or other explanations?

Eijkman and Vorderman's study is significant in part because of its large scope. But imagine for a moment the native Javanese perspective. Why are so many people in prison available for scientific study? The Dutch are managing over a quarter million prisoners on one island! Java is one of the most densely populated areas in the world. Still, almost 1 percent of the population is in prison. From the local perspective, the Dutch colonials are invading foreigners. The prisons, all military prisons, reflect how the Dutch deal with Javanese opposition to their occupation—that is, when they do not rely on mass executions. Vorderman's survey takes advantage of that exercise of colonial power.

In addition, although more Javanese than Dutch suffer from the disease, the Dutch colonials have more at stake than simply aiding the local population. The disease takes its toll on the local armies and work force. The Dutch thus value a cure primarily for military and economic reasons. Likewise, no one has offered the Javanese the tools or resources to study the disease on their own. Although Eijkman and Vorderman are addressing fundamental biological questions, their research is also motivated by the Dutch economic interests and facilitated by the military presence.

BERIBERI AFTER EIJKMAN

Eijkman leaves Java just as his collaboration with Vorderman is ending—for a second time because of illness. Back in the Netherlands, he continues briefly his studies on beriberi. Unsuccessful in his efforts to isolate the bacterium, he focuses on the cure instead. He shows that water and alcohol extracts of the rice cuticle can cure the disease as effectively as the polishings themselves. He confirms that the curative factor is destroyed when heated above 120°C. It can also pass through a membrane, such as the cell membranes of an intestine. He then turns to other research projects on metabolism, seasons, and climate, leaving others to pursue the remaining problems about beriberi.

Beriberi is important enough that research has been occurring in several places besides Java. There are major efforts in Japan, Malaya, and the

Philippine Islands. (In Japan's war with Russia in 1904–1905, four thousand soldiers die of beriberi.)

THINK [8]: How will Eijkman's and Vorderman's dramatic results become known to others? If you are working elsewhere in Asia, how will you know if someone has been studying beriberi nearby? If you are aware of such work, how do you find out about the results? What about differences in language?

Between 1885 and 1906, inspired by Eijkman's conclusions, many researchers search actively for the bacterium or toxin present in rice and try to identify the curative factor in the rice cuticle. Seventeen different researchers claim that they have found the microorganism that causes beriberi. Other researchers, including Koch, search for the infectious agent and fail to find one. They conclude, by contrast, that beriberi is not bacterial at all.

THINK [9]: From the perspective of someone who thinks that beriberi is infectious, why might Eijkman, Koch, and others have failed to isolate the bacterium? Is the failure to find a pathogen definitive in this case? Where should the burden of proof lie?

THINK [10]: Consider the conflicting claims about the causes of beriberi in 1900.

(a) If you are a researcher with limited time and resources for investigation, will you focus on infection or diet as a cause of beriberi? Why?

(b) If you are a public administrator in Java, with a limited budget, what programs will you support to control the incidence of beriberi? How will you justify to potential critics whether you inform the public about consumption of half-polished rice, improve sanitation of rice storage and transport, wait, or do something else?

On the basis of your responses, how does scientific uncertainty seem to affect decision making in different contexts?

In Java, another Dutch doctor, Gerrit Grijns, succeeds Eijkman at his laboratory. Grijns disagrees with how Eijkman has interpreted his results, however. For Grijns, it is not the rice that is toxic, nor the polishings that effect a cure. Rather, something vital seems to be missing from the rice once it has been polished. The rice cuticle must contain a critical nutrient. In other words, Grijns sees beriberi as a nutrient deficiency, not the result of some germ.

THINK [11]: How can Grijns explain Eijkman's and Vorderman's data? How would you try to confirm Grijns's theory experimentally?

Grijns finds that other foods can effectively treat beriberi—notably *kachang-*

ijo, or mung bean. In addition, starchy diets of tapioca root or sago palm can also produce the disease. Rice alone is not responsible. Even a diet without starch—of overcooked meat—can cause beriberi. Grijns's results dramatically undermine and reverse many of Eijkman's conclusions. Beriberi patients do not suffer from something in their diet, but from something missing from it. Beriberi is a deficiency disease, based on the absence of some essential nutrient present in the rice cuticle.

> **THINK [12]**: How could Vorderman's conclusions have been significant and mistaken at the same time? More generally, what can we conclude about both the value and the limits of a controlled experiment?

The work on beriberi by medical researchers eventually intersects with independent investigations of nutrition by biochemists in Europe. In England, in 1910–1912, one researcher, Frederick Gowland Hopkins, feeds young rats highly purified forms of the basic ingredients known to be essential for any diet: proteins, fats, carbohydrates, water, and salts. Though apparently fully nourished, the mice cease to grow. When given as little as 2 or 3 cubic centimeters of milk per day, they begin to grow again. Such amounts are insignificant in terms of their protein or energy. Hopkins concludes that there are "accessory factors" in the milk that are necessary, though only in extremely small amounts.

During the same period, several individuals working independently around the globe—Casimir Funk, a Pole working in London; E. S. Edie, also in England; and Umetaro Suzuki, in Japan—each isolate an anti-beriberi chemical. They recognize more clearly how beriberi and similar diseases are linked to the work on dietary requirements. Scurvy, pellagra, and beriberi are all deficiency diseases. That is, they result from something essential that is not present in the diet. Because the vital missing elements seem to include substantial nitrogen, Funk calls them "vital amines," or vitamins. Later, the specific factors are labeled: vitamin C is associated with scurvy, vitamin B_1 with beriberi, niacin (also in the B complex) with pellagra, and vitamin D with rickets. Ironically, Eijkman does not accept these conclusions when they are first introduced.

The beriberi vitamin, named thiamine, is isolated in 1925 by a pair of Dutchmen, Jansen and Donath, again working in Java. From 300 kilograms of rice polishings, they are able to extract a mere 100 milligrams of thiamine. Even in the rice cuticle—which can prevent beriberi—the vitamin is present in only a few parts per million. Vitamins, they learn, are not typical nutrients.

The significance of Eijkman's work in opening the study of vitamins is marked by a Nobel Prize in Medicine in 1929, awarded jointly to Hopkins and Eijkman, then age eighty-one.

> **THINK [13]**: Who discovered vitamins? When? What does it mean to make a discovery in science? As a member of the Nobel

Prize Committee, how would you advise giving an award on this occasion?

Why had beriberi suddenly become more prevalent in the early 1870s? During that period, Westerners introduced steam-driven mills to the East. The mills replaced more traditional methods of hand-pounding rice. The highly effective milling process stripped the essential vitamins from the rice with increased efficiency. As steam-milled white rice became more common, so too did the occurrence of beriberi.

> THINK [14]: What was the cause of beriberi in Java in the 1880s? Was it a vitamin deficiency? A white-rice diet? Economic conditions that led to poor diet? The introduction of steam mills by the Dutch? Or the whole system of colonialism that established these conditions? How does each view (biochemical, dietary, social, cultural) imply an alternative way to reduce the frequency of beriberi? What does this tell us about the nature of causation?

THINK: NOS REFLECTION QUESTIONS

What does the case of Christiaan Eijkman and the cause of beriberi show about the following features of the nature of science?
- the role of chance or accident
- theoretical perspectives in interpreting data
- the distinction between causation and correlation
- growth of knowledge through small cumulative additions versus through major conceptual reinterpretations
- the role of the individual versus groups in making a discovery
- scientific communication and communities of researchers
- the cultural and economic contexts of science

Further Reading

Carpenter, Kenneth. (2000). *Beriberi, white rice, and vitamin B: A disease, a cause, and a cure.* Berkeley: University of California Press.

Funk, Casimir. (1912). The etiology of the deficiency diseases. *Journal of State Medicine, 20,* 341–368.

Vedder, Edward. (1913). *Beriberi.* New York, NY: W. Wood.

Williams, Robert. (1961). *Toward the conquest of beriberi.* Cambridge, MA: Harvard University Press.

Images

http://ships.umn.edu/modules/biol/beriberi.htm

Christiaan Eijkman and the Cause of Beriberi
Teaching Notes

THINK EXERCISES

The primary purpose of the THINK questions is for students to develop scientific thinking skills and to *reflect explicitly* on the nature of science. The questions are open ended. The notes here are only guides about the possible diversity of responses. In many cases, there is actual history as a benchmark (which can be shared after the students' own work), but by no means does it indicate an exclusively correct answer. Accordingly, the teacher may strive to avoid overt clues, fishing for answers, or implying that a particular response is expected or considered "more right." Again, the case study should illustrate the partly blind process of science-in-the-making. To help promote thinking skills, the teacher should encourage (and reward) thoughtful responses, well-articulated reasoning, and respectful dialogue among students with different ideas or perspectives.

Student inquiry often leads to requests for more information, and the teacher equipped with a deeper perspective is better prepared to provide guidance. The additional information below addresses these pedagogical demands, while also allowing the teacher to extend discussion, once students have found what they consider to be good solutions to the questions in the text.

THINK [1]: WHAT CAUSES DISEASE / THE NATURE OF CONCEPTUAL FRAMEWORKS

There is an epistemic conundrum common in disease research: How do you know what to observe or look for if you do not yet know what it is that you are going to find? Certain types of diseases may have diagnostic patterns, but one must be aware of those characters in advance if one is to search for them as clues. That is, a spectrum of possible hypotheses must precede any target hypothesis about the cause of the disease. This works against finding new disease types, of course.

In the late 1800s, various researchers, both Asians and Europeans working in Asia, explained the cause of beriberi differently. Some insisted that beriberi was not a specific disease at all, but a combination of other known diseases. Others claimed it was a form of poisoning. They disagreed about which toxin was responsible, however. Was it arsenic, oxalate, carbon dioxide, or some compound produced by a microorganism? Later, some viewed beriberi as an infection—but they disagreed about whether it was a protozoan, a tiny worm, or a bacterium. Another blamed moldy rice. Yet other researchers implicated diet. But while some concluded that the cause of beriberi was a deficiency of fats, others thought it was lack of phosphorus or proteins. For one researcher it was insufficient nitrogen, for another an improper balance of nitrogen. How does the scientist determine which of these many reported ideas to trust?

An outbreak of beriberi just a few years earlier (1880–1881) in Japan had been well studied by a doctor in the Japanese Navy, Kanehiro Takaki. He collected data about the patients' clothing, living quarters, diet, occupation, economic status, and geographic region, and about seasonal frequency, hoping to find clues. Each, in a sense, represented a hypothesis, in the form of a question, about what might have been a causal factor. Indeed, his methods largely reflect epidemiological methods today. Takaki found that

- Cases of beriberi were most frequent from the end of spring into summer but were not isolated to those seasons.
- The frequency of disease also varied considerably from one ship to another, and from one station to another within a ship.
- Upper-class individuals suffered less than sailors, soldiers, policemen, students, and shop boys.
- The disease was more prevalent in large cities, but even people living in the same area did not suffer equally.

Do these data provide valuable clues? Why?/Why not?

THINK [2]: UNEXPECTED CURES / THE ROLE OF CHANCE

This is an excellent occasion to discuss the role of chance, or contingency, in science. Unplanned events are far more important than the conventional image of "the scientific method" implies. Eijkman could well have decided that chickens were too unpredictable to use. But he saw the potential for tracking an unknown variable. Students may consider what features are necessary to *notice* chance events as significant. Louis Pasteur, for example, is noted for suggesting that "chance favors only the prepared mind." Molecular biologist Max Delbrück coined "the principle of limited sloppiness," suggesting that laboratories should operate with enough informality that careless errors or chance events were likely to surface occasionally. This was certainly true of Alexander Fleming's habits—and contributed to his famous findings about penicillin. The series of chance events in Eijkman's case also shows how difficult it can be to proceed without any obvious clues, previous theories, or working hypothesis.

THINK [3]: ISOLATING CAUSES / CONTROLLED EXPERIMENTS

The key difference is apparently the rice diet. But it is not clear which factor in the diet is important. As Eijkman himself considers, it could be the storage of the cooked rice, the cooking (heating), the water used in cooking, the rice type, the absence of a protective covering on the rice grain. Each factor must be isolated and tested in turn. The key is to compare the presence and absence of a single factor under parallel circumstances: the essence of a controlled experiment.

THINK [4]: Eijkman's Interpretation of Chicken Diet / Model Organisms

Many researchers failed to accept Eijkman's conclusions because they refused to believe that the disease in chickens was the same as the human disease beriberi. This highlights the role of animal models or model organisms in studying human diseases. Chickens are cheaper and can be used in ways that would be morally unacceptable for human subjects. Yet one must ensure that the conclusions can be transferred from one organism to another. Was chicken polyneuritis indeed equivalent to human beriberi?

In response to criticism, Eijkman characterized the disease more fully. He examined the chickens' tissues and noted the same degeneration of the nerves that the medical commission had identified in human beriberi. Eijkman also tried to show the connection by transferring the disease from humans to chickens with injections of blood or other body fluids from beriberi patients—but with no luck. This could be explained by the intermediate role of a toxin.

Students have suggested other plausible hypotheses. For example, the milling (polishing) of the rice may have been unhygienic and introduced a germ into the starchy white rice, whereas unmilled rice would remain germ free. That is, there would be no antitoxin, only a physical protective barrier.

THINK [5]: Human Experiments / Research Ethics

This can be an occasion to underscore the concerns of using human subjects in research—although ethical standards in the late 1800s were very different from today's. An obvious approach to investigating beriberi in humans is to control the diet of two groups of people. Indeed, on the basis of Eijkman's work, in 1906 two researchers (Fraser and Stanton) took a healthy workforce to a previously isolated area of Javanese forest. They fed one half of the workers white rice, the others a more complete diet. They continued until the workers that were fed only rice became ill with beriberi. They then switched diets between the two groups. The first group was cured, and the second group became ill. Reversing the diet of the same two groups is an elegant example of the use of control. From today's perspective, there are also obvious ethical problems with Fraser and Stanton's study in deliberately exposing people to harm (in this case, moreover, likely without their consent). Also see subsequent comments in the main text on the Javanese perspective.

THINK [6]: Vorderman's Statistics / Natural Experiments and Controls

Each part of the supplemental investigation represents a variable that Vorderman wished to rule out as a possible cause. There were other ways a bacterium might spread: water, air, contagion, unhygienic environment. The Eijkman/Vorderman study is classic in exhibiting the idea of a controlled study—not "controlling" the variables in the lab, but comparing two parallel sets of data that differ by a single variable: what is commonly called a *natural experiment*. This is an important illustration of how scientists can sometimes secure the relevant empirical information (limited to one variable) without performing an experiment.

THINK [7]: VORDERMAN'S RESULTS / SAMPLE SIZE

Vorderman's results addressed many objections about the relevance of chicken polyneuritis to human beriberi (THINK 4). Its large scope—sufficient samples of both prisons and prisoners—helped rule out the role of mere coincidence in the results. Once the results became widely known in the early 1900s, more research began to focus on the rice diet. Large-scale studies, like Vorderman's, continued through 1912—in each case confirming the findings on rice. Between 1905 and 1910, major institutions—armies, navies, prisons, insane asylums, and leper colonies(!)—finally began to change their primarily white-rice diets.

Nevertheless, as noted in the text, several researchers also continued to search for the bacterium responsible for something in the rice.

THINK [8]: COMMUNICATING EIJKMAN'S FINDINGS / SCIENTIFIC COMMUNICATION

This case underscores the importance of communication among members of a scientific community. Journals, of course, are the primary channels for formally reporting results. Most journals in the late 1800s were European, and even the work of researchers from colonial powers working in Southeast Asia were typically published in these European journals. In addition, Eijkman chose to publish his article in his native Dutch—hardly a language used universally, even at the time.

Students might imagine the various forms by which scientists communicate today (e-mail, telephone, correspondence, local and international conferences) and contrast these with what was available in Eijkman's era. This can highlight further the general cultural and technological contexts of science on a mundane but clearly influential level.

THINK [9]: BACTERIAL CAUSES / BURDEN OF PROOF

As suggested in the question, this case is an excellent occasion to discuss asymmetries in experimental reasoning and the corresponding notion of the *burden of proof*. Philosopher Karl Popper is widely known among scientists for his idea of falsification: that we can never logically prove a theory in all cases, but that we can rule it out on the basis of a single counterexample or falsifying instance. This may be true logically, but the beriberi case demonstrates the additional experimental dimensions of the problem: how does one know that one has a definitive falsifying instance? If one is searching for an unknown (of unknown properties—and, hence, hard to find or identify), for example, how exhaustive must one make the search?

In some cases, by contrast, single specimens or events have been influential scientifically ("golden events" in particle physics; fossils; phylogenetically unprecedented animals). In these cases, an individual piece of evidence can demonstrate the plausibility of a previously improbable hypothesis or prove the existence of an important class of previously unknown phenomena. Both cases—falsification and demonstration—raise the question of expectations, null hypotheses, and burden of proof. This can be especially important in cases of social decision making under scientific uncertainty.

THINK [10]: MAKING DECISIONS ABOUT BERIBERI IN RESEARCH AND POLICY / UNCERTAINTY IN SCIENTIFIC AND SOCIAL CONTEXTS

The first decision underscores that scientists do not have unlimited resources for pursuing various investigations. Scientists must make choices about which problems to pursue or which hypotheses to test. Further, they must make these choices without the advantage of hindsight—that is, they cannot know which path is the "right" path to pursue in advance. The beriberi case illustrates that a scientific community, through its diversity, might be able to hedge it bets, by pursuing several different lines of investigation simultaneously. If so, then disagreement in a scientific community may be a productive force, rather than a sign of weakness.

The second decision highlights how public policy must often be decided in contexts of scientific uncertainty. Scientists may have the luxury of withholding judgment; public policy-makers, generally, cannot "wait and see." They have many factors to consider and, therefore, may not always follow the weight of the scientific evidence. They may equally need to consider avoiding the consequences of possible error. Such cases of uncertainty confront us today. Even professional scientists may disagree over interpretations of the evidence. How do we decide the best policy in the mean time?

THINK [11]: GRIJNS'S INTERPRETATION / ALTERNATIVE HYPOTHESES AND CONCEPTUAL CHANGE

Grijns's notion of a deficiency rather than an active cause required a conceptual gestalt switch—seeing background as foreground. The disease was caused by the absence of an essential nutrient, rather than the presence of a disease-causing agent.

For further discussion, consider how Grijns might have interpreted Takaki's findings, if he had known about them (notes for THINK 1).

THINK [12]: THE SIGNIFICANCE OF VORDERMAN'S STUDY / LIMITS OF CONTROL

It is hard to find a better example that demonstrates both the power and the limit of scientific investigation. Correlation does not necessarily document causation. The conclusions of a controlled experiment or controlled study are only as good as the controls investigated. Yet this does not invalidate a study in the context where the controls apply. One may contrast the dramatic decrease in beriberi throughout Asia, based on dietary changes, with the later discovery of thiamine itself.

THINK [13]: DISCOVERY AND NOBEL PRIZES / SCIENTIFIC CREDIT

Historian Thomas Kuhn has argued that discovery is a fuzzy concept and that we cannot pinpoint a specific date, time, or place for most discoveries. They are not single events, but complex shifts in conceptual understanding. For example, can Eijkman be credited with discovering vitamins? He could cure beriberi, but at first he rejected the explanation of deficiency diseases. Does Hopkins earn recognition, even though he did not connect specific molecules with specific diseases, as Funk and Suzuki did? Jansen and Donath were the first to actually isolate thiamine, but would they have done so without the earlier findings? Nobel Prizes tend to reinforce a common notion that science relies on genius and individuals of exceptional talent. How does the case of beriberi fit with this image of science?

Credit in science is often portrayed as a significant motivating factor, if not for research, then for publicizing findings as soon as they are publicly defensible. How does the system of credit motivate scientists? What other motivations might exist? Why do we credit only the first person to publish a discovery? Are there any disadvantages to our system of credit? Should we give prizes or awards in science?—If so, on what basis?

THINK [14]: "THE" CAUSE OF BERIBERI / NATURE OF CAUSALITY

Here, causes seem to operate on at least four levels simultaneously. This challenges many conventional notions of causality as single, linear, and deterministic—proceeding in billiard-ball-like fashion, from one cause to the next.

Reductionistic thinking further leads us to consider the lack of vitamin B_1 as "the" cause of beriberi. Many were able to cure beriberi using Eijkman's (erroneous?) conclusions, long before anyone understood the concept of a vitamin. Why might we tend to privilege one explanation over another?

THINK: NOS REFLECTION QUESTIONS

The reflection on NOS themes functions partly for recall and review but also to help consolidate and, thus, complete the central NOS lessons of the case study. It is essential to closing the lessons and making the NOS thinking explicit and articulate. This reflection includes the following (with relevant earlier discussions noted):

- the role of chance or contingent events (THINK 2)
- theoretical perspectives in interpreting data (THINK 1, 4, 9, 11, 12)

- the role and limit of controlled experiments: distinction between causation and correlation (THINK 3, 6, 7, 12)
- conceptual change (reinterpretations versus cumulative growth of knowledge) (THINK 11)
- collective nature of discovery (THINK 13; and students can be invited to list all the individuals who contributed something significant to the outcome: the medical commission, Eijkman, his critics, Vorderman, Grijns, and Hopkins, at least)
- scientific communication (THINK 8, and the transfer of Koch's methods by the commission)
- the cultural and economic contexts of science (THINK 5, 10)

11 | Rekindling Phlogiston

The notoriety of phlogiston • a primer on phlogiston • orientation and classroom strategy • introduction for students • inquiry into reduction • inquiry into calcination and combustion • inquiry into coupled calcination and reduction • combustion revisited and student synthesis • reflecting on the nature of science • other inquiry opportunities • assessment • commentary

Oxidation and reduction reactions (ox–redox) are notorious as among the most challenging subjects in chemistry to teach. Like the distinction between metals and non-metals, they require careful attention to electrons and valences, at a subatomic level that is neither familiar nor relevant to—nor easily visualized by—students. Here, history offers a potential remedy (Chapter 2). That is, chemists clearly understood such reactions long before the discovery of electrons. They conceptualized the relationships among the reduction of ores to metals, the calcination of metals to their non-metallic calxes, the burning of wood or charcoal, and also, eventually, photosynthesis. They also recognized the connections with the generation of heat and light, the common properties of metals, and, later, electrochemistry. Just as one can conceive electrical current or electricity as a thing without reference to electrons, or heat as a thing (latent heat, specific heat, heat capacity) without reference to molecular movement, chemists were able to conceptualize reducing potential as a thing at a macroscopic level. Eighteenth-century chemists referred to it as *phlogiston* (pronounced FLOW-*JIST*-ON). In this chapter, I profile how the historical concept of phlogiston was used to organize a series of inquiry lessons on the nature of metals, their formation and oxidation, and coupled oxidation–reduction reactions—all without referring to electrons. At the end, the history of the phlogiston concept also served as an occasion to discuss scientific models and their meaning, and to reflect on the nature of conceptual change in science.

Phlogiston, for the typical eighteenth-century chemist, was the material stuff of fire. It's what allowed things to burn. It produced heat and light. It was also the substance in charcoal that was transferred to ores, transforming them to their metals (reduction), and that was released again when the metals were roasted (calcination). It's also what made metals metal: shiny, malleable, and conductive. Phlogiston was a powerful concept for unifying combustion, reduction, calcination, and, later, the chemical composition of fuel in sunlit plants.

Today, phlogiston has a mixed reputation. For many, perhaps most, phlogiston does not exist. It is imaginary, not real. We explain combustion now by oxygen, not phlogiston. Likewise, heat is movement, not a material substance. For these persons, phlogiston is appropriately relegated to the scrap heap of erroneous—and even embarrassing—ideas in the history of science.[1] John Herschel epitomized this view in his virulent 1830 criticism:

> The phlogistic doctrine impeded the progress of science, as far as science of experiment can be impeded by a false theory, by perplexing its cultivators with the appearance of contradictions…and by involving the subject in a mist of visionary and hypothetical causes in place of true and acting principles.[2]

Others, however, see the concept as more fruitful, even in a modern perspective. For them, phlogiston reflects, even if crudely, a simple model of reducing potential.[3] While we now have a more sophisticated view of combustion, phlogiston still functions well at a basic level as a unifying concept, as it once did. The noted chemist Alexander Crum Brown nicely expressed the more accommodating view in 1864:

> There can be no doubt that this [potential energy] is what the chemists of the seventeenth century meant when they spoke of phlogiston.
>
> We have only to regret that the valuable truth embodied in it [the phlogistic theory] should have been lost sight of; that the antiphlogistonistic chemists, like other reformers, destroyed so much of what was good in the old system.[4]

Accordingly, several science educators have advised a place for phlogiston as a simple concept or model in the modern classroom.[5] The approach profiled here adopts this perspective, using history as a tool, while also explicitly acknowledging the tension between simple and complex models (Chapter 8). It underscores the value of a simple conceptual scheme for understanding the close relationship of the processes involved—and of articulating the incomplete and contextual nature of all scientific concepts.

The emphasis in this series of lessons was on inquiry learning. That is, while we were inspired by and guided by history—and commented periodically on historical events or views—the students were not explicitly engaged in the historical context. That is, this was not a case study with a guiding historical narrative, following the work of a famous scientist from the past (as illustrated in Chapter 10). Rather, the scope of the historical concept circumscribed a bounded investigative space for student inquiry—an excellent example of history guiding lessons behind the scenes (Chapter 2). Four classes were each free to explore this problem-space independently. Each explored all the relevant topics, yet followed different trajectories (discussed further below). Ultimately, we capitalized on the historical *perspective* without adhering strictly to historical chronology (Chapter 4). For example, we introduced the thermite reaction, although it was discovered decades after the concept of phlogiston had largely been abandoned. We also realized that using phlogiston conceptually, one could delve into electrochemistry, as suggested by the views of several chemists, although their work waned and was eclipsed by later approaches (see concluding Commentary).

Here, the NOS reflections occurred mostly at the end of the unit, with the introduction of additional historical information and discussion. The NOS content was based in part, however, on the students' inquiry experience, on having actively engaged the concept of phlogiston in interpreting phenomena and in predicting and manipulating events in the lab. In our case, the whole unit spanned seven to eight class days.

A PRIMER ON PHLOGISTON

A fully developed notion of oxidation and reduction reactions as we now conceive them did not emerge until after an atomic model provided a framework for characterizing the reactions in terms of electron transfer. But knowledge of reduction has ancient roots. The reactions were known to the first miners and metallurgists who reduced ores to their corresponding metals. Originally, they sometimes attributed the metallic property to a substance from the fire—a conclusion that emerged, no doubt, from the resemblance of the reflective metal surface to the light of the fire. Oxidation reactions were familiar, of course, to anyone who built a fire. By the early eighteenth century, these phenomena had become linked by the notion of a material principle of fire or inflammability: phlogiston. Using the concept of phlogiston, chemists could explain why things burned and why they emitted heat and light when they did. Wood, oils, alcohol, charcoal, metals, sulfur, and phosphorus were rich in phlogiston. Metals also contained phlogiston, which was released when they were burned, or calcined. Combustion (of organic material) and calcination (of metals)—both oxidations in today's terms—each involved the release of phlogiston. Phlogiston thus powerfully unified the mineral kingdom with the plant and animal kingdoms, earlier considered wholly distinct, by using a shared chemical principle. Chemists also related reduction to its reverse reaction, calcination. Metals lost phlogiston in becoming an earthy material, the metal's calx, whereas ores or calxes gained phlogiston to yield metals. The basic notion of phlogiston may be summarized in the more familiar form of modern chemical equations:

Combustion:
charcoal   →   earthy residue   +   light, heat
(rich in phlogiston)

Calcination:
metal   →   calx   +   light, heat
(compound of calx   (for us, the   (released
 + phlogiston)   metal oxide)   phlogiston)

Reduction:

calx (or ore)   +   phlogiston   →   metal
                       (from charcoal)

In today's terms, phlogistonists had identified something akin to chemical energy or reducing potential. That is, phlogistonists had identified the importance of energy relations in these reactions and then reified them as the gain or loss of a material substance.[6]

Phlogiston also intersected with interpretations of acidity. For example, sulfur was considered a compound of acid and phlogiston. So, when an acid reacted with a metal, phlogiston from the metal was transferred to the acid, yielding the metal's corresponding calx and (in this case) sulfur. In addition, phlogiston was considered to contribute to the unique properties of metals: their luster, malleability, and ductility. Late in the 1700s, chemists saw an analogy between fire and electric sparks and explored the relationship between phlogiston and electricity—even predicting, and then demonstrating, that electricity could reduce calxes to their metals.[7] That is, the concept of phlogiston exhibited successful predictive power. These detailed conceptions became peripheral in our relatively simple classroom project.

As the eighteenth century progressed, chemists began to collect gases, or various "airs." They realized that they could be involved in chemical reactions—as products or reactants. Pneumatic chemistry flourished. Without knowing fully about oxygen, for example, they realized that combustion in a closed vessel was limited by the amount of air. In addition, burning "fouled" the air for further burning. For breathing, too. A burning candle reduced the time a mouse could live when placed in the same vessel. Drawing on the image of smoke leaving a burning substance, some chemists extended the notion of phlogiston. During combustion, air would become "phlogisticated." When the air became saturated with phlogiston, it failed to support further burning. Likewise, "dephlogisticated air"—oxygen, in today's terms—could support extended combustion.

Another "air" was produced by the reaction of acids with metals. The new "air" (hydrogen, in today's terminology) burned remarkably well and earned the name "inflammable air." Given that metals lost their metallic properties, the gas surely contained the lost phlogiston. For some, it perhaps *was* phlogiston! One of the most productive pneumatic chemists, Joseph Priestley, envisioned the reverse reaction: shouldn't this new "air" be able to reduce calxes back to their metals? And this he subsequently demonstrated. Again, a prediction based on the concept of phlogiston proved correct and opened further investigation.[8]

In the late 1700s, Antoine Lavoisier identified oxygen as a distinct elemental gas, able to combine with other elements in solid compounds. For Lavoisier, explanations using oxygen made the role of phlogiston unnecessary. Metals were simple substances, not compounds with phlogiston. Calxes were compounds with oxygen, not simple substances. Lavoisier began crafting what he called a "revolution" in chemistry. He developed a new nomenclature for the elements, which now included the gases hydrogen, nitrogen, and oxygen, while eliminating water and air. That system helped fruitfully reorganize thinking. Lavoisier's view of combustion (and of heat and light, which were also now chemical elements, as well!) was inscribed in the new terminology and was widely adopted with it. In the following decades, the concept of phlogiston waned. Today, many histories (shaped by Lavoisier's perspective) portray phlogistic theory as "overthrown" or "supplanted" by the oxygen theory of combustion.[9] Yet, by today's reckoning, Lavoisier failed to adequately explain aspects of heat, light, and why things can burn at all. Even our simplest notions of fire include a role for both oxygen and fuel. Late phlogistonists criticized these very deficits in Lavoisier's scheme. They underscored the original strengths in the concept of phlogiston (not related to "airs"). Thus, many accepted the discovery of oxygen, while still maintaining the original role for phlogiston—say, in explaining the heat and light of burning.[10] Understanding this historical overlap allows one to approach phlogiston and its role in metals without addressing its separate, and less secure, role in pneumatic chemistry. Indeed, because phlogiston helped guide early understanding of combustion and related phenomena, apart from any concerns about gases, it seems ideally suited—despite its sometimes maligned reputation—for conceptualizing oxidation and reduction at an introductory, macroscopic level.

ORIENTATION AND CLASSROOM STRATEGY

We rekindled phlogiston for teaching chemistry in a modern classroom. The original motivation was to teach about metals and to lay the foundation for oxidation–reduction reactions. The teacher in this project, while experienced, had found teaching these topics (especially motivating students) particularly challenging. Our approach to the material was structured by the organization in Figure 11.1. That is, we wanted the students to address and find relationships among the following:

1. Reduction of "ores" (metal oxides, chlorides, etc.) to metals
2. Oxidation of metals (calcination)
3. Calcination and reduction coupled together
4. Combustion

This provided a fairly well-bounded problem-space that each class could explore in a context of inquiry. Students would easily be able to generate and draw on a set of observations largely available in the early and mid-eighteenth century, when the concept of phlogiston flourished.

Pedagogically, then, our strategy was to situate the students as a group of investigators in a historically informed scenario, while not making the

	Phlogiston lost	Phlogiston gained
METALS	**calcination** (corrosion, rusting, and tarnishing)	**reduction** ("reverse calcination")
ORGANICS (carbon/wood)	**combustion**	?

FIGURE 11.1. Relationship of reduction, calcination, and combustion.

history explicit or dominant (Chapter 2). We guided four classes in finding their own way through the problems and their solutions, occasionally asking simple questions to help them see adjacent areas of the problem-space. The students conducted simple experiments and demonstrations, punctuated by sometimes quite extended discussions in which they collaborated to interpret their results and to map out successive phases of their inquiry. Each class was free to pursue its own path, and each eventually tackled the material in a different sequence.

An inquiry curriculum is, by its very nature, unscripted. Applications are highly contextual and contingent on local features, such as teacher strengths, student abilities and class profiles, curricular setting, institutional resources, time, and so on. Our context was a tenth-grade chemistry class in a relatively affluent college-prep school. Interest in chemistry itself was relatively low, but the teacher enjoyed a good rapport with the students. The following account may thus be viewed as just one instance of the activity. I try to profile local factors that shaped its development and implementation. Other teachers may thus find here a scaffold or flexible model only, to be adapted to their own local circumstances.

INTRODUCTION FOR STUDENTS

We wanted to motivate students at the outset to reflect on the relevance of metals—their cultural role and historical emergence. That would ideally lead to questions about industrial processes and, of course, metals as substances subject to chemical reactions. We used a video ("The Age of Metals: Can It Last?" from the public television series *Out of the Fiery Furnace*). We might equally have discussed the local mining industry (here, taconite, a form of iron ore—but metal ores of some sort are mined nearly everywhere). Or one might compile a list of all the metal objects found just in the classroom and note their importance, perhaps typically taken for granted. Or one might profile the historical impact of metals. For example, in a dramatic battle at Cajamarca, Spanish conquistador Francisco Pizarro and 168 soldiers were able to subdue more than 80,000 Incas in one afternoon, in part because of their steel swords, armor, and guns.[11] In our classes, we also noted the centuries-old problem arising from the economic value of metals: how does one transform an ore into its corresponding metal?

We then allowed students to observe and record in lab the differences between metals and their "ores" (or their oxides—their *calxes* in eighteenth-century terminology; we refrained from providing any chemical formulae). We primed the problem further by showing how a metal can apparently burn, producing once again its chalky calx (our surrogate ore). First, we burned steel wool. That raised a few eyebrows. Then, for sheer spectacle, we burned magnesium ribbon. The desired effect was achieved. These observations served to guide subsequent work, organized around some key questions: What is the nature of the difference between a metal and its ore? How does one interpret the transformation from one to the other? And where did all that light from burning the magnesium come from?

INQUIRY INTO REDUCTION

One class was particularly intrigued by the smelting process. They started off by researching this for homework. Fortunately for the students, we could direct them to a convenient chapter section in their text. One could well have used this opportunity, though we did not, to delve deeper into the history of metal technology—elaborating on discoveries in the Iron and Bronze Ages, their implications for civilization, etc.

Once students had acquainted themselves with the critical role of coke or charcoal, they were ready to reduce their own "ores." They used partially covered crucibles as mini-furnaces. Different groups used $CuCl_2$ (as tolbachite or eriochalcite) and CuO (cuprite) to provide some variation in trials—and to establish the benefit of sharing results. Using their makeshift smelting apparatus, students successfully produced small granules of metal. In some cases, they found a thin but unmistakably colored lamina of copper on the outside of their crucible (from vapors that had rolled over the crucible lip and condensed, we concluded). They confirmed the presence of their products through observable traits and tests for conductivity, which they had learned on the first day of the unit. The charcoal had been able to confer some metallic properties to the ores. In this laboratory exercise, then, they had established one piece of the overall picture, summarized in their reaction equation:

$$\text{calx / "ore" + carbon} \rightarrow \text{metal} \qquad (1)$$

The remaining question was, of course, "What is carbon's role?" They would need to pursue other inquiries, especially about the burning of coal, before being able to answer this directly. In subsequent investigation, in fact, this class would also learn that they had inadvertently omitted a key product from their equation: where had the carbon gone, or was it part of the metal?

INQUIRY INTO CALCINATION AND COMBUSTION

The initial demonstration of burning metals prompted other classes to focus first on the role of the heat of the fire. We provided samples of metal powder for them to roast, or *calcine*, in simple crucible setups. Again, they noted the recognizable features of the metal–calx change. In group discussion, we posed the question whether there was a basis for comparison: did the transformation between metals and their ore-like versions occur elsewhere? Their observations of different samples now provided clues, and some cases were forthcoming; others we teased out by suggestion:

- Tarnishing—of brass candlesticks or doorknobs, silver jewelry or cups, silverware or other cutlery
- Corrosion—of bronze statues or copper roofs and pipes, each signaled by distinctive color changes
- Rusting—of iron nails or fenceposts, cast-iron frying pans, etc., also marked by color change

Our intent was to link the science with more familiar phenomena and to use that familiarity as a channel for introducing their emerging theory into a vernacular perspective. By comparing these cases with their own, students were able to notice that heat was not exclusively responsible, though it did seem to hasten the process. Most students seemed satisfied by this conclusion, and we passed over the opportunity to construct a more carefully controlled experiment. In all cases, however, the metals were exposed to the air or rain or water.

When students reconsidered calcination as burning, they were able to draw on their prior knowledge about the combustion of wood or coal as possibly applicable to metal-burning. They speculated about the role of oxygen as a reactive agent or catalyst. (Addressing such common prior knowledge was one reason we had not endeavored to follow a strict historical scenario.) Students were able collectively to design experiments to test their hypothesis: compare metals in pure oxygen test chambers with those in oxygen-deprived atmospheres. We recalled that during the early eighteenth century, chemists had discovered different "airs" and devised the equipment to collect and manipulate them. So we challenged our modern students to create on their own any special lab materials: how did one obtain this sample of pure oxygen, for example? (How might a modern chemical supplier get it?) The students were forthcoming in suggesting that burning a candle in a closed system would create the requisite oxygen-free atmosphere. They were a bit more baffled about the oxygen. But some remembered a reaction they had done with red mercury powder in chemistry class some time earlier.

Others suggested gathering the "air" from plants or seaweed. Again, we had the opportunity to pursue a tangent—here, into pneumatic chemistry—but we opted for a narrower focus, acknowledged the fruitfulness of their experimental design, and assured them that their tests would confirm their expectations. Another piece was thus added to the puzzle, summarized again in their equation:

$$\text{metal} + \text{oxygen} [+?] \rightarrow \text{calx/"ore"} [= \text{an oxide?}] \qquad (2)$$

This still left open the questions of the light, so dramatically exhibited by the magnesium fire, and of the heat known to accompany burning. Many were ready to speculate that something—akin to smoke, perhaps—was given off during the burning process. (Here, they re-expressed the naive chemical views documented in many cognitive studies, and we did nothing immediately to suppress the misconception. We trusted their own investigations and discussions as a route to a proper conception.) One may note, here, that a role for phlogiston was being established, even as they spoke about oxygen.

INQUIRY INTO COUPLED CALCINATION AND REDUCTION

Given these preliminaries (and, in one case, even prior to completing them), we were ready to introduce formally the notion of phlogiston. We asked rhetorically whether it might not be possible to transfer the metallic qualities from one metal to another, producing a new metal from its ore without charcoal. We then proceeded with what was perhaps the theatrical highlight of the unit, a demonstration of the thermite reaction.[12] In this reaction, aluminum reduces iron ore to iron, while it is oxidized to aluminum oxide. The reaction produces spectacular fireworks and enough heat to melt the iron product. In our demonstration, performed on the school's baseball infield, the molten iron exhibited an impressive orange glow as it dripped nicely out of our flower-pot reaction vessel.) The pyrotechnics were calculated to have an effect—and the students did not disappoint us, even given our deliberately hyperbolic promises.

Students confirmed the cooled iron product as metal by measuring conductivity (resistance). We then returned to the blackboard to summarize the reaction:

$$\text{aluminum} + \text{iron "ore"} \rightarrow \text{iron} + \text{aluminum "ore"} + \text{LIGHT, HEAT!} \qquad (3)$$

Although the thermite reaction was not discovered until 1893 (hence unknown to the eighteenth-century phlogistonists), we confidently interpreted the results in terms of phlogiston. We announced to the students that they had witnessed the transfer of an inflammable (and metal-conferring) substance from the aluminum to the iron ore, while some was lost, accounting for the light and heat. We called the substance *phlogiston*. This, clearly, was what had allowed carbon to reduce their metal "ores" earlier: carbon was a rich source of phlogiston (witness, for example, the combustibility of coal). While many students remained wary, the demonstration allowed them to recognize and address the strong link between calcination (here, of the aluminum) and reduction (of the iron calx). A transfer of properties of some kind had occurred (and we referred back to some of their speculations about things emitted during burning). We sent them home with an excerpt from the 1771 *Encyclopedia Britannica*: "Of the PHLOGISTON."[13] The results of all the investigations were ready to be integrated.

COMBUSTION REVISITED AND STUDENT SYNTHESIS

Through several days of following leads, students had accumulated a wealth of disparate information: about the roles of oxygen in calcination and of carbon in reduction (though this was cast in doubt by the thermite reaction), and about the release of heat and light. To facilitate the synthesis of information from the different contexts, we constructed the table in Figure 11.1. Once the elements of the table had been explored separately, the organization suggested some broader comparisons and posed several questions. Some of the details of this discussion are included here to illustrate the multiple possibilities and to alert the teacher to prepare properly. It may also help to show a modern chemist how to think like a phlogistic chemist.

(a) How did the thermite reaction involving aluminum metal (eq. 3) relate to the reduction exercise involving organic carbon (eq. 1)? (How did top and bottom rows relate?) Can charcoal reduce ore precisely because it can combust—the phlogistic explanation? Does charcoal burn at the same time as it reduces metal ores?

(b) Metals become calxes (top row), but what does wood or coal become when it burns (bottom row)? As one student expressed it—catching us off guard, but delighting us nonetheless: what is the "calx of carbon"? Alternatively, substituting coal for aluminum in the thermite reaction (eq. 3): what is "coal ore" as a product?

(c) If oxygen is acquired during combustion and calcination (left column), is oxygen lost during reduction, as the reverse process (right column)?

(d) If light and/or heat are released in combustion/calcination, is heat or light therefore required for reduction (in addition to whatever carbon does)?

(e) If reduction and calcination are complementary processes, is there a process complementary to combustion? That is, how would one characterize "*reverse combustion*"—the gaping hole in the chart?

Each comparison provided an entry for bringing information together and filling in the holes. Students plumbed their own knowledge to realize that coal ore was the more familiar carbon dioxide: a gas, not an earthy material. It must be formed in reducing metals (revealing an additional, missing product in eq. 1). They concluded that calxes must be metal oxides and that reduction must involve the loss of oxygen. Carbon can be a reducing agent partly because it reacts with the oxygen. Oxygen is lost in the processes in the right column, but oxygen is gained in the processes on the left. Hence, we could provide a new label: *oxidation*. The new term expressed the unity between metallic and organic reactions. Similarly, phlogiston—or some equivalent—must be lost on the left, yielding light and heat, and gained (or required) on the right. The reactions also seem to be coupled: every loss entails a gain somewhere else, and vice versa. But what was reverse combustion? They reconstructed the necessary equation:

$$? + \text{phlogiston (light)} \rightarrow \text{carbon fuel} + O_2 \qquad (4)$$

Again, they did not disappoint us: they identified this as photosynthesis: familiar, but now with a new and deeper meaning in the context of similar reactions with metals.

All the reasoning here was straightforward, but the students needed time to cross the observational ground several times and notice and talk through all the connections. But this itself can be a lesson about how real research in science proceeds: as a balance between blind groping, reasoned guesses about where to go next, and empirical confirmations. We were satisfied that patience on our part was rewarded with the appreciation by students that they had largely reached their conclusions on their own.

Ironically, our students had already learned atomic theory. Thus, they were well prepared to appreciate further lessons about what some perceived as the fuzzy concept of phlogiston. Phlogiston, they could interpret in modern terms, was a form of energy that one might construe crudely as chemical bonds. But they also knew that light, produced by the release of phlogiston under one conceptual system, was to them associated with the release and capture of electrons (associated with spectral lines). They thus re-mapped what they had just learned about phlogiston onto what they already knew about atoms. They could easily reinterpret oxidation (calcination and combustion) and reduction in terms of a more sophisticated or more deeply articulated notion of electron transfer. This reinterpretation is significant, because it demonstrates that one need not follow a strict historical sequence across the curriculum. Historical episodes can be used intermittently, even perhaps anachronistically, to great effect.

REFLECTING ON THE NATURE OF SCIENCE

The students at this point were well positioned to appreciate the observations of renowned chemist Alexander Crum Brown in 1864:

> There can be no doubt that this [potential energy] is what the chemists of the seventeenth century meant when they spoke of phlogiston.
>
> The truth which [Lavoisier] established, alike with that he sub-verted, is now recognizable as partial truth only; and the merit of his generalization is now perceived to consist in its addition to—its demerit to consist in its suppression of—the not less grand gener-

alization established by his scarcely remembered predecessors....
Accordingly, the phlogistic theory and antiphlogistic theory are in
reality complementary and not, as suggested by their names and
usually maintained, antagonistic to one another.

The occasion was ripe to pose deeper questions about the nature of scientific
theories. Is phlogiston real? In particular, is it any more or less real than
electrons? For many of our students, phlogiston was not real: it was not in
their textbooks—and it did not appear on the periodic table! Once we had
reintroduced the more familiar concept of electrons, they quickly abandoned
the old concept as outmoded. As an idea from history, it no longer had
currency and could thus be rejected as imaginary, not real. We emphasized
the explanatory adequacy of the concept and how it could guide their own
interactions with the materials, and we asked how else they would judge the
concept. Our strategy was to create a discrepant event *about the nature of
science*. How would one know or prove what is real or not?

Philosopher of science Ian Hacking has proposed that our criteria
for realism should not be evidence for particular *representations*. Rather,
our judgments about what is real and what is not typically rely on our
interventions in the world.[14] In this case, our students (like their eighteenth-
century counterparts) could intervene by transferring phlogiston from one
substance to another. Historical chemists were even able to predict such
interventions—reducing metals with inflammable air (hydrogen) and with
electric sparks. In such a *functional* framework, phlogiston can indeed be
considered real. It has *instrumental* reality.[15]

We also turned the students' skepticism on its head: how would the
students know if electrons (as a preferred explanation) were any more real
than phlogiston? Did phlogiston, like electrons, not explain things in a
definable context? Did phlogiston not help describe and predict the reactions
for them now just as effectively as it had for chemists in the eighteenth
century? Students gradually recognized that phlogiston and electrons
were both humanly developed, possibly limited concepts. Still, they were
grounded empirically and extraordinarily powerful in interpreting the world
around us. Even if a scientific theory is limited, we may still want to construe
it realistically, at least in a particular context. The concept of phlogiston,
they admitted, was certainly reliable and warranted within a prescribed
domain of application—and had been accepted historically within this
domain. It could be "wrong" and "right" at the same time, depending on
context (Chapter 4). Phlogiston was as real for eighteenth-century chemists
as electrons are for us now. Scientific theories are not universal or absolute.
They have contexts and limits (Chapter 8). This, we felt, was a profound
lesson about the history and nature of science.

Not every student was deeply engaged in this level of thinking. But our
success in provoking philosophical reflection may be indicated, perhaps, by

the students who stayed after class to argue about whether the phlogiston concept is still relevant—and who also arrived in class the next day brimming with fresh thoughts.

OTHER INQUIRY OPPORTUNITIES

The encounter with phlogiston opened several tangential excursions that we wanted to pursue more fully but could not because of our particular schedule. Each could extend the phlogiston/ox–redox unit and/or serve to segue to other units of study.

(1) *Sulfuration*. Despite the term 'oxidation', oxygen is not the only element capable of converting metals into earthy ores. The sulfuration of iron was known early in the eighteenth century, but did not become a major component in the debates over combustion and phlogiston. Some, however, used the phenomenon to argue that the oxygen theory of combustion was incomplete, and that it could not therefore fully replace the notion of phlogiston in burning, as many at the time contended.[16] Some students in one class found that they could get a fine-grained metal to lose its metallic properties by heating it with sulfur powder. With creativity, an instructor might expand to chlorine, or to carbonates, sulfates, nitrates, etc.

(2) *Acid–metal reactions, reduction by hydrogen*. Some students, prompted by the similarity of calcination with corrosion, wondered if acid could calcine (corrode) a metal. By heating zinc with hydrochloric acid under the hood, they demonstrated that indeed they could. Had they been able to isolate the gas released (clearly evidenced by the fizzing bubbles), they might have been surprised that it was highly flammable. In fact, hydrogen had been called "inflammable air" when it was first isolated by phlogiston-minded chemists. It seemed to have taken phlogiston from the metal as it was calcined by the acid. With a bit of guesswork, the students might have predicted that a phlogiston-rich gas such as this could reduce calxes—a speculation that could be confirmed by further testing, as it was historically by Joseph Priestley (see Primer above).

(3) *Electrochemistry*. The thermite reaction offers a suggestive model for the reactions in a galvanic pile. Indeed, Humphry Davy used a phlogiston-like concept to interpret some of his early work on electrolysis.[17] For others, phlogiston was the "principle of the negative pole of the galvanic apparatus." One can thus pose a challenge to students: given the concept of phlogiston, and the knowledge that some metals seem to be able to release it to others (as in the thermite reaction), can one generate or harness a flow of phlogiston? The reducing potential of different metals (at two poles in a battery) can certainly be interpreted as reflecting characteristic

levels or amounts of phlogiston. Here, one would be building on work of many chemists in the eighteenth century who saw a connection between phlogiston and electricity. The light of electrical sparks, they speculated, was analogous to combustion and indicated the release of phlogiston. Indeed, some researchers successfully reduced metals with electricity—a prediction students might also make under appropriate circumstances.[18]

(4) *Respiration.* In a classroom prepared to make cross-disciplinary leaps, the discussions of phlogiston, carbon, and photosynthesis could lead to a further pursuit of biological oxidations and reductions. Here, the burning of wood would be an explicit analog of the burning of plant fuel by an animal.

(5) *Mining and metallurgy.* As noted above, the encounter with metals and ores provides an opportunity to linger on the metal industry and the historical developments of the Bronze Age, Iron Age, etc. For example, how does one determine the purity of a metal extracted from its ore, especially of gold or silver, whose values are closely related to their purity?

Assessment

This project was ripe for several alternative modes of assessment (Chapter 9). For example, we asked the students to keep scientific notebooks/journals of their lab results and their thinking along the way. Initially, we planned to have each student summarize his or her observations and interpretations, along with the class's collective reasoning—all in the format of a modest scientific paper, as though they were reporting original research. By the end of this relatively complex project, however, this seemed somewhat daunting. We also passed up the opportunity for peer review—made even more promising by the possible exchange of papers between classes that had reached their own results in slightly different ways. Nor did we ask students to plan a research agenda or experiments that might investigate one of the Opportunities above.

The solution for the teacher, in this case, was to have students write an essay on one of several topics. One option was to write a letter to a phlogistonist and to explain, in terms he could understand, how we now interpret the four processes in Figure 11.1. A more straightforward version asked about the role of light and heat in the same set of reactions, comparing explanations using phlogiston and electrons. Another, more philosophical topic invited the students to comment on the claim that "phlogiston is just as real as the electron." The most challenging (and least selected) question asked students to consider the results of combining of carbon or silicon with (a) coal, (b) iron ore, and (c) an acid. The intention, here, was that they might predict or articulate the properties of a semi-metal.

COMMENTARY

What does one gain from rekindling phlogiston in the chemistry classroom? How might one generalize from this case to others?

Nature of Science

Teachers often adopt elementary theories from history—even those we now construe as "wrong"—as simple models. For example, the Bohr model of the atom is a standard part of teaching atomic theory and spectral lines, though by the next week we often celebrate how it was replaced and shown to be "wrong" by quantum theory. Likewise, textbooks frequently draw on the fluid model of electricity to explain current, resistance, capacitance, etc., especially in complex circuits, although scientists abandoned the notion of an electrical fluid in the mid-nineteenth century. Such models are useful because they can help lead students from simple to complex ideas. We may need to dwell on their simple applications and contrast their adequacy in certain contexts with their inadequacy in others. These models are excellent opportunities for introducing students to the role of models as conceptual maps, and the ways in which any model is selective in what it represents. They can both inform *and* mislead at the same time. We may profitably reflect and comment on the virtues and qualifications of scientific models or theories as we teach them (Chapter 8).

As a result of the explicit reflections, we think that our students learned something about conceptual change, or the history of ideas in science. Their final letters to the historical phlogistonist, in particular, demonstrated that they could see the same data in two ways. That is, when comparing interpretations using phlogiston and electrons, they were able to appreciate the historical change in concepts, along with the merit and context of the original interpretation. They thus understood that ideas can change: even an entry in an encyclopedia (our excerpt from *Encyclopedia Britannica*) can later be considered "wrong" (Chapter 4).

Finally, they were able to realize that the development of knowledge is not the mere accumulation of facts. Sometimes there are dramatic reconceptualizations, exemplified in the long-term shift from phlogiston to electrons. Having used the concept of phlogiston themselves, however, they were less likely to see current ideas as self evident, or to regard as ignorant or foolish those scientists of the past who advocated ideas we now regard as mistaken (Chapter 3). Some students also clearly appreciated, by analogy, the tentative status of our current knowledge (about electrons, etc.). Working with the historical ideas, rather than merely learning about them through a presentation, may have been critical to appreciating their legitimacy and thus understanding the historical moral.

The Role of History in Science Education

We borrowed the concept of phlogiston from history, but ultimately our project used history creatively rather than adhering to it strictly. First, our students had learned about combustion and the role of oxygen—and even about electrons—well before our unit on metals and phlogiston. This clearly contradicted the actual historical unfolding. In addition, we used the thermite reaction anachronistically and interpreted it as phlogistonists might have decades earlier, before it was actually known. By contrast, we entirely omitted discussions about phlogiston and air, and metals and acids (but see section on "Opportunities"). Our lesson was historically inspired, not a historical simulation. Our aim was not to replicate or recapitulate history. Rather, we used history as a tool (Chapter 2). The history sensitized us to initial and simplified impressions about reduction and oxidation reactions. It also highlighted the relevant observations for developing the concepts fully. Our scenario was not so much historical as historically informed. Still, a teacher that delves into this case will ideally develop some fluency in the historical way of thinking, to help converse with students on the same level.

Although we may have taken liberties with historical events, we did not abandon the perhaps more important historical principle: *respecting historical perspective* (Chapter 4). That is, we preserved the context in which talking about phlogiston did—and still does—make sense. Our application of phlogiston to the thermite reaction, anti-historical in one respect, was governed by just this principle. Nor did we corrupt the original phlogiston concept to fit modern ones. Nor did we treat it as a historical precursor to electron theory, or as an elementary or primitive version of potential energy or reduction potential. Thematically, these links exist; historically, they do not. Throughout, we were sensitive to the historical context, its virtues and limitations both.

According to some models, the individual student recapitulates history in learning scientific concepts. Some regard history as *the* model for designing appropriate instructional sequences (Chapter 2). While the notion is intuitively appealing, our experience demonstrates that history need not be the exclusive authority in specifying a conceptual sequence. While we taught oxidation–reduction in a historical context, students had already learned certain relevant concepts developed only later in history. We introduced the eighteenth-century concept of phlogiston *after* the students had learned about oxygen and Lavoisier's system of the elements, *after* they had learned the role of oxygen in combustion, even *after* they had learned about electrons, atomic models, and the electron's role in light and spectral lines. We allowed students to construct their notions on concepts that, historically, were not available at the time of their historical counterparts. Our chronological juggling, nonetheless, yielded a penetrating interpretation that reconciled phlogiston and oxygen.

Managing Inquiry as Science-in-the-Making

As noted above, we allowed students to make errors early in the unit without correcting their misconceptions: about carbon possibly combining with ores to constitute metals, or about smoke indicating the release of something material from the fire. This was essential to inquiry learning, adopted as a model for authentic science-in-the-making (Chapter 2). Withholding comments at such times was difficult and required a fair amount of confidence that the students' ensuing research would isolate such errors and correct them. By the end of the unit, these tensions had indeed resolved themselves. The danger, of course, was that individuals might not recognize these connections or transformations. In many cases, therefore, we tried to reintroduce the early conceptions into their final discussions, so that they could couple them with their later knowledge and reinterpret them explicitly using their more sophisticated conceptual frameworks. Where possible, we directed questions specifically to the student who originally introduced the simpler notion. Here, we emphasized the revisions and the reasons for shifting explanations. These provided good occasions to acknowledge and celebrate the students' own discoveries and learning.

Authentic research, or science-in-the-making, involves uncertainty. Admittedly, the problem-space we selected was carefully bounded. But it was also shaped by historical awareness. Through historical precedent, we expected students to be able to make discoveries by exploring on their own without getting lost. Still, the students faced (for themselves) genuine unknowns. Sometimes, we reluctantly nudged students into further research. But we did so only by adopting the role of fellow investigators and posing questions in context, from the knowledge at hand. Each class seemed to present its own limits or thresholds. The challenge of contextual prompting seems a likely challenge whenever an inquiry activity has a target domain.

History can be a guide for an inquiry scenario. But for inquiry to model science-in-the-making, activities must be based on the students' own understanding, not on trying to re-enact history (Chapter 4). As a result, our four classes found four different ways through the problem-space (described in the Classroom Strategy), pursuing the topics in their own distinctive order:

Class A*	4	→	2	→	5	→	3	→	1
Class B	2	→	1	→	3	→	4		
Class C	2	→	3	→	4	→	1		
Class D	1	→	3	→	2	→	4		

(*One class also considered sulfuration and treatment of metals with acid, labeled 5.) Each class also raised and addressed its final synthesis questions in a different order. Yet we deliberately chose a problem-space that was well cross-linked. We also guided students along the way. Still, the different pathways nicely demonstrate that given a rich collection of phenomena, many different ways to reason to a conclusion are possible. For this style of project, finding such a problem-space may be key.

Our ideal was student-based inquiry. At the same time, as student autonomy and freedom increased, so too did the teacher's workload. To be able to pursue contingencies as they arise, one must be ready to accommodate them. In principle, this is simple. In the classroom, it is not. In planning the unit, we felt the need (confirmed in retrospect) to scout the territory ourselves and be prepared for the many possibilities. Many prospective experiments were not fruitful. Planning our phlogiston unit (which covered seven to eight class days) involved six ninety-minute sessions. Preparing for contingency, we found, is extraordinarily time consuming. In addition, the teacher sometimes needed to prep the lab for four different setups on the same day. This was challenging. Thus, in subsequent years, the teacher opted to guide all classes more strongly through a common sequence.

We were also concerned that students not only learn the content, but have a rewarding experience in scientific investigation as well. That is, we were sensitive to the *affective* dimension of inquiry and wanted to ensure *closure* and *success* in problem solving in an educational setting. We thus carefully selected a problem-space that was relatively simple to explore. We are well aware that students can learn from their own errors. At the same time, students can be enormously discouraged by such perceived failures. We thus invested considerable time in preparing the unit by trying several investigations and materials that the students might likely have selected on their own. For example, we focused on copper compounds after finding that alternatives with magnesium and zinc did not readily yield clear results. We also abandoned intentions to replicate Lavoisier's measuring of the weight in calcining metals after finding how difficult it was experimentally to get a complete reaction. We used that knowledge to shape and limit the students' inquiry. We established boundaries that would allow the students to explore the phenomena successfully without getting lost or bogged down in the complexity of the real world. Again, the effective teaching scenario was, in some ways, student-based in appearance only.

In summary, we rekindled phlogiston. Borrowing from history, we were able to teach a difficult concept in simple terms while also fostering NOS learning about conceptual change and the nature of scientific models and theories (Chapter 2). Because the concepts that we teach in K–12 are typically simplified (Chapters 7 and 8), such lessons seem important in conveying the limits of science and how to interpret the reliability of its claims when relevant to personal and public decision making (Chapter 1).

12 | Debating Galileo's *Dialogue*: The 1633 Trial

In our culture, Galileo's trial is emblematic of an apparently irresolvable rift between science and religion. However, delving into the case more deeply—with students assuming roles in historical perspective—one encounters a startlingly different, yet equally provocative, image. The trial is an excellent historical case for engaging students in discovery about the nature of science and its cultural contexts (Chapter 2).

In popular lore, the trial has become stock melodrama—another example of a scientific myth-conception (Chapter 3). In those widely retold accounts, Galileo is the hero for science, boldly confronting the Church, the anti-science villain. Galileo defends a Copernican view against a Church establishment that is unwilling to forsake a literal, Earth-centered interpretation of the Bible. The implicit moral is that Galileo was right because science and evidence were on his side. The Church was wrong because it clung to religion, even when faced with obvious evidence against it. With that basic scenario established, the scientific story is typically streamlined to yield the expected outcome (Chapter 4), and other historical facts are shaped to sharpen the ideological moral (Chapter 5).

Too much goes unsaid in the standard account, however. There is considerable richness in a fuller story, which is not especially difficult to render or to understand (Chapter 7). Consider how the case unfolds, following the format of the trial in the classroom project outlined below (also online at http://galileotrial.net).[1]

To begin, we situate ourselves in Rome in 1633. We hear about the Pope and the Reformation in northern Europe, and the ensuing drama of shifting international political allegiances and military conflict. We learn about the wealthy elite, such as the Medicis in Florence, and their opulent courtly life, patronage of the arts, and cutthroat social politics. We are introduced to Galileo Galilei, a natural philosopher, and to his renowned skills in entertaining the Medici court in debates over dinner—whether about buoyancy or sunspots, the new star of 1604 or the comet of 1618, or the phosphorescent Bologna stone. Perhaps we hear about sculptor Giovanni Bernini and the fabulous bronze canopy he has designed for St. Peter's Cathedral.

Last year, in 1632, Galileo published his *Dialogue Concerning the Two Chief World Systems*, presenting evidence for the Copernican doctrine of a Sun-centered astronomy in the familiar format of a conversation among friends. But the book has stirred controversy within the Church. It is April, and Galileo has been called to Rome to answer charges of "vehement suspicion of heresy" for

1. advocating Copernican doctrine, contrary to a 1616 Vatican edict, and

2. holding and defending as probable a thesis contrary to the Bible.

Our task is to consider the evidence and decide for ourselves whether to condemn Galileo.

First, we assess the science. We hear from Galileo's supporters that he has created the telescope. He has seen the rough surface of the Moon. He has seen spots on the Sun. The heavens are not perfect, as the Ancients imagined. There are planetary bodies orbiting Jupiter. Hence, the Earth is not alone in having a moon. Venus exhibits phases, indicating that it orbits the Sun. In short, Ptolemy was wrong.

And the Church? They concur. First, their advocates note the credentials of Church astronomers, including Copernicus himself. They have been quite concerned about the timing of Easter, based on celestial events, and recently reformed the calendar. They have done landmark work on comets and sunspots. So yes, indeed, they are competent to judge Galileo's claims. Notably, all the observations also support a mathematically equivalent system envisioned by Tycho Brahe, in which the planets orbit the Sun and the Sun orbits the Earth. This is the position endorsed by the Church astronomers. And Galileo seems to have deliberately sidestepped addressing that alternative in the *Dialogue*. The only way to discriminate observationally between the Copernican and Tychonic systems is through stellar parallax, based on the annual motion of the Earth. But no measurement can confirm that. The heavens would have to be unimaginably vast and empty otherwise. So the failure to detect parallax is pretty conclusive. Advocating Copernicanism thus overstates the astronomical evidence and its uncertainty.

Yet Galileo has argued that there is *physical* proof that the Earth moves. For Galileo, the tides are caused by a combination of contrary motions, due to the Earth's daily rotation and its annual revolution around the Sun. This is not just a mathematical formalism. This is undeniable physical evidence.

The Church readily admits that the *Dialogue*, the focus of the trial, is structured around this very argument. It is clever, to be sure. But as they point out, mariners and others astronomers, such as Kepler, know that the Moon is critical to the timing and strength of the tides. Tides are somehow a lunar phenomenon. Yet Galileo has dismissed Kepler's ideas in the *Dialogue*. Nor does the timing of the tides (twice daily) seem to match Galileo's explanation (only once). Ultimately, it seems that Galileo is misguided if not *wrong* in his central argument.

Galileo has also tried to address the problems of motion that accompany the idea of a rapidly spinning Earth. Why are there not great winds? Why does a ball thrown vertically land directly below, if the Earth has since moved beneath it? Galileo makes many arguments about relative motion and its perception. But again, he seems to offer only plausible explanations, not real proof. His argument about water in ocean basins forming tides, for example, appears not to apply to air in mountain valleys. Galileo's empirical conclusions, ultimately, seem ill informed and inconclusive.

In presenting alternative accounts, Galileo also makes philosophical assumptions about which explanations to prefer. His supporters echo his principle of simplicity. But such notions are not new, the Church notes, even among Christian scholars. The concept of a Sun-centered system, introduced well before Copernicus, has been consistently rejected as contrary to more fundamental and direct ways of viewing the world. Simplicity is merely an aesthetic judgment, they say, not necessarily yielding truth. Galileo's philosophy of knowledge thus seems ad hoc and dangerously counterintuitive.

There is also a religious context to the trial. The Church has maintained that the Copernican doctrine, taken as more than a mere mathematical formalism, is contrary to scripture. Yet Galileo, in a widely circulated letter to his patron, drawing on views of some historical Christian scholars, has suggested how it is possible to reconcile Copernicanism with the Bible. This should satisfy concerns of faith.

But the Church is not impressed. Indeed, this letter was the occasion for an earlier brush with the Church in Rome in 1616. He was reminded then that the responsibility for interpreting scripture belongs exclusively to the Church fathers. Galileo was plainly out of line, although he presents himself now, as then, as an obedient servant of the Church. On that earlier occasion, Cardinal Bellarmine had helped clarify how Copernicanism violated clearly worded sections of the Bible. The Church, he noted, might indeed some day find occasion to reinterpret those sections in more metaphorical terms. But without solid physical proof of the Earth's movements, such radical imaginings were hardly warranted. And as we have just observed, Galileo has failed to provide such incontrovertible demonstration.

So: Galileo seems on exceptionally shaky ground, both empirically *and* religiously. But a full trial cannot conclude fairly without duly considering the *political* dimension. Galileo's supporters now present him as an innocent victim of Machiavellian court politics. Galileo, like other astronomers and natural philosophers, finds support for his leisure pursuits from the court of a prince—in Galileo's case, Cosimo de Medici. Through the court, Galileo became acquainted with Maffeo Barberini. Maffeo supported him on several occasions of the debates-as-entertainment. Galileo dedicated a 1623 book to him, which he dearly loved. In time, of course, Barberini became

Pope. There, he became prince, in a sense, of another court: the Vatican. And Galileo, as a learned friend, earned stature at the Papal Court. Galileo and the Pope shared a special relationship. He granted Galileo several private audiences, and even approved the writing of the very book that has led to the trial.

What is happening now, Galileo's advocates argue, is a tragic but all-too-routine and familiar event: the fall of the great courtier. The Prince of any court, of course, must demonstrate his absolute authority from time to time. He typically does so by banning one of his favorite courtiers, or *favoritti*. Such banishment cannot be arbitrary, however, if the Prince is to also maintain his moral claim to power. Thus, the Prince typically portrays the courtier as having committed some unpardonable act, such as betrayal of the Prince. The Prince can thereby present his exercise of authority as guided by higher principles, obliging him, no less, to suffer the loss of a close friend. Through the "fall of the favorite" he reasserts his power and enhances his image—with unjust misfortune for the favorite. This, they claim, has happened to poor Galileo at the Pope's whim. Teaching Copernicanism is not a serious offense. Yet it makes a convenient ruse for staging a politically favorable, but ultimately unjust, fall of the favorite. The Jesuits and members of the Collegio Romano have long resented Galileo and his fellow members of the independent Accademia Lincei. Perhaps they have spread false rumors about Galileo, prompting his unjust demise by a Pope who seems increasingly isolated and paranoid about threats to his power? Merely cutthroat, backstabbing politics as usual. Hardly justified grounds for heresy.

True enough, Church representatives agree: the Papal Court is full of politics. And Galileo knows the game as well as anyone, reflected in the history of his shrewd patronage with the Medicis. That is why Galileo should have tread more carefully. Had the Pope not granted Galileo the extraordinary approval to write this book? Had he not been advised to discuss Copernicanism as merely hypothetical? Had he not been reminded to include the notion—introduced by Galileo himself in the book dedicated to the Pope, which the latter so admired—of the inherent uncertainty of our senses and the limits of all human knowledge? What was Galileo's purpose in putting this cherished idea in the voice of the simpleton in the conclusion of the *Dialogue*? How could he not intend to mock the Pope and his absolute authority, at a time when most of Europe is at war and the Catholic Church is threatened on all sides by the Reformation? Is this how he shows appreciation for having been granted special privileges? Galileo's own politics seem deeply flawed, not the Pope's.

Finally, there are the technicalities to consider. What precisely are the charges, and do Galileo's actions fall strictly within the prohibitions, rigorously defined? After all, the Church has not banned discussion of Copernicanism. It has only stipulated that it must be treated hypothetically. And this, say his supporters, is all that Galileo has done. His *Dialogue* is fictitious, a drama to amuse. He uses both proponents and skeptics to present all perspectives, without advocating any particular position. Numerous quotes from the book show this indecisive posture. Finally, there is the Imprimatur of the Inquisition itself, on the page following the title page. Prior to publication, the Church's censors themselves reviewed the manuscript—both in Rome and again in Florence! If there were any substantive flaws in it, surely they would have found them then? If they did not, Galileo is surely not to blame now for any oversight.

Nevertheless, the voices of the Church respond, Galileo's treatment was not at all hypothetical in spirit. Numerous quotes show this, as well—in particular, the consistent ridicule and humiliation of the critic of Copernicanism. Moreover, as noted earlier, Galileo is selective in presenting the evidence, obscuring the Tychonic system and alternative explanations for the tides. The issue is not as conclusive as he pretends. Again, he has not given proper voice, under the Pope's own guidance, to the critical principle of human uncertainty. In his rhetoric and style, Galileo has *advocated* (not merely considered) Copernicanism and defended *as probable* a thesis contrary to the Bible. This is the core of his heresy, as defined by the heritage of the Church.

This, then, is the available evidence. Astronomical. Motion-related. Philosophical. Religious. Political. And procedural. In every case, it seems, Galileo presents an apparently justified perspective. But the Church seems also to have shown such perspectives to be premature, ill founded, limited, or biased. Is Galileo guilty of heresy? As a member of the Inquisition, you must now cast your vote.

Having reached a verdict, we may return to the present day, centuries later, to reflect on the actual outcome. Rather than face the ultimate verdict of heresy in 1633, Galileo was permitted to formally renounce Copernicanism, which he did. He lived the remainder of his life under the watchful eye of the Church, in his pastoral villa outside Florence, on a hill overlooking the convent where his two (illegitimate) daughters resided.

——————————— .

All of the factors discussed above contributed significantly to the historical events. Not all were formally addressed by the Inquisition in 1633, of course. Nor was it a trial in a modern sense. Rather, it was more of a tribunal or inquiry, conducted under the unilateral power of the Church. But staging a trial in this way allows a more comprehensive assessment of the episode, while adopting our modern sense of fairness. The primary aim here is to open understanding about the nature of science and its cultural contexts. With retrospect, and also having relived the experience in context, one can appreciate the complexities of scientific patronage, alternative theories, assumptions and scientific uncertainty, the role of philosophical perspectives about knowledge, how religion can engage science, and interpersonal politics, too. Ultimately, the case seems less about science and religion and more about science and *power*.

The complexity of the nature of science in such cases can be easily and effectively conveyed in inquiry mode through a role-play simulation (also see next chapter). Each student is given a particular topic and position, specializing on a unique aspect of the case. The details, which are so important to understanding the texture and realism of such cases, are distributed among many students. Those details then come together again in the final trial, as each student presents their work to the whole class. As profiled below, there are potentially thirty-four separate roles here.

The themes also allow for collaborative work. While students are responsible for their own presentation, they must also work together with team members to integrate individual topics into a thematic conclusion. This format helps students develop skills for working in groups, while maintaining individual accountability, an important balance in educational contexts.

The final presentations are organized to generally delve ever more deeply as the trial unfolds. They begin with the basic astronomical observations and proceed through problems of explaining motion and interpreting observations, to interpreting scripture, court politics, and the procedural technicalities of the charges against Galileo. This provides different levels of challenge for students with varying abilities. It also helps generate increasing drama as the trial progresses (as described above).

To avoid undue emphasis on historical research skills, I provide information about appropriate sources. This project requires a level of detailed information that is not available exclusively on the Internet, except for some of the more elementary topics. Many of the sources listed in the bibliography below are typically available only through university libraries. Still, an adequate level of detail may generally be achieved by investing in copies of Dava Sobel's *Galileo's Daughter* and Mario Biagioli's *Galileo, Courtier* (which is essential for addressing the political dimensions of the episode).

For assessment, individual teachers will undoubtedly find their own style to meet their particular needs. When working with college students, I require a written position paper. I also require a short, three-minute in-class presentation at the trial to foster public-speaking skills, although this is not graded for quality. Students must also speak from their knowledge. They may not read a statement or rely on note cards. In addition, I require a presentation image, to promote skills in visualizing ideas. I collect their various position papers in advance, so that I can provide feedback before the actual in-class trial, and to discover in advance where presentations may potentially be weak. I also use a post-trial essay. Students explain the reasoning of their verdict or write open-ended reflections on what they learned from the experience.

In presenting the trial, the teacher ideally acts as a moderator interested in hearing all perspectives fully. As teams change to take their position in front of the class, the teacher can neatly summarize the previous team's presentation. This can help highlight the major elements and fill in as needed, where students' presentations are not polished. The trial with all the roles easily fills a 2.5-hour class, or three class periods, with minimal discussion. The ultimate lessons are worth every minute.

Making a judgment at the end of the trial is difficult for most students, who, from a modern standpoint, want Galileo to be wholly "right" and thus fully exonerated (the same tendency towards rational reconstruction discussed in Chapter 4). With the actual evidence in historical context, however, such a judgment becomes problematic to justify. By borrowing conveniently from the actual history, one can provide a compromise option of sorts. As the arguments conclude, the leader of the trial can report a special missive from the Pope, who has been following the proceedings. He offers Galileo one last opportunity. To affirm his professed allegiance to the Church, he may publicly renounce Copernicanism. Failing that, he will be found guilty. This can be offered to students as a third possibility for the Inquisition's decision. (Historically, this situation was likely coupled to threat of torture—leaving Galileo, age sixty-nine and in ill health, with little choice.) This option provides a softer middle ground for students, and many make it their choice. Teachers should also note that students tend to base their decision primarily on their own work, not on the evidence as a whole presented during the trial. This reflects a common cognitive bias, whereby individuals tend to give more importance to information that is familiar or well known. This cognitive tendency can also be a powerful point for discussion following the trial decision.

On every occasion that I have led this project (now more than ten times), the participants have found Galileo guilty (or conditionally guilty, pending his potential renunciation). They greet their own judgment with collective dismay. For me, however, it reflects learning something profound about the nature of science and history.

The particulars for conducting the classroom trial and some resources are listed below and are also available online at

http://galileotrial.net

Debating Galileo's *Dialogue*: The 1633 Trial

PROJECT PROFILE

In 1632, Galileo presented evidence for Copernicanism in his *Dialogue Concerning the Two Chief World Systems*. The following year, he appeared before the Inquisition for violating a 1616 Vatican edict forbidding such teachings. We will situate ourselves in 1633, consider the evidence for ourselves, and decide whether to condemn Galileo. Each person will work with a team to present to the class the case supporting either the Church or Galileo on one particular issue. Afterward, we will act as members of the Inquisition and decide Galileo's fate. Galileo is under "vehement suspicion of heresy" for

1. advocating Copernican doctrine and
2. holding and defending as probable a thesis contrary to the Bible.

RESEARCH AND TRIAL PRESENTATIONS

Teams will specialize and focus on the following issues:

1. **Astronomical Observations**—What evidence does Galileo present for the Copernican view? How do others (notably, Tycho Brahe) interpret the same evidence?
2. **Interpreting Motion**—What other observational problems do Copernican views create? How does Galileo try to solve these, and does he do so adequately?
3. **Interpreting Alternative Theories**—When observations fit two alternative hypotheses, how should we decide between them?
4. **Interpreting Scripture**—What posture should the Church adopt when observations seem to conflict with the scripture?
5. **Politics**—Is Galileo a victim of court politics, or has he failed to respect them?
6. **Procedural Details**—What exactly was Galileo told in 1616 and since? Has he respected Church precepts?

BACKGROUND

Read John Heilbron's "Introduction" to the *Dialogue* (pages xi–xx), a fine overview of the trial episode and its historical context. For more extensive discussion, see Dava Sobel's *Galileo's Daughter* and Mario Biagioli's *Galileo, Courtier*.

GALILEO TEAM NO. 1—ASTRONOMICAL OBSERVATIONS FOR COPERNICANISM

Discuss how Galileo's claims about Copernicanism are well justified by astronomical observations (and thus that the 1616 decision declaring Copernicanism heretical is unjustified). Each person will focus on profiling the significance of one set of evidence:

- **Moon surface, nova,** and **sunspots**
- the **moons of Jupiter** and the **phases and size of Venus**
- **planetary retrograde**
- the **motion of sunspots** and the **annual motion of the Sun**

For further depth, you may also discuss:

- the role of the **telescope** and its reliability
- the validation of observations by **astronomers within the Church,** and/or potentially favorable evidence that Galileo rejects as inadequate (for example, see his treatment of William Gilbert's ideas in Day 3 of the *Dialogue*)
- Galileo's explanation for the lack of observed **stellar parallax** (a potential objection)

Bibliography

For the first three topics, read

- [essential] Galileo's earlier *Starry Messenger* (*Siderius Nuncius*) (1610)
- relevant sections in the *Dialogue* (use the index and/or word search!)
- [for background] Crowe's *Theories of the World from Antiquity to the Copernican Revolution*
- [for background] Swerdlow's chapter 7 in Machamer's *Cambridge Companion to Galileo*
- [for background] sections in Fantoli's *Galileo: For Copernicanism and the Church*

For the claims about sunspots, review

- Galileo's *Dialogue*, Day 3 (esp. section starting on p. 400)
- Fantoli's *Galileo: For Copernicanism and the Church* (on Day 3)

CHURCH TEAM NO. 1—ASTRONOMICAL OBSERVATIONS AND THE TYCHONIC SYSTEM

Discuss how astronomical observations do not necessitate accepting Copernicanism—and that the 1616 edict to regard it as hypothetical and a mathematical formalism is warranted scientifically.

- Imagine yourself, perhaps, as **Christopher Scheiner**. Discuss the Church's support of astronomy. Show that the Church includes skilled astronomers (such as Christoph Clavius and yourself) who can credibly disagree with Galileo's claims. Discuss your own work on and interpretation of **sunspots**. (For further depth, discuss how Galileo treats you in the *Dialogue*.)

- Show how **Tycho Brahe** provides an important alternative for accommodating Galileo's observations (of the Moon's surface, the moons of Jupiter and the phases of Venus, etc.). (For further depth, discuss how Galileo addresses—or fails to address?—the Tychonic system in the *Dialogue*.)
- Imagine yourself, perhaps, as **Orazio (Horatio) Grassi**. Discuss your work on **comets**, their importance to the Tychonic system, and Galileo's apparent errors in an earlier publication.
- Discuss the significance of **stellar parallax** in deciding between Tychonic and Copernican systems—and Galileo's failure to provide evidence for it.

Bibliography

On the role of the Church in astronomy, see

- Lerner's chapter 1 and Kelter's chapter 2 in McMullin's *The Church and Galileo*
- Feldhay's *Galileo and the Church* (esp. chapter 11)
- Heilbron's *The Sun in the Church*

On Scheiner, see

- Blackwell's *Behind the Scenes at Galileo's Trial*, chapter 4
- Feldhay's *Galileo and the Church*

On the Tychonic system, see

- Kuhn's *The Copernican Revolution*, chapter 6, or
- Crowe's *Theories of the World from Antiquity to the Copernican Revolution*, chapter 7
- Margolis's "Tycho's system and Galileo's *Dialogue*"

On Grassi, see

- Biagioli's *Galileo, Courtier*, pp. 267–301 on "Courtly comets"
- Drake & O'Malley's *The Controversy on the Comets of 1618*

On parallax, see

- overview at The Galileo Project website
- relevant sections of the *Dialogue* (esp. Day 3; see the book's index)
- Casanovas's "The problem of the annual parallax in Galileo's time"
- Siebert's "The early search for stellar parallax: Galileo, Castelli, and Ramponi"
- Graney's "But still, it moves: Tides, stellar parallax, and Galileo's commitment to the Copernican theory"

GALILEO TEAM NO. 2—INTERPRETING MOTION

Defend Copernicanism further by showing how Galileo addresses the problems of motion coupled with the Earth's rotation (thereby implying further that the 1616 edict listing Copernicanism as heretical is unjustified).

- If the Earth is moving, why (according to Galileo) do we not feel it? Why are there no great winds? Why are objects not hurled outward from the rotating Earth? Why do objects dropped from a great height fall directly underneath where they are dropped? Present the various problems and explain how Galileo solves them through the **concept of relative motion**.
- Discuss Galileo's **theory of the tides** and why he considers it so important (for example, why he first wrote about it to Cardinal Orsini in 1616).
- Discuss how Galileo further explains the **timing of the tides**, and their monthly and annual variations.

Bibliography

On relative motion, see

- Day 2 of the *Dialogue*, starting on p. 132
- Finocchiaro's *Galileo on the World Systems* (his commentary on how Galileo addressed objections, pp. 323–325, 332, and footnotes for pp. 142–212)
- Swerdlow's chapter 7 in Machamer's *Cambridge Companion to Galileo*

On the tides, see

- Heilbron's "Introduction" to the *Dialogue*, pp. xii–xv
- [essential] Day 4 of the *Dialogue*
- Galileo's "Discourse on the flux and reflux of the sea," reprinted in Finocchiaro's *Galileo Affair*
- Fantoli's *Galileo: For Copernicanism and the Church*
- Naylor's "Galileo's tidal theory"

CHURCH TEAM NO. 2—INTERPRETING MOTION

In continuing Team no. 1's criticism of Copernicanism, show that Galileo does not decisively solve the problems of motion implied by a rotating Earth.

- Discuss the weaknesses in **Galileo's concept of falling bodies**.
- Discuss how Galileo ultimately fails to solve the problem of motion with his **concept of relative motion**. Underscore the common sense of direct observations (for example, perhaps calculate the proposed velocity of the Earth in Rome).
- Discuss the weaknesses in **Galileo's theory of the tides** (especially their timing) and Kepler's alternative (1609).

For further depth, discuss

- [in coordination with Team 4] Discuss Galileo's desperate **effort to provide physical evidence** related to the Earth's motion, exemplified in his unsuccessful effort to enlist Gilbert's arguments (end of Day 3) and/or in his dismissal of Kepler's theory of the tides.

Bibliography

On falling bodies, see

- Day 2 of the *Dialogue*, esp. fig. 8 on p. 192 (and Galileo's discussion of it)
- Heilbron's *The Sun in the Church*, pp. 176–180

On motion, see

- Day 2 of the *Dialogue*
- McMullin's chapter 8 in Machamer's *Cambridge Companion to Galileo*
- Feldhay's *Galileo and the Church*

On tides, see

- Day 4 of the *Dialogue*
- The *Dialogue:* use the index or word search to find discussion of Kepler
- Drake's "History of science and tide theories"
- Cartwright's *Tides: A Scientific History*
- Conner's *A People's History of Science*, pp. 206–209
- Finocchiaro's footnote commentary in *Galileo and the World Systems*, pp. 282ff
- Shea's *Galileo's Intellectual Revolution*, pp. 172–189
- McMullin's *Galileo: Man of Science*, pp. 31–42
- Kepler's *Somnium* [However, note that this was not *published* until 1634]

GALILEO TEAM NO. 3—INTERPRETING ALTERNATIVE THEORIES

Discuss how Galileo resolves the problem of interpreting observations when there are alternative explanations that each concur with the evidence (on planetary systems, consult Team no. 1, and on relative motion, consult Team no. 2).

- Discuss the role of **simplicity**, especially regarding retrograde motion of the planets and the multiple motions of the Sun (sunspots).
- Discuss how to decide between alternative explanations, if both describe the observations. (How does one know when to trust one's senses and when to reason—for example, if we do not sense a rotating Earth?) In particular, discuss methods for reconciling **"appearances" versus "understanding."**

Note: On retrograde and sunspot motion, consult Team no. 1. Ideally, become familiar with the arguments of Team no. 2 on the concept of relative motion. Focus instead on the philosophical principles for deciding between alternative interpretations of our observations. For additional depth, consider

- how Galileo differs philosophically with **Aristotle** (Day 1) and why this is important in 1632.

Bibliography

- Heilbron's "Introduction" to the *Dialogue*, pp. xvii–xviii
- Finocchiaro's "Methodological reflection" in *Galileo on the World Systems*, pp. 335–356
- De Santillana's "Philosophical intermezzo" in *The Crime of Galileo*, pp. 55–77

On simplicity, see

- *Dialogue*, pp. 132–144, 396–400, 401–416, 453; Finocchiaro's *Galileo on the World Systems*, pp. 28–31; Kuhn's *The Copernican Revolution*

On deciding between alternative explanations, review especially

- *Dialogue*, pp. 123–144, 295–298, 492–494, conclusion on p. 536

On appearances vs. understanding, review especially

- Day 1, pp. 116–121

Church Team no. 3—Interpreting Alternative Theories

Identify what principles are appropriate in interpreting observations, especially when they fit alternative explanations.

- Summarize the arguments of **Nicholas of Cusa** (1400–1464) and others in the Medieval tradition about **regulating the use of reason**.
- Summarize the views of **Nicole Oresme** (1320–1382), who once considered, but then rejected, Sun-centered astronomy. Focus on the **principles of observations** (noting that Team no. 4 will address principles of interpreting scripture).

For context, consult Team no. 2 on the dilemma of the indeterminacy of relative motion and Team no. 1 on the alternative planetary systems.

Bibliography

- Review Day 1 of the *Dialogue*; on deception of senses, see pp. ~288–298 (Day 2)
- On the concept of "saving appearances," see McMullin's chapter 6 in McMullin's *The Church and Galileo*

On Nicholas of Cusa and the Medieval tradition, see

- Grant's *God and Reason in the Middle Ages*
- Nicholas of Cusa's *On Learned Ignorance* (online)

On Oresme, see

- Oresme's "The compatibility of the Earth's diurnal rotation with astronomical phenomena and terrestrial physics," in Grant's *Sourcebook on Medieval Science*, pp. 503–510; or in Clagett's *Science of Mechanics in the Middle Ages*, pp. 600–606

GALILEO TEAM NO. 4—INTERPRETING SCRIPTURE

Show how, according to Galileo, one may interpret scripture so as to reconcile it with Copernicanism (astronomical observations from Team 1 and concepts of motion from Team 2) (thereby indicating that it should not be heretical).

- Show that Galileo is a **faithful servant of the Church**. For more depth, profile the Church's support of astronomy and the defense of Galileo's ideas within the Church (esp. by Castelli and Campanella).
- Profile the context and content of **Galileo's 1615 letters** to Benedetto Castelli, Piero Dini, and, later, the Grand Duchess Christina.

Bibliography

For background and commentary, see

- McMullin's chapter 4 in McMullin's *The Church and Galileo*
- McMullin's chapter 8 in Machamer's *Cambridge Companion to Galileo*
- Feldhay's *Galileo and the Church*, chapter 2
- Blackwell's *Galileo, Bellarmine, and the Bible*, chapter 3 and appendix IX

On Galileo's support of the Church, see

- Sobel's *Galileo's Daughter*
- Pedersen's chapter in Coyne and Zycinski's *The Galileo Affair*

On the Church's support of astronomy, see

- Heilbron's *The Sun in the Church*

On Campanella, see

- Campanella's *Defense of Galileo*

On Castelli, see

- Galileo's "Letter to Castelli"

On Galileo's letters (to Castelli, Dini, and Christina), see

- Finocchiaro's *The Galileo Affair*, Drake's *Discoveries and Opinions of Galileo*, or Blackwell's *Galileo, Bellarmine, and the Bible*

CHURCH TEAM NO. 4—INTERPRETING SCRIPTURE

Discuss the Church's position on interpreting scripture with respect to astronomical observations and on who has the authority to interpret the Bible.

- Review and explain the **1616 decision of the Council of Trent** on Copernicanism. Consider especially the views of **Cardinal Bellarmine**, his authority to interpret Church doctrine, and his openness to Copernicanism and studies of nature.
- Imagine yourself, perhaps, as **Melchior Inchofer**. Summarize and explain your (1633) report.

Familiarize yourself with the arguments of Team no. 3. For more depth, also consider

- **Tommaso Caccini**'s views and their relevance in 1614–1616, and
- the implications of Galileo's materialistic interpretations for **atomism and the doctrine of the Eucharist**.

Bibliography

- Bellarmine's "Letter to Paolo Foscarini" in Finocchario's *The Galileo Affair* or Blackwell's *Galileo, Bellarmine, and the Bible*, appendix VIII (also online)
- Inchofer's report, in appendix 1 of Blackwell's *Behind the Scenes at Galileo's Trial*; also see commentary in chapters 2–3 [reprinted in Finocchiaro's *The Galileo Affair*]

For background, see

- on Bellarmine, McMullin's *The Church and Galileo*, chapter 6 and Redondi's *Galileo, Heretic*, chapter 2
- Fantoli's *Galileo: For Copernicanism and the Church*, sections 3.2–3.5
- De Santillana's *Crime of Galileo*, chapter 6
- Feldhay's *Galileo and the Church*
- Pera's chapter 10 in Machamer's *Cambridge Companion to Galileo*
- Blackwell's *Galileo, Bellarmine, and the Bible*, chapter 2 and appendix III

On atomism, see

- Redondi's *Galileo: Heretic*, esp. chapters 7 (pp. 203–226), 8 (pp. 232–240, 258), and 9 (pp. 283–284, 301)
- Galileo's *The Assayer* in Drake's *Discoveries and Opinions of Galileo*, pp. 308–314
- McMullin's *The Church and Galileo*, chapters 4 and 8
- Artigas, Martínez, and Shea's "New light on the Galileo affair?" in McMullin's *The Church and Galileo*, pp. 213–233. Also retrievable from http://www.unav.es/cryf/english/newlightongalileo.html

GALILEO TEAM NO. 5—POLITICS

Present a case that this trial is based primarily on interpersonal politics, not Church doctrine—and thus that all charges should be dismissed.

- Discuss (on a general level) the system of patronage and professional recognition and the structure of court politics. Discuss, in particular, the series of events known as **"the fall of the favorite"** as background for understanding Galileo's circumstances in the Papal court.
- Discuss how the general framework articulated above applies, in particular, to the **relationship between Galileo and Pope Urban VIII** (formerly Maffeo Barberini). Include the historical events (through 1624) that led to the writing of the *Dialogue*.
- Discuss **Jesuit antipathies** towards Galileo and how they are relevant to the current charges against him. Note, in particular, the controversies involving Christopher Scheiner (on sunspots) and Horatio Grassi (on comets and interpreting nature). For further depth, coordinate with the instructor and Team no. 6 on "the false injunction."

Note that Team no. 6 will be addressing the Inquisition's process of approving the final book manuscript. For further depth, consider

- the vulnerable **political stature of Pope Urban VIII** in 1632–1633,
- the questionable behavior and **mental state of Pope Urban VIII** in late 1632 and early 1633, and
- the **case of Giovanni Ciampoli** as a parallel.

Bibliography

- [essential] Biagioli's *Galileo, Courtier* (esp. chapter 6 on "the fall of the favorite")
- Shea and Artigas's *Galileo in Rome*
- Feldhay's *Galileo and the Church*, chapters 6, 9, 10–13
- Sobel's *Galileo's Daughter*

On the Jesuits, see

- Fantoli's *Galileo: For Copernicanism and the Church*, sections 2.5–2.6, 4.1–4.5
- Westfall's essay on the Jesuits in his *Essays on the Trial of Galileo*
- Biagioli's *Galileo, Courtier*, chapter 5 (on Grassi)
- Feldhay's *Galileo and the Church* (esp. chapter 9 and 12–13, and passages on Scheiner)
- Blackwell's *Behind the Scenes at Galileo's Trial*, chapters 1 and 4 (esp. on Scheiner)
- Freedburg's *Eye of the Lynx*

On Urban VIII, see

- Sobel's *Galileo's Daughter*

On Ciampoli, see

- De Santillana's *Crime of Galileo* (use index)

CHURCH TEAM NO. 5—POLITICS

Show how Galileo's interpersonal and court politics have been misguided and unduly challenge the Church's authority.

- Discuss precedent **cases of Giordano Bruno** (1600) **and Paolo Foscarini** (1616). Discuss the Church's exceptional support for Galileo on earlier occasions, in 1616 and 1624.
- Discuss the **relationship between Galileo and Pope Urban VIII** (formerly Maffeo Barberini). Show how (Copernicanism aside) Galileo failed Urban's trust, especially in revising the *Dialogue*. Be sure to discuss the critical relevance of Galileo's earlier work, *The Assayer*.
- Discuss the context of **political challenges for the Catholic Church in Europe** [events later known as the Thirty Years' War]. Note the role of Spanish Cardinal Borgia in Rome. For further depth, be aware of Urban's state of mind during this period (partly evident in events to follow).

For further depth, consider

- the relevance of the **case of Ciampoli**, and
- the threat of a potential case of deeper heresy (now suppressed) concerning Galileo's alleged **atomism and its implications for the doctrine of the Eucharist**.

Bibliography

For orientation and background, see

- Sobel's *Galileo's Daughter*, chapter 20
- Blackwell's *Behind the Scenes at Galileo's Trial*, chapter 1

On Bruno, see

- Yates's *Giordano Bruno and the Hermetic Tradition*
- Gatti's *Giordano Bruno and Renaissance Science*
- Blackwell's *Galileo, Bellarmine, and the Bible*, chapter 2

On Foscarini, see

- Fantoli's *Galileo: For Copernicanism and the Church*
- Bellarmine's "Letter to Paolo Foscarini" (also central to Team no. 4)
- Blackwell's *Galileo, Bellarmine, and the Bible*, chapter 4
- De Santillana's *Crime of Galileo*, chapter 5

On Galileo and Barberini/Urban, see

- [essential] Heilbron's "Introduction" to the *Dialogue*(!), and footnote to p. 538.
- [essential] Biagioli's *Galileo: Courtier* (esp. chapters 5 and 6)
- Sobel's *Galileo's Daughter*
- Fantoli's chapter 5 in McMullin's *The Church and Galileo*
- Feldhay's *Galileo and the Church*
- Galileo's *The Assayer* in Drake's *Discoveries and Opinions of Galileo*

On Urban and the Thirty Years' War, see

- Sobel's *Galileo's Daughter*, pp. 223–226
- De Santillana's *Crime of Galileo*, pp. 204–214, 222, 229–231
- Biagioli's *Galileo, Courtier*, pp. 333–340
- general histories of Europe and accounts of Urban VIII's papacy

On Ciampoli, see

- Biagioli's *Galileo, Courtier*
- De Santillana's *Crime of Galileo*

On atomism, see references for Team no. 4.

GALILEO TEAM NO. 6—PROCEDURAL DETAILS

Show that, in terms of procedural details, Galileo cannot be held accountable. Discuss how, in approaching an extremely delicate topic, Galileo heeded all the guidance and instructions given to him by Church officials. (If any flaws remain, the Church—not Galileo himself—is to blame!)

- Present the **1616 certificate** of Cardinal Bellarmine, discuss its context, and show (through quotes and analysis of the *Dialogue*) how Galileo has honored it and neither explicitly "defends" nor "holds" Copernicanism.
- Show (with a healthy use of quotes) that Galileo honored the 1616 Congregation of the Index by treating Copernicanism only **hypothetically** in the *Dialogue*.
- Profile the procedure for the **Inquisition's review** of books and their examination and approval of Galileo's manuscripts. Show, in particular (again, through healthy use of quotes), how Galileo revised his work as instructed (by Riccardi) before receiving a final imprimatur.

For further depth:

- Coordinate with Team no. 4 on Galileo's overall support of the Church.
- Coordinate with Team no. 5 on the context that makes procedural details so essential.
- Coordinate with the instructor and Team no. 5 on "**the false injunction**."

Bibliography

For orientation, see

- Blackwell's *Behind the Scenes at Galileo's Trial*, chapter 1
- Fantoli's chapter 5 in McMullin's *The Church and Galileo*
- [essential] 1616 documents in Finocchiaro's *The Galileo Affair* (esp. p. 153) or online
- Feldhay's *Galileo and the Church*

On hypothetical rhetoric, see
- Shea's chapter 6 in Machamer's *Cambridge Companion to Galileo*
- Moss's *Novelties in the Heavens: Rhetoric and Science in the Copernican Controversy*
- Moss and Wallace's *Rhetoric and Dialectic in the Time of Galileo*
- on Galileo's rhetoric, Finocchiaro's commentary in *Galileo on the World Systems*, appendix 3

On the review and approval process, see
- [for orientation] Sobel's *Galileo's Daughter*
- Fantoli's *Galileo: For Copernicanism and the Church*, chapters 5 and 6
- "miscellaneous" 1631–1633 documents in Finocchiaro's *The Galileo Affair* or Burke's *Science and Culture in the Western Tradition*

CHURCH TEAM NO. 6—PROCEDURAL DETAILS

Show that, in terms of procedural details and technicalities, Galileo has honored neither the spirit nor the explicit guidance offered to him (through the grace of the Church) in 1616 and again in the review and revision of his manuscript in 1632. (Your discussion should extend beyond Team no. 4's discussion of interpreting scripture by highlighting Galileo's rhetoric and intent.)

- Present the **1616 injunction**, and show that Galileo has failed to honor it by his continuing to teach Copernicanism. Show further (through a healthy use of quotes and analysis of the structure of the book's primary arguments) that, rhetorically speaking, Galileo's treatment of Copernicanism is **not merely hypothetical**, in violation of the 1616 Index. Ideally, as background, consult with Team no. 1 and Team no. 2 on the limits of Galileo's evidence. For further depth, consult with the instructor about preparing for discrepancies between "the injunction" and "the certificate."
- Show (again, through a healthy use of quotes) that Galileo exhibits a selectively biased view—not fairly balanced—and thus, through his rhetoric and style, **defends as probable a thesis contrary to the Bible**.
- **"Bellarmine's Ghost"**: Although Cardinal Bellarmine has died, his testimony is critical to interpreting events of 16 years ago. Using available documents as clues, speak on Bellarmine's behalf regarding what he (you) communicated to Galileo in 1616. Note Bellarmine's background in astronomy and his general sympathy with the aims of Galileo's work. Note also that you will be the final speaker in the trial and, as a respected member of the Church, your message will carry great authority: it should exhibit balance and moderate reasoning, appealing to all. Consult with Team no. 4 on Bellarmine's theological views and Team no. 3 on other Catholic views of Copernicanism.

Bibliography

For orientation, see

- Blackwell's *Behind the Scenes at Galileo's Trial*, chapters 1 and 3
- Sobel's *Galileo's Daughter*
- 1616 and 1632–1633 trial documents in Finocchiaro's *The Galileo Affair* or Burke's *Science and Culture in the Western Tradition*
- footnote to p. 119 in the *Dialogue*
- Fantoli's *Galileo: For Copernicanism and the Church*, sections 6.3–6.4
- Inchofer's report in Blackwell's *Behind the Scenes at Galileo's Trial*, appendix 1, or Finocchiaro's *The Galileo Affair*

On rhetoric, see

- Shea's chapter 6 in Machamer's *Cambridge Companion to Galileo*
- Moss's *Novelties in the Heavens: Rhetoric and Science in the Copernican Controversy*
- Moss and Wallace's *Rhetoric and Dialectic in the Time of Galileo*
- Finocchiaro's commentary in *Galileo on the World Systems*, appendix 3
- the opening and closing sections (especially) of each Day of Galileo's *Dialogue*

On the "injunction," see

- Fantoli's chapter 5 in McMullin's *The Church and Galileo*

On Bellarmine, see

- Blackwell's *Galileo, Bellarmine, and the Bible* (esp. for more translated documents)
- McMullin's chapter 6 in McMullin's *The Church and Galileo*
- Bellarmine's "Letter to Paolo Foscarini"

RESOURCES

Core Books

Biagioli, Mario. (1993). *Galileo, Courtier: The Practice of Science in the Culture of Absolutism*. Chicago, IL: University of Chicago Press.
Galilei, Galileo. (1632/2001). *Dialogue Concerning the Two Chief World Systems* (Stillman Drake, Trans.). New York, NY: Modern Library. —*This version is used for the page references cited above. Also see the electronic version retrievable from* http://galileotrial.net/library/Dialog/index.html *(abridged for Day 2, missing pp. 126–265).*
Sobel, Dava. (2000). *Galileo's Daughter: A Historical Memoir of Science, Faith, and Love*. New York, NY: Penguin.

Websites

The Galileo Project [Albert van Helden, Rice University]—http://galileo.rice.edu
Includes photos and biographies of key persons, images related to astronomical arguments, more
The Trial of Galileo [Douglas Linder, University of Missouri, Kansas City]— http://law2.umkc.edu/faculty/projects/ftrials/galileo/galileo.html
Includes a chronology and many historical documents from the original trial

Seventeenth-Century Texts [find links at http://galileotrial.net/resource.htm]

Bellarmine, Roberto. (1615). Letter to Paolo Foscarini. New York, NY: Fordham University. Retrievable from http://www.fordham.edu/halsall/mod/1615bellarmine-letter.asp
Campanella, Thomas. (1622/1976). *The Defense of Galileo of Thomas Campanella* (Grant McColley, Trans.). Merrick, NY: Richwood.
Campanella, Thomas. (1622/1994). *A Defense of Galileo* (Richard J. Blackwell, Trans.). Notre Dame, IN: University of Notre Dame Press.
Documents from the 1616 Inquisition proceedings. Retrievable from http://galileotrial.net/library/1616docs.htm (includes the "injunction" and the "certificate," and the statement of the Index banning Copernicanism)
Elert, Glenn. (2008). The Scriptural basis for a geocentric cosmology. E-world. Retrievable from http://hypertextbook.com/eworld/geocentric.shtml
Galilei, Galileo. (1957). *Discoveries and Opinions of Galileo* (Stillman Drake, Trans.). Garden City, NY: Doubleday [includes *The Assayer*].
Galilei, Galileo. (1610). *Siderius Nuncius [Starry Messenger]*. http://books.google.com/books?id=jGNz4X1UmkUC
Galilei, Galileo. (1615). Letter to the Grand Duchess Christina of Tuscany. New York, NY: Fordham University. Retrievable from http://www.fordham.edu/halsall/mod/galileo-tuscany.asp

Kepler, Johannes. (1634/1967). *Somnium; the dream, or posthumous work on lunar astronomy* (Edward Rosen, Trans.). Madison: University of Wisconsin Press.

Nicholas of Cusa. (1440/1981). *On Learned Ignorance* (Jasper Hopkins, Trans.). Minneapolis, MN: Arthur J. Banning Press. Retrieved from http://jasper-hopkins.info/DI-I-12-2000.pdf (Also see Finocchiaro's *The Galileo Affair: A Documentary History.*)

Specialized Books and Articles

Blackwell, Richard. (1991). *Galileo, Bellarmine, and the Bible.* Notre Dame, IN: University of Notre Dame Press.

Blackwell, Richard. (2006). *Behind the Scenes at Galileo's Trial: Including the First English Translation of Melchior Inchofer's Tractatus Syllepticus.* Notre Dame, IN: University of Notre Dame Press.

Burke, John. (1987). *Science and Culture in the Western Tradition.* Scottsdale, AZ: Gorsuch Scarisbrick.

Cartwright, David. (1999). *Tides: A Scientific History.* New York, NY: Cambridge University Press.

Casanovas, John. (1985). The problem of the annual parallax in Galileo's time. In G. V. Coyne, M. Heller, & J. Zycinski (Eds.), *The Galileo Affair: A Meeting of Faith and Science.* Vatican City: Specola Vaticana.

Clagett, Marshall. (1959). *The Science of Mechanics in the Middle Ages.* Madison: University of Wisconsin Press.

Conner, Clifford. (2005). *A People's History of Science: Miners, Midwives, and "Low Mechanicks."* New York, NY: Nation Books.

Coyne, G. V., Heller, M., & Zycinski, J. (Eds.). (1985). *The Galileo Affair: A Meeting of Faith and Science.* Vatican City: Specola Vaticana.

Crowe, Michael. (2001). *Theories of the World from Antiquity to the Copernican Revolution* (2nd ed.). New York, NY: Dover.

De Santillana, Giorgio. (1955). *The Crime of Galileo.* Chicago, IL: University of Chicago Press.

Drake, Stillman. (1979). History of science and tide theories. *Physis, 21,* 61–69.

Drake, Stillman, & O'Malley, Charles. (1960). *The Controversy on the Comets of 1618: Galileo Galilei, Horatio Grassi, Mario Guiducci, Johann Kepler.* Philadelphia: University of Pennsylvania Press.

Fantoli, Annibale. (1994). *Galileo: For Copernicanism and the Church.* Vatican City: Vatican Press.

Feldhay, Rivka. (1995). *Galileo and the Church: Political Inquisition or Critical Dialogue?* Cambridge, England: Cambridge University Press.

Finocchiaro, Maurice. (1989). *The Galileo Affair: A Documentary History.* Berkeley: University of California Press.

Finocchiaro, Maurice (Ed.). (1997). *Galileo on the World Systems.* Berkeley: University of California Press.

Freedberg, David. (2002). *The Eye of the Lynx: Galileo, His Friends, and the Beginnings of Modern Natural History*. Chicago, IL: University of Chicago Press.

Gatti, Hilary. (1999). *Giordano Bruno and Renaissance Science*. Ithaca, NY: Cornell University Press.

Graney, Chris. (2008). But still, it moves: Tides, stellar parallax, and Galileo's commitment to the Copernican theory. *Physics in Perspective, 10*, 258–268.

Grant, Edward. (2001). *God and Reason in the Middle Ages*. Cambridge, England: Cambridge University Press.

Heilbron, John. (1999). *The Sun in the Church: Cathedrals as Solar Observatories*. Cambridge, Massachusetts: Harvard University Press.

Kuhn, Thomas. (1985). *The Copernican Revolution*. New York, NY: MJF Books.

Machamer, Peter (Ed.). (1998). *The Cambridge Companion to Galileo*. Cambridge, England: Cambridge University Press.

Margolis, Howard. (1991). Tycho's system and Galileo's *Dialogue*. *Studies in the History and Philosophy of Science, 22*, 259–275.

McMullin, Ernan. (1968). *Galileo: Man of Science*. New York, NY: Basic Books.

McMullin, Ernan (Ed.). (2005). *The Church and Galileo*. Notre Dame, IN: University of Notre Dame Press.

Moss, Jean Dietz. (1993). *Novelties in the Heavens: Rhetoric and Science in the Copernican Controversy*. Chicago, IL: University of Chicago Press.

Moss, Jean Dietz, & Wallace, William. (2003). *Rhetoric and Dialectic in the Time of Galileo*. Washington, DC: Catholic University of America Press.

Naylor, Ron. (2007). Galileo's tidal theory. *Isis, 98*, 1–22.

Redondi, Pietro. (1987) *Galileo, Heretic* (Raymond Rosenthal, Trans.). Princeton, NJ: Princeton University Press.

Shea, William. (1972). *Galileo's Intellectual Revolution*. London, England: Macmillan.

Shea, William, & Artigas, Mariano. (2003). *Galileo in Rome: The Rise and Fall of a Troublesome Genius*. Oxford, England: Oxford University Press.

Siebert, Harald. (2005). The early search for stellar parallax: Galileo, Castelli, and Ramponi. *Journal for the History of Astronomy, 36*, 251–271.

Westfall, Richard. (1989). *Essays on the Trial of Galileo*. Notre Dame, IN: University of Notre Dame Press.

Yates, Frances. (1964). *Giordano Bruno and the Hermetic Tradition*. Chicago, IL: University of Chicago Press.

13 | Debating Rachel Carson's *Silent Spring*, 1963

Teaching about Silent Spring • *NOS issues* • *scientific concepts* • *simulation profile* • *roles* • *resources* • *teaching notes*

In 1962, Rachel Carson published *Silent Spring*, criticizing the indiscriminate use of synthetic pesticides and advocating respect for the integrity of nature. With its vivid images and dramatic evidence, it helped spark growing concerns about pollution and protecting wildlife into the modern environmental movement. It is frequently cited among the most influential books of the twentieth century. At the time, Carson's book also generated considerable controversy. The public debates about both scientific claims and values epitomize the need for scientific literacy and NOS education (Chapter 1). The case thus offers a prime opportunity for teaching the nature of science.

One teaching strategy is to follow Carson's story in historical context, highlighting the NOS dimensions and posing reflective questions to students, as illustrated in the beriberi case in Chapter 10.[1] Another, fuller and more engaging, option is profiled here: guiding students through a role-play simulation, similar in format to the Galileo trial profiled in Chapter 12. The scenario here is based on President Kennedy's Science Advisory Committee in 1963. The committee was charged to assess Carson's claims and recommend any appropriate policy or legislation. The panel's review provides an occasion to introduce the perspectives of multiple stakeholders and to work towards a consensual solution based on the available evidence. An additional question—should Carson receive a Presidential Medal of Freedom for her public service?—provides an opportunity to reflect on the role of Carson's writing style and methods of promoting the public understanding of science.

The prepared guidance for the activity includes up to 26 roles, ranging from ecologists and economic entomologists to representatives of the chemical pesticide industry and the U.S. Secretary of Agriculture. By design, the roles reflect the spectrum of stakeholder perspectives in 1963. They do not faithfully replicate historically the actual members of the committee. This is a historical *simulation*, an open-ended exercise for exploring the nature of science by posing a particular problem in a particular historical context.

Although based on real historical characters and conceptual positions, the goal is neither to re-enact history nor to "correct" it (see Chapter 4). Still, a large number of original historical documents have been assembled on the Internet (http://pesticides1963.net). They help clarify what was known at the time and how the information was interpreted from various perspectives.

Included in the online resources are reviews of Carson's book, hard-to-find scientific studies, items from the chemical industry magazines, commentaries from *Audubon* magazine, and more. There are also various historical images and documents to set the scene. The ultimate aim is to allow participants to grapple with the problem themselves and experience the uncertainty and contingent dynamics of history-in-the-making (Chapter 2). Such simulations are especially valuable for exploring how many alternative perspectives interact. Students typically learn to appreciate the complexities of the nature of science and the challenges of resolving problems that, in retrospect, may seem quite plain and simple.

The episode surrounding *Silent Spring* became an important historical landmark. Investigations about the adverse effects of pesticides were also launched by journalists and the U.S. Congress, leading to calls for reform. In the years to follow, other environmental concerns entered public debate. Within a decade, the United States had a new Environmental Protection Agency, a nascent Earth Day, and major legislation regulating DDT, endangered species, and air and water pollution. Carson's role in helping modern environmentalism take root has since come to be celebrated (and sometimes romanticized, as in *Time*'s account in its *100 Persons of the Century*) (Chapter 3).

NATURE OF SCIENCE ISSUES

The case highlights several features of the nature of science, each discussed briefly below:

- scientific credibility
- science and values
- scientific uncertainty
- public understanding of scientific issues
- science and gender

First, how does the public assess scientific credibility? A central feature of this episode is not just the science itself, but the public's understanding of science and its role in public policy. The case highlights the difference between what "is known" scientifically and *who knows it*. How do non-scientists come to have scientific knowledge, and how do they distinguish it from inauthentic science presented as scientific by various political or economic interests? Carson's expertise was certainly widely challenged. Others endorsed her claims. Were pesticides indeed poisoning our food? Were they carcinogenic? Did use of pesticides prompt the evolution of resistance, ultimately making them ineffective in controlling the insects that ravaged crops or carried disease? Such issues touched the ordinary citizen. Yet not everyone can be an expert. Who, then, can one trust? How does one establish an effective system of credibility? Who makes decisions: the experts or others who may or not be well informed by them?

Second, how does one integrate values and science? Carson coupled her specific argument about pesticides with a general critique of human attitudes towards controlling nature. Pesticides were thus also cast as damaging wildlife and upsetting a purported balance of nature. Were those views relevant in public discourse? Were we to care (as a society) about the loss of birds, dramatized in a prospective "silent spring"? How does one make a public decision involving both moral perspectives and scientific information? How does one debate values when facts are also involved? How does one debate facts—or the interpretation of facts—when they seem laden with values?

Third, *Silent Spring* introduces the challenge of public policy under scientific uncertainty. Science is not always complete and does not always yield ideal information to resolve public controversy. The evidence for toxicity of food and for the carcinogenicity of pesticides was particularly problematic at this time. What is the appropriate response? Does one use a scientifically guarded posture, even if it might later prove wrong? How does one hedge against possible alternatives? Does one entertain worst-case scenarios and apply a precautionary principle? One can always appeal to the need for further research. But how much will it cost? Who will pay for it? What is to be done in the mean time? In 1962, how was policy to be established without knowing more and having only limited alternative technologies at hand?

Fourth, the case raises striking questions about how the public becomes fully informed on scientific issues relevant to our common welfare. The role of *Silent Spring*'s publication is especially dramatic when framed in historical context. The discovery of the insecticidal properties of DDT— and its role in controlling typhoid and malaria in World War II—earned a Nobel Prize in 1948. For the next decade, the culture continued to bask in the triumph of DDT and the promise of "better living through chemistry." Studies about the adverse effects of DDT had been published almost from the outset but had little effect on public policy. How does the public develop understanding of such issues? Carson's emotional imagery was surely important in raising awareness and shaping public opinion. Were her persuasive tactics appropriate, or possibly even necessary? Did Carson or other scientists have an ethical responsibility to inform the public or, more deeply, to effect political change? How does one balance the ideal of scientific objectivity and political advocacy or activism? The question of whether Carson deserves special recognition highlights questions about this broader context of science in society.

Finally, the episode also raises questions about science and gender. Carson's credibility was challenged in part by portraying her as "a woman" (at a time when the culture largely peripheralized women as voices of authority in science and politics). In addition, one may consider how Carson's views about the control of nature were shaped by values fostered more deeply among women. How important was Carson's gender to her interpretation of science, her perceived role in communicating it to the public, and her writing style?

Many of these general questions about the nature of science enter into the concrete discussions of the President's Committee on Pesticides as it makes its decisions. Other aspects are ripe for explicit reflection and exchange after the simulation is complete.

SCIENTIFIC CONCEPTS

The case is also an occasion to explore several scientific concepts:

- ecology of agriculture and the role of pest control
- ecology of disease transmission and the role of pest control
- natural selection of resistance to pesticides
- concentration of elements in the food chain
- predator–prey and parasite–host interactions (as a basis for biological control)
- cellular biochemistry and the mechanisms of pesticide action
- "balance of nature" and interactions in complex systems

These are familiar lessons for biology teachers, so no further comment on them seems needed here. One may note, however, that these concepts provide a curricular context for situating the simulation in a standard biology class. Moreover, the case shows how such concepts are not isolated, but are highly integrated in practice. In addition, the case provides a concrete context that typically enhances student learning of such abstract principles.

The use of pesticides also raises important questions about ecosystem stability and/or the unpredictable behavior of complex systems. Carson appealed to the "balance of nature," a concept now widely discredited by ecologists. One may discuss the status of the scientific basis for this concept and, thus, its prospective relevance in a policy context.

Equally important here, perhaps, is probing the limits of science in guiding public policy. Science cannot justify values, even when providing information relevant to forming value judgments. Assessments of risk, for example, may be quantified, but the level of risk that is deemed safe expresses a value. This case illustrates well how science may inform policy decisions, but does not eclipse other considerations.

The President's Advisory Committee on Pesticides, 1963

PROJECT PROFILE

Rachel Carson published *Silent Spring* in 1962, advocating limitations on the use of synthetic pesticides. With its vivid images and dramatic evidence, it sparked emerging environmental concerns into a major public controversy. President Kennedy directed his Science Advisory Committee to investigate the claims and make recommendations. Our challenge is to situate ourselves in late 1963 and respond to Kennedy's charge. Each person will assume a different role and give testimony and initial policy proposals. As a group, everyone will then debate the arguments in *Silent Spring*, the evidence, the individual proposals, and make a final set of recommendations to the president. In addition, the committee will decide whether to recommend Rachel Carson for the new Presidential Medal of Freedom for her public service.

RESEARCH, TESTIMONY, AND DISCUSSION

Orient your research and writing around the following major issues:

1. **Carson's credibility**—Can the facts in *Silent Spring* be rigorously documented, and are the sources reliable? Is the presentation balanced and complete? How should Carson's emotive rhetoric and style affect our interpretation of her claims?

2. **Benefits of pesticides | Harm to non-target species**—What evidence indicates negative dimensions of pesticide use—to insect pollinators, to other organisms in the habitat, to the food chain? What are the nature and scope of the benefits that pesticides provide? How ought we to maximize benefits while minimizing any risks or harm?

3. **Indiscriminate use**—Should pesticide use be regulated? In what way, to what extent?

4. **Safety**—Is our food safe? Do pesticides cause cancer? Are the health and safety of workers threatened?

5. **Insecticide resistance**—Do insects evolve tolerance to pesticides? If so, what should be our response?

6. **Alternatives**—Are there any alternatives to chemical pesticides? If so, what are their relative costs and effectiveness?

7. **Control of nature, balance of nature**—In what ways does use of pesticides reflect larger issues about human relationships with nature that ought to be addressed?

ROLES

Each person will adopt the role of one historical figure, prepare a statement, and give testimony to the committee. See the webpages (http://pesticides1963.net) for guidance on each person's perspective and relevant sources:

- [optional] Jerome Wiesner, President's Science Advisor and Committee Chair
- P. Rothberg, President of Montrose Chemical Corporation (manufacturers of DDT)
- LaMont Cole, ecologist, Cornell University
- Edwin Diamond, science writer
- Robert L. Rudd, zoologist
- Robert White-Stevens, chemist, Assistant Director of Research and Development, American Cyanamid
- George J. Wallace, zoologist, Michigan State University
- Roland C. Clement, staff biologist, National Audubon Society
- Clarence Cottam, Director, Welder Wildlife Foundation
- Walter H. Larrimer, U.S. Department of Agriculture Insect Control Division; member of the NRC Committee on Pest Control and Wildlife Relations (1960–1963)
- Murray Bookchin, author, libertarian, socialist
- Mitchell R. Zavon, Department of Industrial Medicine, University of Cincinatti, consultant to industry, and member of the NRC Subcommittee on Research Needs for Pest Control and Wildlife Relations
- George Larrick, Commissioner of the Food & Drug Administration
- Ira L. Baldwin, Chair, National Academy of Science–NRC Committee on Pest Control and Wildlife Relations (1960–1963)
- Orville Freeman, Secretary of the U.S. Department of Agriculture

Extension Roles

- Robert Bushman Murphy, ornithologist, American Museum of Natural History
- William A. Brown, Jr., economic entomologist
- Samuel W. Simmons, U.S. Public Health Service
- George Decker, Illinois Natural History Survey and member, NRC Committee
- Roy Barker, Illinois Natural History Survey
- Ira Gabrielson, member of the NRC Committee; chair of Subcommittee on Research Needs
- Tom Gill, forestry, member of the NRC Committee
- Luther Terry, Surgeon General
- Thomas Jukes, chemist, American Cyanamid Company
- William J. Darby, Vanderbilt School of Medicine

RESOURCES

Carson, Rachel. (1962). *Silent Spring*. New York, NY: Houghton Mifflin.

General Histories

Graham, Frank, Jr. (1970). *Since Silent Spring*. Boston, MA: Houghton Mifflin.

Bosso, Christopher. (1987). *Pesticides and Politics: The Life Cycle of a Public Issue*. Pittsburgh, PA: University of Pittsburgh Press.

Brooks, Paul. (1972). *The House of Life: Rachel Carson at Work*. Boston, MA: Houghton Mifflin.

Dunlap, Thomas. (1981). *DDT: Scientists, Citizens and Public Policy*. Princeton, NJ: Princeton University Press.

Lear, Linda. (1997). *Rachel Carson: Witness for Nature*. New York, NY: Henry Holt.

McWilliams, James. (2008). *American Pests: The Losing War on Insects from Colonial Times to DDT*. New York, NY: Columbia University Press.

Mellanby, Kenneth. (1992). *The DDT Story*. Woking, England: Unwin Brothers.

Murphy, Priscilla Coit. (2005). *What a Book Can Do: The Publication and Reception of Silent Spring*. Amherst: University of Massachusetts Press.

Russell, Edmund. (2001). *War and Nature: Fighting Humans and Insects with Chemicals from World War I to Silent Spring*. Cambridge, England: Cambridge University Press.

Whorton, James. (1975). *Before "Silent Spring": Pesticides and Public Health in Pre-DDT America*. Princeton, NJ: Princeton University Press.

Historical Sources (available in 1963)

Agricultural Chemicals

- "Pesticide sales increase for eighth straight year" (Aug., 1962)
- "Pesticides and the industry are target of one-sided attack" (Aug., 1962)
- Editorial: "A public relations crisis" (Sept., 1962)
- "For the defense: Pesticide authorities speak out" (Sept., 1962)
- "NACA speakers emphasize industry's role in pesticide safety" (Oct., 1962)
- "Communications create understanding" (Nov., 1962)
- Editorial (Nov., 1962)
- "Rachel Carson influence gains momentum as book appears" (Nov., 1962)
- "Scientists score 'Silent Spring'" (Dec., 1962)
- "Bias, misinformation, half-truths reduce usefulness of 'Silent Spring'" (Feb., 1963)
- "Natural insect control measures are grossly inadequate for today's farming" (March, 1963)

- "The impact of pesticides on public health and recreation" (May, 1963)
- Baldwin, I. L. (1962). Chemicals and pests (book review). *Science, 137,* 1042–1043.

Chemical Week

- "The chemicals around us" [viewpoint] (July 14, 1962), p. 5
- "Nature is for the birds" [viewpoint] (July 28, 1962), p. 5
- "Response to criticism" (Aug. 11, 1962), p. 42
- "Bracing for broadside" (Oct. 6, 1962), p. 23

Clement, Roland. (1962a, Jan.–Feb.). The failure of common sense. *Audubon, 64,* 41.

Clement, Roland. (1962b, July–Aug.). How far can man go in 'controlling' nature? *Audubon, 64,* 186–189, 224.

Clement, Roland. (1962c). *Silent Spring* (book review). *Audubon, 64,* 356.

Cole, Lamont. (1964). The impending emergence of ecological thought. *BioScience, 14*(7), 30–32.

Cole, Lamont. (1964). Pesticides: A hazard to nature's equilibrium. *American Journal of Public Health, 54*(Supplement), 24–31.

Committee on Pest Control and Wildlife Relationships. (1961). *Pest control and wildlife relationships: A symposium by George C. Decker [and others] March 10, 1961.* (National Research Council Pub. no. 897). Washington, DC: National Academy of Sciences, 1961. Retrievable from http://books.google.com/books?id=1mIrAAAAYAAJ.

Committee on Pest Control and Wildlife Relationships. (1962–1963). *Pest Control and Wildlife Relationships* (National Research Council Pub. no. 920-A, B, C). Washington, DC: National Academy of Sciences.

CBS Reports. (1963). The silent spring of Rachel Carson (TV episode). Video excerpts at http://www.youtube.com/watch?v=8WS4GwIXv24. Audio (Windows Media) at http://pesticides1963.net/library/CBSreports1.wma and http://pesticides1963.net/library/CBSreports2.wma. Guide to interviews at http://pesticides1963.net/CBS-cd.htm

Cottam, Clarence. (1963). A noisy reaction to *Silent Spring. Sierra Club Bulletin, 48,* 4–5, 14–15.

Darby, William. (1962, Oct. 1). Silence, Miss Carson (book review). *Chemical & Engineering News, 40,* 62–63.

Diamond, E. (1963, Sept. 28). The myth of the pesticide menace (book review). *Saturday Evening Post, 236*(33), 16–18.

Dunlap, Thomas (Ed.). (2008). *DDT, Silent Spring, and the Rise of Environmentalism: Classic Texts.* Seattle: University of Washington Press.

Geigy Company. (1944). Now it can be told (press release).

Jukes, Thomas. (1962, Aug. 18). A town in harmony. *Chemical Week, 40,* 5.

Lavine, Irvin, & Zimmerman, O. T. (1946). *DDT: Killer of Killers.* Rochester, NH: Record Press.

Leary, James, Fishbein, William, & Salter, Lawrence. (1946). *DDT and the Insect Problem*. New York, NY: McGraw-Hill.

McLean, Louis. (1967). Pesticides and the environment. *BioScience, 17*(September), 613–617.

Monsanto Magazine. (1962, Oct. 4). The desolate year, *42*(4), 4–9.

Nobel Foundation. (1948). The Nobel Prize in Medicine or Physiology [to Paul Muller]. Retrievable from http://www.nobelprize.org/nobel_prizes/medicine/laureates/1948

Rudd, Robert. (1963). The chemical countryside: A view of Rachel Carson's *Silent Spring* (book review). *Pacific Discovery, 16*, 10–11.

Time. (1962, Sept. 28). Pesticides: the price of progress (book review), *80*(13), 46–48. Retrievable from http://www.time.com/time/magazine/article/0,9171,940091,00.html

Time. (1963, July 5). The pest-ridden spring, *82*(1), 65. Retrievable from http://www.time.com/time/magazine/article/0,9171,875035,00.html.

U.S. Fish and Wildlife Service. News releases on DDT, 1945–1962. http://www.fws.gov/contaminants/Info/DDTNews.html
- August 22, 1945
- May 18, 1946

U.S. News and World Report. (1962, Nov. 26). Are weed killers, bug sprays killing the country? *49*, 86–94.

Wallace, George, & Bernard, Richard. (1963, July/Aug.). Tests show 40 species of birds poisoned by DDT. *Audubon, 64*, 198–199.

Wallace, George, Nickell, Walter, & Bernard, Richard. (1961). *Bird Mortality in the Dutch Elm Disease Program in Michigan*. Bloomfield Hills, MI: Cranbrook Institute of Science [Bulletin 41].

Other Relevant Articles

Buhs, Joshua. (2002). Dead cows on a Georgia field: Mapping the cultural landscape of the post-World War II American pesticide controversies. *Environmental History, 7*, 99–121.

Buhs, Joshua. (2002). The fire ant wars: Nature and science in the pesticide controversies of the late twentieth century. *Isis, 93*, 377–400.

Buhs, Joshua. (2004). *The Fire Ant Wars*. Chicago, IL: University of Chicago Press.

Gunter, Valerie, & Harris, Craig. (1998). Noisy winter: The DDT controversy in the years before *Silent Spring*. *Rural Sociology, 63*, 179–198.

Jukes, Thomas. (1971). D.D.T., human health and the environment. *Environmental Affairs, 1*(Nov.), 534–564.

Lear, Linda. (1992). Bombshell in Beltsville: The USDA and the challenge of *Silent Spring*. *Agricultural History, 66*, 151–170.

MacIntyre, Angus. (1987). Why pesticides received extensive use in America: A political economy of pest management to 1970. *National Resources Journal, 27*, 534–577.

Russell, Edmund. (1999). The strange career of DDT: Experts, federal capacity and environmentalism in World War II. *Technology and Culture, 40,* 770–796.

Wang, Zuoyue. (1997). Responding to *Silent Spring*: Scientists, popular science communication and environmental policy in the Kennedy years. *Science Communication, 19,* 141–163.

Historical Supplements

"News," 1963. http://pesticides1963.net/news.htm

Time magazine: review of *Silent Spring* and other items in the news (Sept. 28, 1962); letters in response (Oct. 5, 1962)

1963 Billboard Top Pop Hits. Rhino Records no. 27158.

For further details on roles and resources, see

http://pesticides1963.net

The President's Advisory Committee on Pesticides, 1963 Teaching Notes

FORMAT AND DAY PLANS

The historical resources assembled for this project may be used in many ways, of course. For example, one may simply discuss the history of the episode or the various perspectives or use documents related to some particular ethical or scientific issue. However, a simulation emphasizes experience as essential. By adopting a role, one understands a particular perspective in depth, while also coming to appreciate how and why other perspectives may differ. Each participant must be provided time to prepare his or her role, to be able to represent that particular perspective in a creative, open-ended exercise.

Everyone reads chapters 2 and 7 of *Silent Spring*. For each role, there is particular guidance, identifying essential chapters in *Silent Spring* and background articles—especially published reviews of *Silent Spring*. Almost all the characters are mentioned in Frank Graham's *Since Silent Spring*. There is also a list of common resources, including some Internet resources. A background essay—which can also be the basis for an instructor's presentation—introduces the history of DDT, Rachel Carson, and her book. Additional information details news items, popular music, and cartoons. These, too, can be used by an instructor to help set the scene in 1963 (say, at the beginning of class).

Like any case study, activities may expand to include more detail or context. The following is a guide to scheduling, with optional extensions noted.

Introduction (one-half to one class period)

(a) *History of DDT*. (~15 mins.) Possibly use text/visuals from the background essay. This may be expanded with more information about agriculture as monoculture, the problems of disease and crop pests, and pesticides used before 1945.

(b) *Optional*: Set the scene by re-enacting the 1948 Nobel Ceremony. The instructor takes the role of Prof. G. Fischer and presents the Nobel Prize to Paul Müller (possibly played by a teaching assistant or designated student). Use the Nobel presentation speech (online), possibly edited. *Optional*: musical fanfare.

(c) *History of Rachel Carson and* Silent Spring. (~15–20 mins.) Possibly use text/visuals from the background essay.

(d) Alternate to (a),(c): Assign background essay to be read independently by students. The aim is not to provide an exhaustive analytical history or biography, but to present information that an ordinary citizen might know about Carson and pesticides in 1962.

(e) *Task Charge*. (~15 mins.) Present the simulation scenario and the responsibilities of the committee (see project profile). Assign roles, discuss reading and writing assignments and available resources.

Optional: The instructor may adopt the role of President Kennedy and issue the charge directly to students as the committee. This begins to establish the spirit of the simulation and demonstrates for students how role playing works. Note the brief recording of Kennedy on *CBS Reports*.

Preliminary Discussion (optional; one class period)

For added depth in reading *Silent Spring*, allow students to discuss their personal responses to Carson's book, outside a historical context. This may be based on the whole book or selected chapters. Chapter 2, "The Obligation to Endure," features many of Carson's themes about control of nature. Chapter 7, "And No Birds Sing," is about harm to birdlife, echoing the book's title. Such discussion might be used to identify or highlight the themes that can guide the committee's later discussion. A short reaction essay may be required.

Testimony (one class period, depending on class size)

In this phase, each student gives a presentation to the committee based on his or her role. I have students prepare a written position statement in advance, which also becomes the basis for evaluating their work. The paper is to provide an assessment of Carson's claims from each role's perspective (each focusing on certain issues, as highlighted in the role descriptions). They should include any policy recommendations (new administrative rules or actions, new laws, funding requirements, etc.). I typically preview the statements, so that students have feedback comments before presenting information to class.

I limit presentations to three minutes and require a visual to foster development of presentation skills (no note cards allowed).

Optional: Alternatively, such papers may be posted on a shared website and serve as either required background reading (with no presentations) or a reference.

Optional: Students may be allowed, as members of the committee, to ask questions.

To lead the committee, the instructor may adopt the role of Jerome Wiesner (PSAC chair) or it may be assigned to a student as a role (to foster leadership skills).

Proposals and Discussion (one to two class periods)

If not included as part of the testimony phase, students present concrete proposals or recommendations to include in the report to the president. (More recently, to economize on time, I have delved into this activity without extended testimony.) Discussion may be more formally organized—using the key issues identified in the project profile and listed below as a structure. For example, discussing Carson's overall credibility is an appropriate opening. Alternatively, one may consider specific proposals in turn. The structure, or agenda, may also be established by the person playing Jerome Wiesner, if that role is assigned.

Students may need to decide whether they will work towards consensus (ideal, for working through all conflicts) or some other form of reaching a group decision. The most challenging topic, if adopted for discussion, is Carson's claims about "control of nature": is environmental action beyond pesticides warranted? If so, what?

Some instructors may want students to work on the language and wording of the proposals and, hence, of a final joint report. If so, segments taken from individual position statements (including justification) may facilitate group writing.

See Discussion Guide below.

Presidential Medal of Freedom (optional; one-half to one class period)

The presidential charge may include considering whether Carson should receive recognition (such as the Medal of Freedom) for her public service. This discussion can highlight more dramatically the role of responsible voices and communication style in public understanding of science. Carson's information all came from published sources, yet her emotive style influenced public opinion. Is her work especially significant or deserving of merit for this reason?

Epilogue (one-half to one class period)

When positions in the simulation are well researched, the participants typically echo the findings of the actual 1963 committee. That is, they likely validate most of Carson's claims, but also reaffirm the role of pesticides in modern agriculture, hardly entertaining a ban on DDT or other pesticides. You may refer to the actual report of the President's Science Advisory Committee from May 1963 (http://pesticides1963.net/library/psac1963.pdf).

Equally important, perhaps, may be the fact that despite such recommendations, little action was taken. The political power of agricultural businesses managed to suppress major action until the late 1960s. See

- MacIntyre, Angus. (1987). Why pesticides received extensive use in America: A political economy of pest management to 1970. *Natural Resources Journal, 27,* 534–577.
- Wang, Zuoyue. (1997). Responding to *Silent Spring*: Scientists, popular science communication and environmental policy in the Kennedy years. *Science Communication, 19,* 141–163.

The lesson about science and politics is part of learning about the nature of science, too. An epilogue may also be an occasion to reflect on several warnings by scientists and scientific organizations, some in the popular press, in the late 1940s and early 1950s. See

- Russell, Edmund. (1997). Testing insecticides and repellents in World War II. In M. Roe Smith & G. Clancey (Eds.), *Major Problems in the History of American Technology* (pp. 399–409). Boston, MA: Houghton Mifflin.

One may discuss why they did not have a cautionary effect at that time.

A retrospective is also a good occasion for discussing gender. Carson's gender was sometimes portrayed as relevant to her credibility, as exemplified in the review in *Time* magazine or other references to "Miss" Carson. One may also discuss whether Carson's perspective was gendered in a way critical to her effectiveness.

One may also wish to view and comment on many of the political cartoons inspired by Carson's work, some included on the website. Also see Paul Brooks's *House of Life*.

Optional: One may also discuss current controversies over (1) the use of DDT in developing nations for control of malaria and (2) the unregulated use of pesticides for individual residences.

Optional Supplements

Various elements may help set the scene in 1963 (for example, as a prelude or opening to class):

- *1963 Billboard Top Pop Hits* (Rhino Records)
- Tom Lehrer's satirical song, "Pollution." See online video (http://www.youtube.com/watch?v=nz_-KNNl-no or http://www.youtube.com/watch?v=XCojBngA--s).
- News headlines

DISCUSSION GUIDE

As noted elsewhere, discussion may be led by the instructor or by a student in the role of Committee Chair Jerome Wiesner.

The aim is for participants to not merely express their views, but to develop a joint recommendation. Reasoning is central. Thus, one major role of the discussion leader is to ensure that comments clearly articulate the reasons for a particular claim, based on evidence, ethical principles, or other shared values. A helpful standard for facilitators is the tone of a journalistic interviewer, interested in bringing information to light and seeking clarity and elaboration.

Another role of the discussion leader is to ensure that all stakeholder voices are addressed in developing the final decision(s). See Figure 13.1 as a general guide for where to expect, and possibly draw out, particular positions. Students often exhibit a tendency to "correct" history to reflect modern interpretations. The discussion leader can help amplify the contemporary disagreement in 1963. Similarly, the leader can help clarify

TABLE 13.1. Rachel Carson's supporters and critics, by topic.

	Carson Supporter	Carson Critic

credibility
• documentation of
  sources
• reliability of facts
• bias/balance of
  presentation
• rhetoric/tone/style
	Cole (science)
	Brown
	Cottam
	Bookchin
		Diamond
		Stevens
		Larrimer?
		Baldwin (bias)
		[Jukes, Darby]

harm (non-target species) • insect pollinators • shared habitat • food chain (*fish, birds) • other	Cole Clement Cottam [Wallace, Barker, Rudd]	Stevens Larrimer Baldwin?
benefits (target species!) • crops: food/fiber/ forest • disease control • nuisance/comfort		Rothberg Cole? Diamond Stevens Larrimer Zavon Baldwin Freeman? [Decker, Gill, Jukes, Darby, Simmons]
safety • food (carcinogens?) • worker safety	Cole? Bookchin (Baldwin)	Rothberg? Stevens Zavon Baldwin (workers) [Bean]
insecticide resistance	Brown Cottam Cole?	White-Stevens (alternatives)
indiscriminate use • excess/runoff	Cole Cottam (aerial) Freeman [Gill]	Rothberg [Decker, Gill]
alternatives: • biological (+) • earlier organics (-) • pesticide persistence • research	*Freeman (biol.) [Gill]	Baldwin (wary) [Decker]
balance of nature	Clement	White-Stevens
control of nature	Clement	~Bookchin Cole White-Stevens [Decker]

items of disagreement and actively engage those with contrary perspectives in fruitful exchange.

The leader may wish to clarify, possibly through group discussion, the standards for agreement—consensus, simple majority, two-thirds majority, or other.

The other major challenge for leading discussion in this simulation is helping to deepen the level of discussion, especially where participants may be underprepared. For example, policy proposals may be vague—advocating a position, not concrete actions or remedies that embody that position. Participants may need to be encouraged and supported in developing specifics.

Also, students tend to appeal to easy solutions, such as "we need more research"—thereby avoiding the real issues of managing any current problem. Decisions must often be made under circumstances of scientific uncertainty or incomplete knowledge. Again, one may need to offer further guidance:

- Who pays for the research? How much? (This raises ethical questions about who benefits and who bears risks and costs. It may also help expose the historical problem of incentives.)
- Are biological controls truly practical? (Their specificity and cost, in contrast to cheap, broadbased pesticides, may be easily overlooked.)
- How long will the development of alternatives take? What do we do in the mean time?
- Assuming that participants document errors in the past (such as exceeding recommended dosages, or human error in safe application), why did such errors occur? What problems (and prospective regulations) should be considered at this deeper level? For example, are the problems with the system of agriculture/forestry or insect-pest control? (Here, Carson's arguments about control of nature become relevant. Bookchin may present a case for the economic system and the regulation of industry, echoing some of Carson's statements about big business.)

The accompanying table (Figure 13.1) lists the various issues and maps them to the particular roles.

The Extension Activity, on whether Carson should receive the Medal of Freedom, allows deeper discussion of Carson's rhetoric and persuasive methods, the relationship of her scientific claims to her emotional imagery, the relevance of arguments about the control of nature, and her scientifically unfounded claims about the "balance of nature."

14 | Collecting Cases

The challenge of developing cases • assessing cases • writing cases

NOS education flows naturally when appropriate materials are in hand. The human and cultural contexts motivate students. Human characters and a narrative format render scientific practice—and scientific concepts, too—in approachable terms. The NOS dimensions become vivid and meaningful, often posing healthy interpretive puzzles to solve. Well-framed NOS questions provoke fruitful reflection and discussion. Students and teacher are all engaged in effective learning.

Currently, a chief challenge remains developing a more extensive repertoire of such cases, which can all meet the many standards profiled in the "Perspectives" section. One may consider the cases in Chapters 10–13 as good models. Still, more are needed. As more cases become available, teachers will need to separate the wheat from the chaff, by learning how to discern those case studies that are workable and effective—and render NOS well. Once familiar with using case studies, some teachers may feel inspired to delve into their own favorite historical discovery and assemble a case study themselves. This final chapter addresses these practical goals.

WHAT MARKS AN EFFECTIVE NOS CASE STUDY?

Effective NOS case studies come in many formats, reflecting the diverse personalities and styles of individual teachers and authors. Yet ideal case studies will all share some central features. For example, as noted in Chapter 1, they will draw the student into explicit NOS lessons and engage them in NOS reflection. Indeed, NOS analytical skills may well be exhibited through the ability to find and interpret relevant NOS dimensions and ask the right kinds of questions, rather than to simply enunciate or echo a set of prescribed NOS principles (Chapter 9).

For the teacher assessing prospective NOS lessons, two features are paramount. First, do they effectively engage the students in fruitful NOS reflection, in the spirit of inquiry learning and science-in-the-making (Chapter 2)? Second, do they express authentic NOS, or—like so many cookbook labs and popular myths—some imagined or contrived School Science NOS (Chapters 3–8)? Each benchmark is addressed below.

NOS Inquiry and Problematizing NOS

Those who appreciate the importance of inquiry in learning science will readily understand its importance for learning NOS as well. That is, effective NOS lessons will combine NOS with the strategies of inquiry learning, making the inquiry as much about the nature of science as the scientific concepts themselves. How does science work? How do we know that the conclusions are reliable? Students do not learn merely by watching videos, hearing stories, or even doing experiments. They need to participate actively in their own learning. Nor can one expect lecturing about NOS tenets to be fully effective, even if illustrated with vivid historical examples. Activities and discussions should *actively engage students in thinking about NOS problems and articulating their maturing perspectives.* The effectiveness of active learning has already been widely acknowledged throughout all types of education:[1] NOS education is no exception.

Most notably, in the model of inquiry learning, instruction ideally becomes student centered. A primary challenge is to motivate student engagement. Adopting this orientation, one might shift the characterization of NOS from tenets to be learned, to questions to be answered by students through informed reflection.[2] That is, declarative NOS knowledge needs to be *problematized.* One should hope to find NOS tenets—now familiar through several stock lists—reframed as *unsolved NOS problems.* In reviewing a potential NOS lesson for potential classroom use, the prominence and nature of student NOS problems should be an initial benchmark.

What kinds of NOS questions or problems guide students towards fruitful reflection on NOS? One may, of course, ask students plainly to comment on NOS interpretations of particular scientific episodes or narratives of research. However, such analysis is likely to be remote, given the student's role as a spectator. They are not fully immersed in the scenario. They are not reasoning about and generating the prospective answers. As observed in other fields of education (including science), students learn most effectively when addressing and solving problems on their own.

History is valuable because it provides a context for framing, or motivating, the NOS problems (Chapter 2). Ideally, the NOS questions emerge naturally from following the science *in practice.* They are not forced upon the case by the author or instructor. That is, many problems appear as decisions or choices for individuals within the case, ensuring that the problems—and the features of the nature of science—are on a human scale and, thus, accessible to learners. They parallel the kinds of questions about the reliability of scientific claims or the conduct of science that students will encounter as consumers and citizens in today's society.

Further, to foster fruitful reflection and skill development, the problems will be open ended (versus closed).[3] That is, a case study should not be designed to expect a single target solution or one exclusively "correct" answer, such as one may find hidden from students and divulged to instructors in special teaching notes. The aim is to explore the nature of science and foster thoughtful analysis and questioning, not to master particular concepts. It should be clear that interpreting the nature of science is rooted in the case itself, not provided by an external authority. For example, consider the case of Robert Hooke and the invention of the watch spring. Another scientist had claimed priority to the discovery, which Hooke believed to be his own. Here, the student can adopt Hooke's perspective and consider how to address the problem. This underscores how priority, credit, and social class contribute to scientific practice, without dictating particular tenets or conclusions.[4] In another case, students might imagine themselves as members of a new academic society in the early 1600s and decide whether to document a set of new organisms through drawings, descriptions, sample specimens, or some other means. Here, they develop an appreciation for the role of visualization in scientific communication, without being *told so* explicitly.[5] Another case poses an ethical question about Thomas Willis's surgical experiments with dogs while researching the brain in the 1600s. Research ethics is embodied in doing science.[6] NOS is coupled with science-in-the-making (Chapter 2).

NOS problems may well emerge through *NOS anomalies*, or *NOS discrepant events*. That is, events from the case may directly engage and address preconceptions about how science works. There are many popular but ill-informed impressions about science. Concrete cases can help challenge such views and begin to expose their deficits. Occasions of cognitive dissonance can, with appropriate guidance, help motivate and orient inquiry specifically on NOS themes. Accordingly, NOS teaching resources will be enhanced when they articulate common student *NOS* misconceptions—and perhaps profile how they may be found in history, too. For example, one case on psychologist B. F. Skinner underscores the gradual nature of his discovery of operant conditioning—at odds with the popular notion of "eureka!" moments of scientific insight.[7] In another, discussion of the competition between Westinghouse's alternating current (AC) and Edison's direct current (DC) challenges the student to consider whether science is always pure and how research may intersect with commercial interests and can exhibit bitter rivalry.[8]

NOS problems may also focus on errors or missteps in science, because they highlight methodological questions and lead to discussion about how to remedy the mistakes and/or avoid them in other cases (Chapter 5).[9] For example, despite eventually winning a Nobel Prize for his work, Christiaan Eijkman originally conceived of beriberi as caused by a germ, when it later proved to be a vitamin deficiency (Chapter 10). His theoretical outlook led both to useful controlled experiments that helped isolate the cause to diet, and also to discounting alternative explanations based on a nutrient deficiency. Eighteenth-century chemists used phlogiston as a concept to interpret combustion, calcination, reduction, and photosynthesis. But the model, like all models, was also limited (Chapters 8 and 11). Such cases also help contribute to rendering science as contingent, or provisional, a major NOS aim.

With the focus on problems, one may consider NOS education in the context of case-based and problem-based learning, already well established and researched in other contexts. While such teaching formats are often framed in narrowly programmatic ways, the extensive literature on these modes of teaching may inform how to approach NOS cases oriented to NOS problem solving.[10] For example, the level of student autonomy, or the amount of guidance needed by students, will vary. It depends, of course, on the level of student skills and a teacher's own skills in facilitating discussion. A case may need to be adapted to the individual classroom context. A corresponding challenge is framing problems for students at an appropriate developmental level.

One popular format is the interrupted case study.[11] That is, as illustrated in the beriberi case (Chapter 10), a narrative is interspersed with scientific and meta-scientific problems. The periodic questions allow the class to alternate between the planned guidance of the instructor and the open-ended problem solving of students at work. The teacher can monitor student responses at each step and build on them. Teachers with little experience in teaching cases often find this format a comfortable way to begin, rather than expect students to organize too much work independently.

The role of problems and active reflection is especially important where the goal is developing analytical skills. Scientific literacy implies not merely recognizing that "science is tentative" or "observations are theory laden," but being able to assess particular claims encountered in everyday life (Chapters 1 and 9). Skill development needs modeling and practice, not just stories or commentaries. Problems should encourage students to probe how science works (or does not work!) and provide them ample time, background, and tools to do so.

In summary, then, the NOS teacher is well advised to preview prospective NOS case studies and lessons in the context of inquiry learning. Are there provocative NOS questions or problems? Are they framed by human and cultural context to motivate student engagement? Are they clearly expressed to orient work? Can they be adapted to the developmental level of a particular group of learners (as needed)? Are common preconceptions identified, and engaged through appropriate NOS discrepant events? Does the case study include the relevant background information or historical context for addressing the questions fully? A good NOS case study leads students through the NOS-style thinking and analysis. It helps develop functional scientific literacy for our contemporary culture.

Authentic Science, Authentic History, Authentic NOS

When reviewing prepared NOS lessons for use, the second benchmark standard should be authentic NOS. That may seem obvious. But many lessons focus on a list of abstracted NOS principles as an end in itself, rather than use such a list to interpret and highlight aspects of real scientific practice. Namely, the proximal goal upstages and displaces the ultimate goal. The resulting understanding tends to be derivative and simplistic and, thus, fails to fulfill the ultimate purpose of informing scientific issues in social and personal decision making (Chapter 7). Instead, the lessons are relatively peripheral, about science in an idealized or purely metaphysical world (Chapter 8). NOS case studies should reflect real scientific practice.[12] They may draw from history,[13] contemporary research,[14] or a student's own investigative experience[15]—or from all combined.[16]

That is, first, authentic NOS lessons will be rooted in authentic science. For example, any unknowns should be genuinely unknown and the investigation meaningful. The questions may surely be reduced in scale and the methods basic to accommodate student abilities. Still, the nature of science should be fully *contextualized* in the process of creating genuine knowledge. In many available lessons, however, the NOS is *decontextualized*. For example, one common activity presents students with a "black box" (or sealed tube or cannister) filled with unknown items. The objective is to reason about what is inside the box without opening it. Students build a model based on their investigation of the box's observable properties. The exercise is often a metaphor for thinking about atoms or the Earth's interior or a cell's genes, or other phenomena that cannot be observed directly. As a metaphor for indirect reasoning, the activity can be quite useful. However, as a model for the nature of scientific practice, it can be grossly misleading. First, intrinsic student motivation for inquiry is rarely considered. Most important, the student is well aware that the teacher knows the expected answer. The activity can easily become a guessing game, trying to succeed in a school setting, not an authentic investigation. This posture is reinforced if the box is opened at the end, or if the teacher uses hidden knowledge of the box's contents either for evaluating student work or even for providing leading hints to the student. Such approaches portray knowledge as already known, rather than assembled solely by reasoning from available evidence.

A variant of the black box activity is "the unfinished cube"—or what I call a "green box" activity (when I first encountered this lesson, the cube was made of green cardboard). Here, all the relevant evidence is plainly visible. The box has information on five sides, and the sixth side is blank. Students are to infer and discuss what might belong on the blank face. This can be a fun little puzzle-solving activity for developing skills in recognizing patterns and generating hypotheses. But in one demonstration I witnessed, the instructor insisted that there was one "right" answer, and was totally deaf to entertaining reasoning about other alternatives—the very purpose of the exercise. Worse, there is no opportunity for *testing* the hypotheses. Ironically, the instructor declared that this was not necessary because the "correct" answer was "obvious." Here, scientific investigation was modeled as an activity in pure logic, with answers validated by an external authority, absent any empirical confirmation. Taught this way, the green box activity exemplifies inauthentic NOS.

By design, many teachers deliberately endeavor to clarify NOS through simple, but also artificial, investigations. That is, the lesson is built around the NOS concept, rather than the NOS being highlighted in an authentic case. For example, in "Tricky Tracks," students observe a series of hand-drawn animal tracks unfold over time. They are to reason that one animal must have eaten the other, without actually observing the event, demonstrating the powerful role of inference.[17] In another, students are asked to predict what happens to drops of food coloring when a probe with detergent is inserted into a bowl of milk. They are to appreciate the role of prediction in science. But students readily recognize that these activities are contrived classroom activities. They are nothing more than a game, typically a guessing game. They see the aim as appeasing the teacher, the local authority, not to satisfy innate curiosity or solve a compelling problem. Students do not regard them as simplified versions of real science—namely, what they see on CSI or other television shows, in movies, or in the news about fracking, climate change, or voyages to Mars. They also fail by not being meaningful or properly motivated, authentic investigations. Even where there may be potentially valid lessons, students do not transfer the conclusions to real science in concrete settings, with all their complexity. (Or, if they do, they tend to transfer it with a false sense of simplicity.) Such is the commonly overstated status of decontextualized NOS activities.

In another popular activity, the discovery of the periodic table is modeled through students trying to arrange a set of cards by color, notches, holes, and symbol shapes, rather than by the relevant atomic properties (boiling point, reactivity, density, etc.). They are also told in advance to expect holes and make predictions about the properties of missing cards. Teachers readily appreciate the parallels to the actual history of Mendeleev's discovery. But these parallels can mostly be fully appreciated only in retrospect, knowing the answer. Mendeleev did not—could not—know in advance whether there would be holes in the pattern. Science-in-the-making is blind to such clues. Nor are researchers told which properties will be relevant in forming the pattern. Indeed, that is largely an outcome of the research. This activity, designed to help students through its carefully planned structure, is not authentic science. It pre-solves the most important problems for students. It is just another puzzle, with a hidden answer. And students see through it easily. The same applies to many mock forensic activities, when the data are mindfully simplified to make the project clear for the students. Again, the exercise excises precisely the work essential for understanding NOS. Neither activity effectively conveys the intended NOS lessons.

"Black box" and "green box" activities, mock forensics, algorithmic "discovery" exercises, and other decontextualized NOS lessons, while not wholly unhelpful, have limited effectiveness. If used at all, they must be carefully scaffolded and linked to the examples of real science they are intended to inform.[18] The NOS lesson ultimately emerges from the authentic case itself. The activity becomes merely a supplemental analogy or metaphor. Given that the ultimate aim is to inform science in real-life contexts, the classroom will contextualize the science at the outset, using authentic cases, even if the examples are simple or truncated, with a level of challenge scaled to the students. The human and cultural contexts are important for making the science both meaningful and worth pursuing (Chapter 6).

A second critical feature for authentic NOS is respect for authentic history. The history may be selective, of course, appropriate to the lesson. But simplistic history is no better than a simplistic decontextualized classroom activity. Where informing analysis of socioscientific issues is the goal, idealized science cannot substitute for real science (Chapter 6). Students need to understand scientific practice as it is, not just as someone imagines it ought to be (Chapter 5). This applies especially to history, which is readily susceptible to Whiggish interpretations and myth-conceptions as it is converted to classroom lessons (Chapter 3). The teacher should expect to find some trial and error, and some unanticipated twists and turns, accidents and other contingencies that indicate the history has not been rationally reconstructed (Chapter 4). For example, a historical case that announces the discovery or conclusion in advance forsakes the core NOS perspective that science-in-the-making is necessarily blind to the outcome. Such prior announcements preempt exploring the role of evidence and reasoning. They convey how to defend conclusions, rather than how to reach them from a position of initial uncertainty. Such case reconstructions should be cast aside as having forsaken authentic nature of science. Adopting a posture of ignorance may be difficult, of course, where the scientific discovery or its discoverer are already well known. Thus, less famous cases may well be more appropriate for learning NOS.

Fidelity to authentic science and authentic history as a second fundamental NOS teaching standard adds yet further challenges to finding and developing effective resources. Foremost, case studies must be historically *and* philosophically *and* sociologically well informed. Otherwise, rather than convey well-informed NOS, they present a distorted caricature, readily susceptible to naive NOS preconceptions (Chapter 3). That is, they can easily perpetuate the grossly under- or ill-informed NOS views promoted decades ago, which current instruction hopes to remedy. In particular, the process of scientific discovery must be properly contextualized, not rationally reconstructed or romanticized (Chapters 3–5). NOS problems will likewise be situated in concrete contexts, not abstract metaphysical space. Accordingly, resources worth using or worth sharing on a wide scale will reflect the participation and expert review of professional historians, philosophers, and sociologists of science. Complementary skills are needed and require cross-disciplinary collaboration. NOS education thus requires integrating expertise beyond the experience of most science educators, just as it requires more educational expertise than most historians, philosophers, or sociologists of science can provide unassisted.

As easy as it is to articulate this principle, it is not so easy to find resources that embody it. Respecting history entails a bit more work, at least in researching and writing a case study. It involves consulting informed, scholarly histories. It may involve perusing original scientific papers or other documents, such as letters, journals, or speeches. It may involve collaborating with a professional historian of science. Not all case-study writers see the need for the complete investment. So the consumer of historical case studies needs to be discriminating. One should seek and adopt only those case studies that have been cowritten by or reviewed by professional historians of science, or show the author's serious investment with informed historical material. Some websites that can serve as trusted sources are listed in Figure 14.1.

Other Principles

Effective NOS resources will reflect other principles as well. For example, NOS should not be divorced from the science itself. NOS learning should integrate seamlessly with science content and process-of-science skills: Whole Science (Chapter 1). Special units that aim to teach NOS once and for all typically falter. NOS should emerge as a part of science itself, not as a peripheral—and thus dispensable—adjunct. When NOS lessons are designed as sidebar vignettes for textbooks, or as anecdotal asides, or possibly even as specialized lessons, one conveys significant messages to students about the (ir)relevance of NOS. A supplemental guiding principle, then, is an appreciation for teaching science, process-of-science skills, and NOS reflection as an integrated ensemble.[19] Such a broad-scope approach readily accommodates other widespread objectives, as well, such as profiling science as a human endeavor. Laboratory activities or investigations can be integrated as opportunity allows. For example, students might classify organisms while learning about patronage for science and the first scientific society in Italy

---

FIGURE 14.1. Websites with collections of historical NOS lessons.
- Doing Biology — doingbiology.net
- SHiPS Resource Center — ships.umn.edu/modules
- HIPST — hipstwiki.wetpaint.com/page/hipst+developed+cases
- The Story Behind the Science — storybehindthescience.org
- Contextual Science Teaching — sci-ed.org

in the early 1600s.[20] Or they may weigh gases while exploring the initially unfavorable reception to Avogadro's hypothesis on the volumes of gases.[21] Or they might explore heat as a variable in dyeing while considering the cultural context of indigo and its historical significance.[22] Standard labs, such as ones on spectroscopy or exploring light and magnification of the microscope, can also be coupled with and contextualized by the historical narratives.[23]

More Good Exemplars

The multiple demands of ideal NOS lessons—historically informed inquiry cases that foster NOS reflection and engagement with faithfully rendered Whole Science—may seem to severely limit the development of good resources. Still, many such resources are already available. Some notable examples are listed in Figure 14.2.

Notably, over a decade ago, *Doing Biology* offered a set of 17 historical case studies in biology, focusing on such standard concepts as natural selection (peppered moth), homeostasis, sex-linked inheritance, endosymbiosis of mitochondria, the citric acid cycle, production of antibodies, and behavior as an adaptation.[24] These are now online at http://doingbiology.net.

A few other examples might convey a sense of some of the good resources now available. For example, consider a case on *Alfred Russel Wallace & the Origin of New Species*. Most people are familiar with Charles Darwin and his theory of evolution by natural selection. Textbooks often discuss his travels on the *Beagle* and how he developed his ideas. This case focuses, by contrast, on the less familiar work of co-discoverer Alfred Wallace. It highlights his middle-class background, his career as a collector of exotic plant and animal specimens, and the observations and experience in the Amazon and the Malay archipelago that led to his own insights on how new species originate. Wallace, in contrast to Darwin, had to earn a living with his science, and this established a context for how he worked and the kind of observations he made. Yet Wallace shared with Darwin an appreciation of the diversity of life, extensive travel, attention to biogeographic patterns, and reading of Malthus's *Essay on Population*. Both were thus able to develop an understanding of the divergence of species and the role of competition in selecting certain varieties over others. The case touches on many NOS themes, including funding, priority and credit, scientific communication, and the diversity of scientific thinking.[25]

Another fine case focuses on *Charles David Keeling & Measuring Atmospheric CO_2*. The Keeling Curve, an icon of climate science, documents the steady rise of carbon dioxide in the Earth's atmosphere over more than half a century. But how did this famous graph begin, long before anyone knew of its ultimate significance? This case study considers the origin and development of Keeling's measurements, illustrating how science-in-the-making looks very different than it does in retrospect. It begins in a strictly

FIGURE 14.2. A sampling of good historical case studies for teaching NOS.

BIOLOGY

Alfred Russel Wallace & the Origin of New Species	Friedman (2010)
Carleton Gajdusek & Kuru	Gros (2011)
Modeling Mendel's Problems	Johnson & Stewart (1990)
Sickle-Cell Anemia & Levels in Biology, 1910–1966	Howe (2010)
Lady Mary Wortley Montagu & Smallpox Variolation in 18th-Century England	Remillard-Hagen (2010)
King D Carlos, A Naturalist Oceanographer	Faria, Pereira, & Chagas (2011)
Richard Lower & the "Life Force" of the Body	Moran (2009)
Interpreting Native American Herbal Remedies	Leland (2007)
Picture Perfect?: Making Sense of the Vast Diversity of Life on Earth	Carter (2007)
Henry David Thoreau & Forest Succession	Howe (2009)

CHEMISTRY

Determining Atomic Weights: Amodeo Avogadro & His Weight–Volume Hypothesis	Novak (2008)
Splendor of the Spectrum: Bunsen, Kirchoff & the Origin of Spectroscopy	Jayakumar (2006)

PHYSICS

Episodes in the History of Electricity	Henke & Höttecke (2010)

- William Gilbert: Separating Electric from Magnetic Effects
- Otto von Guericke: Analogies, Forces & the Quality of Scientific Instruments
- Charles dú Fay: Explorative Experiments: Describing & Explaining Electrical Phenomena
- Stephen Gray: Electrical Conduction on the Wrong Track
- Traveling Showmen: Electricity, Entertainment & the Construction of Scientificality

Contested Currents: The Race to Electrify America	Walvig (2010)
Robert Hooke, Hooke's Law & the Watch Spring	Horibe (2010)
William Thompson & the Transatlantic Cable	Klassen (2007)
Electromagnetism & the Telegraph	Barbacci, Bugini, Brenni, & Giatti (2011)

EARTH SCIENCE

Charles Keeling & Measuring Atmospheric CO_2	Leaf (2011)
Evolution of the Theory of the Earth	Dolphin (2009)

geological context and passes through several stages as the relevant context shifts, and Keeling faces successive crises in continuing his funding. The case highlights several NOS features: the role of long-term data and funding in science; the role of accuracy, precision, and calibration of instruments in simple measurements; and the cultural and political contexts of science.[26]

As just one more example, peruse *Carleton Gajdusek & Kuru*. In the 1950s, a mysterious neurodegenerative disease called *kuru* appeared among the primitive Fore people of Papua New Guinea. Natives attributed it to sorcery. American D. Carleton Gajdusek conducted epidemiological studies among the remote tribe and later transmission studies in lab animals. He found that kuru was propagated by a "slow virus," a new disease transmission type, a discovery for which he received a Nobel Prize. Gajdusek's interactions with the Fore people—some captured on film now available on YouTube—raise many questions where science, culture, and ethics intersect. Here, again, the NOS features are the basis of inquiry questions posed to students. For example, how do scientists pose problems? How does one articulate a research question about something one does not yet understand? How does one research the cause of a disease when one must, in a sense, assume a cause to collect the relevant information about it? This case is striking because Gajdusek had encountered a new disease transmission type (identified decades later as a prion). There are also research ethics to consider—concretely, not abstractly. Gajdusek needed to conduct autopsies and secure tissue samples, not endorsed by the local culture. Students must decide whether to trade matches and knives with the indigenous people in exchange for scientific evidence. Gajdusek developed a close relationship with the Fore people suffering from kuru. That included several adoptions and, later in his life, allegations of sexual abuse. Scientists are inevitably human: how should we integrate the personal and professional dimensions of their identity? In parallel, what principles should guide how the scientists themselves integrate their private and public lives?[27] Addressing all these questions helps promote an understanding of authentic scientific practice.

One particular NOS feature that has seemed especially challenging for NOS education is presenting science as developmental, changeable, provisional, fallible, and/or contingent—or tentative in the common label. History, however, is an excellent vehicle for this lesson. Case studies can address conceptual change explicitly, sometimes leading students through the experience. For example, one fine case addresses how Berzelius's electrical concept of atoms excluded the possibility of diatomic molecules but was later qualified, opening the way for Avogadro's hypothesis on the volumes of gases.[28] Another conveys how William Harvey's landmark work on circulation promoted erroneous views of the role of the heart in vivifying the blood, and how that was ultimately remedied, decades later, by Richard Lower.[29] Other cases approach the theme of development by underscoring the limitations of scientific thought. For example, the cases on smallpox variation from Turkey and on Native American herbal remedies highlight the roles of credibility and cultural contexts, along with the potential bias they introduce.[30]

Other, more ambitious modules situate students in a richer historical setting suitable for simulation and/or role playing—that is, rehearsing for scientific-literacy-in-practice (see Figure 14.3). These large-scale works exemplify the virtues of work developed over several years: work that is ultimately worth sharing and using widely.

All these cases contrast notably to the short stories or vignettes frequently found in textbooks and elsewhere. First, such anecdotes, while entertaining and memorable, may not lead to the desired, more sophisticated NOS lessons. Everyone knows the story of Newton and the apple, but towards what end? Such anecdotes do not engender the explicit reflection that is needed. Second, these cases are not boilerplate lessons that merely illustrate some prescribed NOS tenets. Rather, they open inquiry into NOS and invite reflective discussion. They use problems to teach NOS. The role of case studies is to provide substantive lessons in NOS, deeper than the superficial treatment of anecdotes and stories. Finally, these cases contribute to NOS understanding for real-life settings. They underscore the need for NOS analytical skills, known as scientific literacy (Chapter 1).

Resources such as those profiled here, while essential, hardly meet all the various challenges in implementing NOS in the classroom. For example, practicing teachers often tend to follow habits and resist departures from content.[31] Another major constraint is the trend of institutional accountability (Chapter 9). But these cases provide healthy exemplars as well as concrete resources from which to start.

How Does One Write a Good Case Study?

Having profited from case studies such as those introduced above, some teachers may wish to assemble a case of their own. Experience with NOS inquiry and NOS in authentic contexts—the two foremost standards profiled above—provides an excellent foundation. Understanding the perspectives presented in earlier chapters is also indispensable. These foundations open

---

FIGURE 14.3. Some historical role-play simulations.

Debating Galileo's *Dialogue*: The 1633 Trial	Chapter 12; galileotrial.net
Debating Rachel Carson's *Silent Spring*: The President's Advisory Committee on Pesticides, 1963	Chapter 13; pesticides1963.net
Darwin, the Copley Medal and the Rise of Naturalism	Driscoll, Dunn, Siems, & Swanson (2009)
Debating Glacial Theory	Montgomery (2010); glacialtheory.net

many possibilities.

Modules may fit many formats. Indeed, a case benefits from an individual's unique insights and style. Here is one effective framework, however, illustrated in the case study in Chapter 10, "Christiaan Eijkman and the Cause of Beriberi":

1. *Select a concept in science.*

   Choose a favorite? But be specific and focus narrowly at first. Ideally, the discovery can be traced to a single individual. While a single central character is not essential, it makes it easier, especially for a novice case-study author, in limiting information and tracing a clear narrative. The aim is to guide the student through this discovery, using the history as a framework. That is, the student will learn the science as the scientists did historically—but with a conveniently collapsed timeline and less work. At the same time, the history helps convey the nature of scientific practice through science-in-the-making, including the cultural context, the process of science (experiments, methods, reasoning), and alternative theories (Chapter 2).

2. *Find high-quality historical sources.*

   With luck, you may find a recent scholarly book that provides an overall script for you to follow. Beware, of course, of sensational popular books and their common myth-conceptions (Chapter 3). For an authoritative overview, a first source might be the multivolume *Dictionary of Scientific Biography* (published by Charles Scribner's Sons). Each entry includes a bibliography of major sources (although possibly not the most up-to-date).

   Because authentic history is integral to rendering authentic NOS (Chapter 5), the quality of historical sources is critical. Also plan to use the FirstSearch academic search engine for History of Science, Technology and Medicine (HistSciTechMed). This is the professional gateway to scholarly articles and books. Consulting a historian of science at a local college or university is well worth the effort. In particular, they may be able to provide access to articles in scholarly journals.

3. *Identify the important features of the nature of science.*

   After you have become generally familiar with the episode, a handful of NOS issues (three to seven) will likely emerge as especially vivid or well rendered, or critical to the central discovery. Ideally, these features appear while delving into the case itself. One does not need to consult some master list of NOS principles. The aim is to render how science works— and the prominent features vary from case to case. The case should inform your interpretation, rather than illustrate some predetermined principles or prior impressions (Chapter 5). Still, it may be helpful to review the spectrum of NOS dimensions, from experimental work, through conceptual interpretations, to social interactions and cultural contexts (see Figure 1.3 and Chapter 6). Reflection helps foster depth, completeness, and balance in the NOS profiled.

If possible at this early stage, identify specific occasions in the episode when the NOS issues become important to the participants themselves. Look for key decisions or choices by the focal scientist, or perhaps critics, or others who applied the scientific results. These often reflect NOS features in a problem-oriented way, conducive to inquiry learning. If there are clear NOS questions to pose to students, these can become early benchmarks for mapping out the narrative. This will help frame the story as a succession of NOS problems, a structure that helps guide the flow and motivate students. The NOS elements may indicate how to divide the narrative into thematic segments or episodes. Questions may also be framed (or reframed) at any later stage in developing the case study.

4. *Contextualize the scientific problem.*

Context is essential to motivating students to participate in inquiry. The question or problem must be meaningful or important. The cultural and human contexts are critical. So: identify the problem the scientist was originally trying to solve. Note that often this is *not* the problem we now regard as ultimately being solved (Chapter 4)! (Eijkman did not set out to discover vitamins.)

Elaborate on the problem's cultural context and significance. Render the motivating elements. One often finds a vivid event introducing or underscoring the problem that can function as an opening scene to the case study, to engage students. The emergence of the scientific problem is thus typically where the case story begins. Prime the inquiry. Having clearly articulated the problem, one is often well situated to invite students to imagine ways to solve it. This may involve some initial hypotheses. It may involve plans for securing more relevant information. It may involve reflection on methods. Strategies, plans, and prospects are more important at this moment than prejudiced solutions. Open the inquiry—setting an initial trajectory for resolving it.

In addressing the problem, document earlier historical efforts to solve the problem. Be generous. Portray early efforts as genuine, not merely short sighted or obviously deficient. Profile how those perspectives seemed well warranted, even if they later proved ill informed. Note that these views may reflect student preconceptions. One needs to acknowledge them as legitimate starting points (Chapter 4).

5. *Delve into the context that proved important in solving the problem.*

The particular resources for solving the scientific problem typically overlap with the scientist's biographical background. But one does not need an exhaustive account of upbringing and every life event. (A biography alone is rarely a good road map for a case study.) Note that many particular, contingent details may be important: fortuitous encounters, hobbies, colleagues with informative perspectives, chance errors, unexpected new instrument technologies, etc. Typically, the discovery is due to context more than any inherent genius. Nor should you try to shoehorn the process into some presumed "scientific method" (Chapter 5).

Use the historical information to set the context for students to address the problem and to apply their own thinking skills. Their tasks may involve interpreting evidence, imagining alternative explanations, or designing tests. Include methodological or ethical issues, too. Reflect that you are teaching process-of-science through example and practice. Use history as a guide (not a rigid script) to highlight important elements. Follow some of the dead ends or errors and where they lead (Chapter 4). Where appropriate, include a lab or introduce original historical data. Get the students active and involved in thinking from the details of the historical scenario.

6. *Profile the idea's historical reception.*

Again, the historical characters may reflect how your students respond. Address alternative interpretations. Frame any debate. Follow subsequent experimental work. Help lead the students through resolving any controversy. Wherever possible, provide opportunities to reflect on how one ensures the reliability of the claims. While simply expressed, this section can sometimes be the bulk of the case study. You may prepare an epilogue that discusses later corrections or important developments beyond the discovery itself.

7. *Consider other special contexts.*

What else makes this episode noteworthy? Research ethics? Social power (race, class, gender, or other)? Special instruments or new methods? Interpersonal dynamics? Unusual publications? Institutional intrigue? Cultural implications of the discovery? How the research was funded? Elaborate on these to enrich the science beyond its narrowest focus. Render scientific practice, not just the scientific thinking (see Figure 1.3 and Chapter 6).

Pause to consider the story as a whole. Everyone tells stories. They also tell them in their own style. So reflect on how the episode is more than just an assemblage of scientific evidence. It unfolds in time. As you begin to assemble all the pieces into a whole, articulate the narrative—possibly outline it, or diagram it in a historical flowchart—and use that overall structure to guide your writing. Find a dramatic or striking opening to capture interest. Where possible, connect it to the central problem to engage the students in the science, not just the story. Also consider how the case resolves, or closes. Ideally, find occasion to celebrate the achievement, however modestly. There may be modern developments that help complement or echo the historical story.

8. *Return to the central NOS features and questions.*

With the case and overall narrative well in hand, review how the NOS features are presented and how you pose questions for student reflection. Sharpen the significance of each NOS question. Shape or revise the surrounding narrative to assist in framing the questions and laying the background for addressing them. Check that you can articulate at least three possible student answers for each question. If you cannot, the question is too narrow or closed. Posing good questions is one of the most challenging elements of writing a case study. It is also the most important to the aim of teaching NOS. Successive revisions may be needed. Adjustments to the NOS questions may require rearranging or reorganizing the narrative. Feedback from colleagues is immensely valuable at this stage.

The case study should conclude with students explicitly reviewing the major NOS features and how the case study contributes to a deeper understanding of each. This is a standard part of consolidating any learning. The written case study should reflect that.

9. *When time is available, enhance the case with more cultural and contemporary context.*

One can help set the scene at the outset with familiar events from the time and place where the case is set: art, music, cultural and political events. This further helps to humanize and contextualize the science. Sometimes, these events reappear in the case itself in interesting ways. News events from the period can help mark the passage of time within the narrative. Images, of course, help students visualize the scientists, what they studied, and where they worked.

10. *Finally, consider how students may effectively demonstrate what they have learned.*

Articulate what you want students to learn from the case study. Be clear and transparent. Perhaps one might ask students to think or act in historical context (or imagine a dialogue between historical characters). Perhaps they might interpret alternative data, to show that they understand patterns of reasoning. Perhaps they might submit their reflections during the case work, recorded in an NOS journal. Perhaps they might formally describe each NOS lesson in a summary at the end.

Of course, these steps need not be followed in a specific order—any more than a scientist follows a guaranteed step-by-step "scientific method." At the same time, all these elements can contribute to a more robust case study. Yes, it involves a substantial investment of effort. It also provides an equal amount of reward when done well.

When the time is ripe, plan to share your case study (and hard work) with colleagues. Assemble a set of teaching notes. Introduce the case, its scientific or historical significance, and the major NOS features that the case highlights. Inform other teachers about the NOS questions, including supplemental information where appropriate. Include handouts and labs. Identify any helpful websites. Package up the images, where copyright is available. Communicate bibliographic sources and other references clearly.

One signal of having written an effective lesson—surely any lesson, not just NOS case studies—is that students become engaged. Not just interested. Or attentive or entertained. Or casually curious. But personally invested. The activity is meaningful. They apply themselves in thinking through the problems or questions. They take pride in the outcome. At such times, the teaching, too, becomes easier. And yields a greater sense of reward. The great virtue of teaching the nature of science through case studies is that when the hard work of writing them is done well, learning seems to happen effortlessly. That, surely, is a touchstone of quality for teaching the nature of science.

| Notes

CHAPTER 1: THE NATURE OF SCIENCE: FROM TEST TUBES TO YOUTUBE

1. Conant (1947)
2. Kelly, Chen, & Crawford (1998); Lederman, Wade, & Bell (1998)
3. American Association for the Advancement of Science [AAAS] (1993); National Research Council [NRC] (1996); Rutherford & Ahlgren (1990)
4. AAAS (2009); Committee on Conceptual Framework for the New K–12 Science Education Standards, NRC (2012); Organization for Economic Cooperation and Development [OECD] (2009)
5. Krajik & Sutherland (2010); OECD (2009, pp. 14, 126); Osborne (2007)
6. Ziman (1978)
7. Millar & Osborne (1998); NRC (1996, p. 22)
8. Daempfle (2013); Feder (1999); Park (2000); Pigliucci (2010)
9. Allchin (2006a)
10. Losee (2005)
11. Handschy (1982); Holton (1988); Lakatos (1978, pp. 73–78); Swenson (1970)
12. Nendick, Scrancher, & Usher (2007)
13. Quoted in Judson (1980, p. 169)
14. For informative historical reviews, see Kelly et al. (1998); Lederman et al. (1998).
15. Analysis by McComas & Olson (1998)
16. AAAS (1993); McComas (1996); McComas & Olson (1998); National Science Teachers Association [NSTA] (2000)
17. Osborne, Collins, Ratclifffe, Millar, & Duschl (2003)
18. Lederman et al. (1998)
19. Juel (2011)
20. http://www.drroyspencer.com/2009/04/some-global-warming-qa-to-consider-in-light-of-the-epa-ruling. Formerly posted on the website www.climatechangefraud.com.
21. Kintisch (2010)
22. Oreskes & Conway (2010)
23. Goodrich (2011)
24. Ford (2008); Rudolph (2000)
25. Kolstø (2001, quote on pp. 292–293)
26. Kolstø (2001)
27. Latour (1987)
28. Allchin (1999b)
29. Ryder (2001)
30. Ryder (2001, table 4)
31. Ryder (2001, p. 26)

32. Ryder (2001, p. 35).
33. Shapin (1994)
34. Goldman (1999, 2002); Hardwig (1991); Selinger & Crease (2006)
35. Gaon & Norris (2001); Norris (1995)
36. Anand (2002)
37. Ihde (1997); L. Thomas (1981)
38. Latour (1987, pp. 195–257)
39. Allchin (2001, 2012b); Guinta (2001); Mermelstein & Young (1995)
40. AAAS (2009); Committee on Conceptual for the New K–12 Science Education Standards Framework (2012); Duschl, Schweingruber, & Shouse (2007); Ford (2008)
41. Latour (1987); Longino (1990); Rudwick (1985); Shapin (1994); M. Solomon (2001)
42. Franklin (1986); Hacking (1983); Kohler (1994); Pickering (1995); Rheinberger (1997)
43. Poincaré (1902)

CHAPTER 2: HISTORY AS A TOOL

1. Allchin (1993)
2. Arora & Kean (1992, 58–60)
3. Dolphin (2009); Johnson & Stewart (1990); Luhl (1990); Shahn (1990); Swanson (1995a, b)
4. Mix, Farber, & King (1996)
5. Aikenhead (1991); Conant & Nash (1957); Hagen, Allchin, & Singer (1996); Klopfer (1964–1966)
6. Barth (1995); Becker (1992); Pi Suñer (1955); Shamos (1987)
7. Collison (1992); Lockhead & Dufresne (1989)
8. e.g., Falkenberg (2009); Randak (1990)
9. Eakin (1975); http://www.lib.berkeley.edu/MRC/berkeleyvideos.html
10. Habben, Mehrle, Heering, Meya, Reiß, Rohlfs, & Sibum (1994); Henke & Höttecke (2010); Kipnis (1993); Reiß (1995)
11. Heering (1992a, b); Sibum (1995)
12. Friedman (2010)
13. Carvalho (1990); Robin & Ohlsson (1989)
14. Clement (1983); Franco (1992); Mas, Perez, & Harris (1968); Steinberg (1992)
15. Monk & Osborne (1997); Wandersee (1986)
16. e.g., Machold (1990) on objections to special relativity theory; Grapi (1992) on chemical reversiblity; also see Nersessian (1989)
17. See Barker (1995) on plant nutrition; Buster (1995) on electricity; Gauld (1995) on Newton's third law; Jensen & Finley (1995) on evolution; and Weck (1995)
18. Gould (1977)
19. DeMeo (1992)
20. Brannigan (1981); Kuhn (1970)
21. Shapin (1988, 1989)
22. Honey (1992, p. 513)
23. Brush (1974); Holton (1978a)
24. Chambers (1983)
25. Broad & Wade (1982); Hull (1988); Jones (1981); Judson (2004); Rudwick (1985); Watson (1968)
26. Barba, Pang, & Tran (1992); Haggerty (1992)

27. Compare J. R. Martin (1989) and Haraway (1989)

28. Conant & Nash (1957, p. ix)

29. Hagen et al. (1996); Herreid (2005)

30. Confrey & Smith (1989); Monk & Osborne (1997)

31. Collins (1992); Galison (1987); Pickering (1995)

32. Reiß (1995)

33. Beveridge (1950, pp. 37–55); Copeland, Larson, & Morton (1996); Judson (1980, pp. 68–750); Kohn (1989); Roberts (1989); Taton (1962)

34. Hagen et al. (1996); B. Martin (1991); Rudwick (1985); Silverman (1992)

35. Kuhn (1970)

36. Laudan (1984, pp. 56–59)

37. Burdett (1989); Collins & Pinch (1993); Guinta (2001); Mermelstein & Young (1995); O'Rafferty (1995); Silva, Silva, Passos, Morais, & Neves (1995)

38. Latour (1987); Monk & Osborne (1997)

39. For excellent overviews, see Hull (1988); Rudwick (1985)

40. Bourdieu (1975)

41. Haraway (1989); Harding (1991); Latour (1987); Longino (1990); Shapin (1994)

42. e.g., Garrison (1995); Keiny & Gorodetsky (1995); Magnusson & Templin (1995)

43. Allchin (2005b)

44. For a concise summary, see Ziman (1976, pp. 240–280)

45. Alcoze & Barrios-Chacon (1993); Aikenhead & Michell (2011); DeKosky & Allchin (2008); Freely (2009); McClellan & Dorn (1999); Ronan (1985); Selin (1997); Teresi (2002)

46. Lunberg, Levin, & Harrington (1999)

47. American Chemical Society (2006); Leonard, Penick, & Speziale (2008); Millar (2000); Postlethwait & Hopson (2003); Schwartz, Bunce, Silberman, Stanitski, Stratton, & Zipp (1997)

48. Dimopoulos & Koulaidis (2003); Elliot (2006); Korpan, Bisanz, & Bisanz (1997); McClune & Jarman (2010); Norris & Phillips (1994); Phillips & Norris (1999); Wellington (1991); Wong, Hodson, Kwan, & Yung (2008)

49. J. Thomas (2000)

50. Dimopoulos & Koulaidis (2003, p. 248)

51. Collins & Pinch (1993)

52. Allchin (2005a)

53. On Millikan, see Franklin (1986, chapter 5); Holton (1978b). On Mendel, see Franklin, Edwards, Fairbanks, Hartl, & Seidenfeld (2008). On mapping chromosomes, see Wimsatt (2007, pp. 94–132). On maps, see Turnbull (1993). On Newton, see M. White (1997). On Murchison, see Rudwick (1985).

54. On pellagra, see Kraut (2003); on beriberi, see Carpenter, (2000). On thalidomide, see Stephens & Brynner (2001); on genetic engineering, see Fredrickson (2001) and Hindmarsh & Gottweis (2005). On spiritualism and the paranormal, see M. Gardner (1957) and Lyons (2009, chapters 4 and 5).

55. R. Gardner (1990). Similar examples are now found easily on the Internet—see Rubin (2011)

56. Latour (1987)

57. Cliff & Nesbitt (2005)

58. Tobin & Roth (1995)

59. Akerson, Abd-El-Khalick, & Lederman (2000); Craven (2002); Khishfe & Abd-El-Khalick (2002); Scharmann, Smith, James, & Jensen (2005); Schwartz, Lederman, &

Crawford (2004); Seker & Welsh (2005)
    60. NRC (1996); Hodson (2008, pp. 155–158)

CHAPTER 3: MYTH-CONCEPTIONS

    1. Allchin (1996); Jayakumar (2006)
    2. Flannery (2001)
    3. Chambers (1983); Finson (2002)
    4. The term 'myth' has many meanings. In the popular mind, myth is *false belief*, often opposed to science (as true belief). When applied to science or science education ("the myth of the scientific method" or "the myth of scientific objectivity"), critics use myth to refer especially to *widespread and unquestioned* false beliefs (MacCormac, 1976; Savan, 1988; Bauer, 1992; Sismondo, 1996; McComas, 1998). In these cases, the term 'myth' serves primarily as a rhetorical device for casting the author as an authority rescuing readers from credulity. I am *not* using the term in these common senses. Others construe myth in science as *a foundational religious metaphor* (MacCormac, 1976; Midgley, 1992). They equate myth with a cultural perspective or cognitive meta-structure (sometimes critiquing it, sometimes endorsing it). While much can be (and has been) said about scientism, I do *not* mean scientific myth in this sense either.
    In Greek, *mythos* means 'telling' or 'story'. In my use, therefore, a myth is always a *narrative*, a literary form, or genre. (I do *not* adopt the adjunct meanings used theoretically by anthropologists or psychologists.) Like parables, myths *function primarily as explanations and/or justifications* (Bauer, 1992; Milne, 1998). They thus generally *contain superhuman elements and/or natural phenomena*, whence they, in part, draw their persuasive power. In many cases, myths embody a worldview by *providing formulae or archetypes for appropriate or sanctioned behavior*, hence their narrative format. In construing myth in this way, my analysis does not focus on false belief itself. Rather, I examine how errors symptomatically reveal the structure of the narratives and how they function as myths.
    5. Sapp (1990)
    6. Witness the hyperbolic defenses of Mendel in the light of the criticism offered here: R. Fritz, "Allchin's many mistakes," *BioQuest Notes, 10* (1999, no. 1), 9–12, and reply by Allchin, pp. 12–14; J. Westerlund & D. Fairbanks, "Gregor Mendel and 'myth-conceptions'," *Science Education, 88* (2004), 754–758, and reply by Allchin, pp. 759–761; A. Lawson, "Allchin's shoehorn, or why science is hypothetico-deductive," *Science & Education, 12* (2003), 331–337, and "A reply to Allchin's 'Pseudohistory and Pseudoscience,'" *Science & Education, 13* (2004), 599–605, and replies by Allchin (2006b); and "Reducing errors to sighs," *Science Education, 90* (2006), 293–295.
    7. Campbell, Reece, & Mitchell (1999, p. 249)
    8. Brannigan (1981); Corcos & Monaghan (1993); Hartl & Orel (1992); Monaghan & Corcos (1990); Olby (1985, 1997); Sapp (1990)
    9. Westerlund & Fairbanks (2010)
    10. Hartl & Orel (1992)
    11. Westerlund & Fairbanks (2010, p. 298)
    12. Endersby (2007); Mendel (1869/1966); Nogler (2006)
    13. Kroeber, Wolff, & Weaver (1969); and the following websites (accessed 21 January 2013): *American Heritage Science Dictionary*, 2005 (http://www.thefreedictionary.com/Mendel's+law); Hamburg University, 2001 (http://www.biologie.uni-hamburg.de/b-online/library/falk/Inherit/Inherit.htm); Darmouth University biology course, 1997 (http://www.dartmouth.edu/~cbbc/courses/bio4/bio4-1997/01-Genetics.html);

Memorial University of Newfoundland, "Principles of Genetics" course, 2012 (http://
www.mun.ca/biology/scarr/2250_Dominance_Segregation_&_Assortment.html);
PinkMonkey Online Study Guide, 2004 (http://www.pinkmonkey.com/studyguides/
subjects/biology-edited/chap7/b0707401.asp); Biology Online http://www.biology-
online.org/dictionary/Law_of_Dominance); Human Genetics for the Social Science,
University of Colorado (http://psych.colorado.edu/~carey/hgss/hgsschapters/HGSS_
Chapter09.pdf)

14. Mendel (1866/1966)
15. Allchin (2005c)
16. Olby (1985)
17. Di Trocchio (1991)
18. Olby (1985); Sapp (1990)
19. Sapp (1990)
20. Ho (1999); WGBH (1998)
21. Macfarlane (1985)
22. Ho (1999)
23. Ho (1999)
24. Judson (1980)
25. Fisher & Lipson (1988, p. 184); Judson (1980, p. 71)
26. Pagel (1967, 1976)
27. Frank (1980)
28. Lawson (2000)
29. Lawson (2000, p. 482)
30. For example: Asimov (1964, pp. 24, 29); Baumel & Berber (1973, pp. 12–13);
Leinhard (1997); Lewis (1988)
31. Lawson (2000, pp. 483, 484; also 2002, p. 16)
32. Harvey (1628/1952, chapter 7, p. 283)
33. Harvey (1628/1952, pp. 283–284, 296; 1649/1952, pp. 308, 322)
34. Young (1929, p. 1); Elkana & Goodfield (1968)
35. Harvey (1649/1952, p. 311)
36. Lawson (2002, p. 16)
37. Malpighi (1661/1929, p. 8)
38. Adelmann (1966, I, pp. 171–198)
39. De Mey (1982, pp. 192–201); quote from Harvey (1628, chapter 13)
40. Lawson (2000, p. 482)
41. Gregory (2001)
42. Allchin (2006b)
43. Johnson (2008); Rivers & Wykes (2008); Uglow (2003)
44. Nash (1957)
45. Matthews (2009b)
46. Matthews (2009b, pp. 931–932)
47. Matthews (2009b, pp. 934, 935, 937–938, 955, all emphasis added)
48. Matthews (2009b), quotes on p. 929, 955–956, and 937, respectively
49. Nash (1957, p. 350)
50. Johnson (2008, pp. 29, 48–51, 62, 70–72, 74–75, 91, 147)
51. Quoted in Nash (1957, pp. 351, 361)
52. Nash (1957, p. 361)
53. Brock (1993, pp. 99–100); Johnson (2008, pp. 71, 85–87, 135–138)
54. Matthews (2009b, pp. 942–943)
55. Nash (1957, pp. 358–369); Schofield (2004, pp. 139, 154–157)

56. Quoted in Nash (1957, p. 360). Matthews similarly tries to rationalize the inconsistencies as a symptom of "the beginning stages of any science" (p. 946). Indeed, he tries to sustain Priestley's virtue by spinning his position into an epitome of intellectual integrity. For the sake of the story, he suggests how Priestley must have thought, contradicting what the available historical documents indicate.

57. McEvoy (1987)

58. Matthews (2009b, pp. 931, 936–937)

59. Nash (1957, p. 328)

60. Matthews (2009b, p. 955)

61. Compare Matthews (2009b, pp. 941–942) to Nash (1957, p. 361).

62. Brock (1993, pp. 99–99, 103–104, 124)

63. Matthews's error on p. 940. See Brock (1993, p. 101).

64. Nash (1957, p. 371)

65. Matthews's error on p. 942. See Johnson (2008, pp. 80–82); Nash (1957, p. 355).

66. Nash (1957, pp. 377–384, 409–419); Schofield (2004, pp. 154–157)

67. Nash (1957, pp. 420–431)

68. Quoted in Nash (1957, p. 351)

69. Matthews (2009b, pp. 929, 937, 943)

70. Quoted in Johnson (2008, p. 91)

71. Johnson (2008, pp. 48–51, 63, 91)

72. Brock (1993, pp. 99, 101, 124); Johnson (2008, pp. 56, 60–63); Levere (2001, pp. 52–54, 58)

73. Johnson (2008, p. 70)

74. Johnson (2008, pp. 37–38, 66, 80–82, 112)

75. Ironically, all the information presented in the analysis here was available in sources cited by the author himself (Matthews 2009a, b). This is not a case of simplification or failure of access to informed and balanced histories. Rather, the case illustrates ideological selective use of facts and corresponding disrespect for both the history and the nature of science.

76. Gilovich (1991, pp. 90–94)

77. Milne (1998)

78. Rutherford & Ahlgren (1990, pp. 9–13); NRC (1996, p. 201)

79. Brush (1974, 2002)

80. e.g., Martin & Brouwer (1991)

81. Milne (1988, p. 184)

82. Butterfield (1959); Kuhn (1970, pp. 136–143)

83. Milne (1998, pp. 178–179)

84. Bauer (1992)

85. Steinbach (1998)

86. Milne (1998, p. 177)

87. Shaffer (1990)

88. Kipling (1902)

89. Allchin (2008)

90. Brush (1995)

CHAPTER 4: HOW *NOT* TO TEACH HISTORY IN SCIENCE

1. Nason & Goldstein (1969, pp. 15–16, 276–277); echoed in Crofton (2010, p. 76)

2. Helmont's experiment is widely told, and subject to myth-conceptions (see Tobin & Dusheck, 2005, pp. 126–127; also Silverstein, Silverstein, & Nunn, 2008, a children's

text now copied on several websites). While commentary is thick, the original account is surprisingly sparse. Here it is, complete (from the 1664 English translation):

> But I have learned by this handicraft-operation, that all Vegetables do immediately, and materially proceed out of the Element of water onely. For I took an Earthen Vessel, in which I put 200 pounds of Earth that had been dried in a Furnace, which I moystened with Rain-water, and I implanted therein the Trunk or Stem of a Willow Tree, weighing five pounds; and at length, five years being finished, the Tree sprung from thence, did weigh 169 pounds, and about three ounces: But I moystened the Earthen Vessel with Rain-water, or distilled water (alwayes when there was need) and it was large, and implanted into the Earth, and least the dust that flew about should be co-mingled with the Earth, I covered the lip or mouth of the Vessel, with an Iron-Plate covered with Tin, and easily passable with many holes. I computed not the weight of the leaves that fell off in the four Autumnes. At length, I again dried the Earth of the Vessel, and there were found the same 200 pounds, wanting about two ounces. Therefore 164 pounds of Wood, Barks, and Roots, arose out of water onely. (Helmont, 1664, p. 109)

It appears in a chapter titled "The fiction of elementary complexions and mixtures" (pp. 104–111), in the midst of a section on the gas of coal being made of water.

3. Gould (1977)

4. Brock (1993, xxi–xxii, 49–53); Ducheyne (2008); Helmont (1664); Pagel (1982)

5. Here is the original:

> For from hence, not onely lice, wall-lice or flies breeding in Wood, Gnats, and Worms, become the guests and neighbours of our misery, and are as it were bred or born of our inner parts, and excrements: but also, if a foul shirt be pressed together within the mouth of a Vessel, wherein Wheat is, within a few dayes (to wit, 21) a ferment being drawn from the shirt, and changed by the odour of the grain, the Wheat it self being incrusted in its own skin, transchangeth into Mice: and it is therefore the more to be wondered at, because such kinde of insects being distinguished by the Signatures of the Sexes, do generate with those which were born of the seed of Parents: That from hence also, the likeness or quality of both the seeds, and a like vitall strength of the ferments may plainly appear: And which is more wonderfull, out of the Bread-corn, and the shirt, do leap forth, not indeed little, or sucking, or very small, or abortive Mice: but those that are wholly or fully formed. (Helmont, 1664, p. 113)

6. Tobin & Dusheck (2005, p. 127); also see Hershey (1991, 2003)

7. Hershey (2003); also Hershey (1991, p. 458)

8. Helmont (1664, treatise no. 15, pp. 70–77; treatise no. 16, pp. 78–81; treatise no. 20, pp. 106–108); Pagel (1982); van Klooster (1947)

9. Helmont (1664, treatise no. 20, p. 106)

10. Kuhn (1970)

11. Gilovich (1991); Kahnemann (2011); Nickerson (1998); Sutherland (1992)

12. Skeptics may examine the bitter comments by D. Hershey, *Journal of College Science Teaching, 31*(May, 2002), 6–7; or *Science & Education, 15* (2006), 121–125; or A. E. Lawson, "On the hypothetico-deductive nature of science—Darwin's finches," *Science & Education, 18* (2009), 119–124.

13. Dobbs (1975); Principe (1998)

14. Hershey (1991, 2003)

15. The concept of replicates emerged from R. A. Fisher's work on statistics in the early 20th century. Measurement error became clear in the late 18th century, statistical analysis in the late 19th (Porter, 1986; Hacking, 1990). The concept of control was articulated in the late 1800s (Boring, 1954). The modern concepts of gas (in contrast to air) and carbon dioxide emerged with Lavoisier's work in the late 1700s (Brock, 1993).

16. Brock (1993); Ducheyne (2005)

17. Duschl (1987, 1993)

18. Duschl (1993, p. 190)

19. Giere (1984)

20. Clough (2006); Posner, Strike, Hewson, & Gertzog (1982); Tyson, Venville, Harrison, & Treagust (1997)

21. Allchin (1993)

22. Gauld (1991); Monk & Osborne (1997, pp. 412–413); Shulman (1995); Stork (1995); Swanson (1995); Wandersee (1992)

23. Burke (1978, 1996, 1999, 2003)

24. Stocklmayer & Treagust (1994)

25. Schiebinger (1993, pp. 75–114)

26. Haraway (1989)

27. Micale (1995); Schiebinger (1993)

CHAPTER 5: PSEUDOHISTORY AND PSEUDOSCIENCE

1. Toumey (1996)
2. Nash (1951, p. 151)
3. Biagioli (1993)
4. Bechtel (2006); Franklin (1986, 1991); Galison (1987); Hacking (1983)
5. Collins (2004); Franklin (1997); Hacking (1983); Rothbart & Slayden (1994)
6. Pitt (2000, pp. 92–96)
7. Lawson (2002), quotes on pp. 21, 2, and 7, respectively (also pp. 9, 19, 20)
8. Biagioli (1993); Finnochiaro (1997); J. D. Moss (1993); Shea (1998)
9. Biagioli (1993, chapter 3)
10. Medawar (1964)
11. Bazerman (1988); Booth (1993)
12. Knorr-Cetina (1981, pp. 94–135)
13. Lawson (2002, pp. 2, 21)
14. Lawson (2002, pp. 2, 4–9, 17, esp. table II, step no. 9)
15. N. Jardine (1994, quotes on p. 280)
16. Sapp (1990); H. White (1987)
17. Gould (1990, pp. 46, 102, 244)
18. Lawson (2003)
19. Lawson (2003, pp. 163, 173)
20. Lawson (2003, pp. 156–157)
21. Judson (1980, chapter 4); Kohn (1989); Merton & Barber (2003); Rheinberger (1997, 2009); Roberts (1989)
22. Glen (1994); Powell (1998); Raup (1992)
23. Lawson (2002, p. 2)
24. Gauld (1992)
25. Whitaker (1979)
26. Gauld (1977)

27. This strategy is advocated, for example, by Matthews (1994, pp. 79–80).
28. Dobbs (1975); Gould (1980, pp. 145–151); Principe (1998)
29. Boyle (1699, Book III, Vol. 2, pp. 275–299)
30. Boyle (1673, pp. 108, 122, 166)
31. Boyle (1699, Book II, p. 249; 1673, chapters V, VI)
32. Gould (1981)
33. Cordingley (2001)
34. Fee (1979, p. 419)
35. Exemplified in many popular books, such as Daempfle (2013); Feder (1999); Fritze (2009); Park (2000); Pigliucci (2010)
36. Klein (1972); Whitaker (1979)
37. Sterling (1994); Wallace (1996)

CHAPTER 6: SOCIOLOGY, TOO

1. Brush (1974); also see Holton (1978a)
2. Knorr-Cetina (1981); Latour & Woolgar (1979); Traweek (1988)
3. Collins (1992)
4. Shapin (1979)
5. Shapin & Schaffer (1985)
6. Irzik & Irzik (2002); Slezak (1994)
7. Gould (1983); Kamin (1974)
8. Hodson (2008, chapter 7); Kelly, Carlsen, & Cunningham (1993); Roth (1997)
9. Osborne et al. (2003, esp. table 2)
10. Hess (1997); Pickering (1992)
11. Cole (1992)
12. Price (1963); Gieryn (1999); Merton (1973); Toumey (1996)
13. Broad & Wade (1982); Judson (2004)
14. Service (2002, 2003); Vogel, Proffitt, & Stone (2004)
15. Committee on Science, Engineering, and Public Policy (2009)
16. Latour (1987); Latour & Woolgar (1979)
17. Bickard (1997); Geelan (1997); Grandy (1997); Longino (2001)
18. Schiebinger (1993, pp. 40–74, quotes on pp. 71, 74)
19. Lewontin, Kamin, & Rose (1984); L. Moss (2002); Rose (1997)
20. Lyons (1998); Buller (2005); Richardson (2007)
21. Michaels (2008); Oreskes & Conway (2010)
22. Dolby (1996); Rousseau (1992)
23. Bechtel & Richardson (1993); Darden (1991); Losee (1972, 2005)
24. Merton (1973)
25. Harding (1991, 1998)
26. Longino (1990, 2001); J. Solomon (1987)
27. Latour & Woolgar (1979); M. Solomon (2001)
28. Latour (1987, 1988); Latour & Woolgar (1979); Pickering (1984)
29. Hull (1988)
30. Goldman (1986, 1999, 2002)
31. Norris (1997)
32. Hardwig (1991); Shapin (1994)
33. For an overview, see Allchin (2012a)
34. Kahneman (2011); Wimsatt (2007)
35. Allchin (1999a)

36. Chalmers (1990); Davson-Galle (2002)
37. Collins & Pinch (1993, 2005)
38. McComas & Olson (1998); Committee on Conceptual Framework for the New K–12 Science Education Standards (2012)
39. Brush (2002)

CHAPTER 7: KETTLEWELL'S MISSING EVIDENCE: A STUDY IN BLACK AND WHITE

1. Robin (1992); Tufte (1997)
2. Starr (1994, pp. 7, 202)
3. Kettlewell (1959)
4. Kettlewell (1973, plate 9.1)
5. Kettlewell (1955, 1956)
6. Kettlewell (1973, pp. 134–136); also Lawrence Cook (personal communication)
7. Kettlewell (1959, p. 51)
8. Kettlewell (1973)
9. Endler (1986)
10. Kettlewell (1959, p. 51)
11. Kettlewell (1973, pp. 106–107)
12. Derry (1999, pp. 69–88); Judson (1980, pp. 112–129); Wimsatt (2007, pp. 94–132)
13. Hagen (1993, 1999); Hagen et al. (1996, pp. 1–10)
14. Rudge (1999)
15. Holmes (2001, p. 368)
16. Holmes (2001, p. 429)
17. Holmes (2001)
18. Scerri (2007); Schaffer (1989); Shapiro (1996)
19. Steinbach (1998)
20. B. Martin (1991); Toumey (1996, pp. 63–80)
21. Toumey (1996)
22. Taylor (2005); Wimsatt (2007)
23. Jungck (1996)

CHAPTER 8: TEACHING LAWLESS SCIENCE

1. Hunter (2009); Sargent (1995, pp. 95–103); Steinle (1995, pp. 334–341)
2. Frank (1980, pp. 128–139); Hunter (2009, pp. 124–138); Shapin (1994, pp. 323–333); Shapin & Schaffer (1985); Webster (1966)
3. Fazio (1992); but also note caveats from de Berg (1990, 1995)
4. Boyle (1662, p. 58)
5. Even the use of $\propto$ to notate proportionality was not introduced until the late 1700s. The main purpose of using the mathematical expressions here is to satisfy those who regard them as more scientifically rigorous. Elsewhere, Boyle describes the hypothesis even more obliquely as "the Air in that degree of density and correspondent measure of resistance to which the weight of the incumbent Atmosphere had brought it, was able to counterbalance and resist the pressure of a Mercurial Cylinder of about 29 Inches" (1662, pp. 56–57).
6. Carroll (1994); Hempel (1966, pp. 54, 58); Kosso (1992, pp. 52–60, 190); Woodward (2003, pp. 167, 236–238, 265–266); Ziman (1978, p. 32)

7. Regnault (1847)
8. Andrews (1869)
9. Gay-Lussac (1802)
10. Waals (1910)
11. e.g., Matthews (1999)
12. Cartwright (1983, pp. 45–47)
13. Kuhn (1970); Toulmin (1960, pp. 31, 63, 78–79, 87)
14. Taylor (2005); Wimsatt (2007)
15. Carroll (1994)
16. Lakoff & Johnson (1980)
17. Milton (1981); Steinle (1995, 2002)
18. Quoted in Milton (1981, p. 192)
19. Cartwright (1999)
20. Besson (2012)
21. Boyle (1662, p. 58); Shapin (1994, pp. 333–338)
22. Boyle (1661); Hunter (2009); Sargent (1995); Shapin (1994, pp. 328–330, 338–350); Steinle (1995, p. 337)
23. Boyle (1660, pp. 65, 123, 133; 1682, p. 50)
24. Matthews (1999, quote on p. 289)
25. L. Jardine (1999); also see Pyenson & Sheets-Pyenson (1999)
26. Daston & Park (1998); Henry (1997, pp. 8–41); Moran (2005); Shapin (1996, pp. 57–64)
27. Aikenhead & Michell (2011); DeKoksy & Allchin (2008); Freely (2009); Ronan (1985); Selin (1997); Teresi (2002); Turnbull(2000)
28. Kuhn (1970, pp. 23–34)
29. Giere (1995, 1999)
30. Franklin (1986); Hacking (1983); Rheinberger (1997)
31. Creager (2002); Creager, Lunbeck, & Wise (2007); Comfort (2001); Kohler (1994); Rader (2004)
32. Creager et al. (2007); Sagoff (2003); Taylor (2005)
33. Dunbar (2002); Nersessian (2002)
34. Turnbull (2000)
35. Rieß (1995)
36. Heering (1992a, b)
37. Cartwright (1983)
38. Feyerabend (2000)
39. Judson (1980); Monmonier (1991); Turnbull (1993)
40. Black (1962); Derry (1999, pp. 69–88); Judson (1980, pp. 112–129); Jackson & Mendoza (1979)
41. Levins (1966); Wimsatt (2007)
42. Giere (1999); Sismondo & Chrisman (2001); Turnbull (1993, 2000); Ziman (1978)
43. Stocklmayer & Treagust (1994)
44. Turnbull (2000, pp. 131–160)
45. Kuhn (1970, pp. 98–99, 101–102)
46. Matthews (1999, p. 262)
47. Conant & Nash (1957, pp. 55–56)
48. Turnbull (1993)
49. Haraway (1989)
50. Höttecke & Silva (2011)

CHAPTER 9: NATURE OF SCIENCE IN AN AGE OF ACCOUNTABILITY

1. Lederman et al. (1998)
2. Elby & Hammer (2001); Osborne et al. (2003, pp. 712–713); Clough & Olson (2008); Ford (2008)
3. Nott & Wellington (1998); Phillips & Norris (1999)
4. Ford (2008); Glynn & Muth (1994); Korpan et al. (1997); Murcia & Schibeci (1999); Philips & Norris (1999); Norris & Phillips (1994); Norris, Phillips, & Korpan (2003)
5. Nickerson (1998); Sadler, Chambers, & Zeidler (2004)
6. Pellegrino, Chudowsky, & Glaser (2001, p. 4)
7. Schiebinger (1993)
8. Kelly et al. (1998); Krüger, Ruhrig, & Höttecke (2013); Nott & Wellington (1998); Ruhrig, Ohlsen, & Höttecke (2013); on journaling, see Henke & Höttecke (2010)

CHAPTER 11: REKINDLING PHLOGISTON

1. Cohen (1985); Conant (1957); McCann (1978); Melhado (1989); Musgrave (1976); Thagard (1990, pp. 184, 201)
2. Herschel (1830/1987, pp. 300–301)
3. Allchin (1992); Chang (2009); Kim (2008); Odling (1871a, b); Scott (1958)
4. Brown (1866, p. 328)
5. Guinta (2001); Mamlok-Naaman, Ben-Zvi, Hofstein, Menis, & Erduran (2005); Scott (1958)
6. Brock (1993); Brown (1866); Chang (2009); Kim (2008); Odling (1871a, b); Partington & McKie (1937–1939); Scott (1958)
7. Allchin (1992); Sudduth (1978)
8. Carrier (1991, pp. 29–30); Partington (1961–1964, III, pp. 268–270)
9. See note 1.
10. Allchin (1992, 1994)
11. Diamond (1997, pp. 67–81)
12. Bodner (2002)
13. *Encyclopedia Britannica* (1771)
14. Hacking (1983)
15. Kim (2008)
16. Allchin (1994)
17. Siegfried (1964)
18. Allchin (1992); Sudduth (1978)

CHAPTER 12: DEBATING GALILEO's *DIALOGUE*: THE 1633 TRIAL

1. The retrial of Galileo is a favorite exercise, available in many versions. The version here is available online at http://galileotrial.net. Other versions include
• Solomon, J. (1989). The retrial of Galileo. In D. E. Herget (Ed.), *The History and Philosophy of Science in Science Teaching* (pp. 332–338). Tallahassee, FL: Florida State University Departments of Science Education and Philosophy. *For elementary students. Eight witnesses with highly prescribed testimony. The aim is "to illustrate how difficult it is to overturn an old theory."*
• Gregory, F. (1995). Science and religion in Western history. Seattle, WA:

History of Science Society.

For college history of science students. Four teams each for Galileo and for the Church.

- Linder, D. 2002. The Trial of Galileo [website]. Kansas City: University of Missouri, Kansas City School of Law. http://law2.umkc.edu/faculty/projects/ftrials/galileo/galileo.html

  Part of a series of famous trials, originally for law students. Aims primarily to "make the cases come alive," and thus provides more documentary material for study than guidance on a simulation with students as participants.

- Purnell, F., Jr., Pettersen, M. S., & Carnes, M. C. (2008). *The Trial of Galileo: Aristotelianism, the "New Cosmology," and the Catholic Church, 1616–1633.* New York, NY: Pearson Longman.

  For college history students. Emphasizes the shift from Aristotelian thinking to more empirical reasoning, with extensive engagement in original historical documents. Set both in 1616 and 1633, firmly within the context of the Church's perspective. Seven roles, plus duplications. Designed as a competitive game, which shifts the student's aim to winning, perhaps at the cost of deep understanding.

The version presented in this chapter differs by focusing on enriching a broad understanding of the nature of science and its cultural contexts. It thus includes a broad cross section of relevant facts surrounding the episode that would not have been part of the original proceedings. A more complete view is also required for a modern interpreter who is reflecting on what it would mean in 1633 to treat Galileo—and his scientific claims—fairly and reasonably. The version here is also designed to provide distinct and independent roles for many students, allowing each to contribute significantly to the outcome. This provides an additional indirect lesson about the need to integrate multiple perspectives for complete understanding and, thus, the virtues of delving into contrary views.

CHAPTER 13: DEBATING RACHEL CARSON'S *SILENT SPRING*, 1963

1. See "Rachel Carson & *Silent Spring*" in Hagen et al. (1996, pp. 185–196), also online at http://doingbiology.net/carson.htm

CHAPTER 14: COLLECTING CASES

1. Bonwell & Eison (1991); Mayer (2004); Michael (2006); NRC Committee on Undergraduate Education (1997)
2. Clough (2007)
3. Cliff & Nesbitt (2005)
4. Horibe (2010)
5. Carter (2007)
6. Stanley (2007)
7. Gallagher (2005)
8. Walvig (2010)
9. Allchin (2012b)
10. Lundberg, Levin, & Harrington (1999); Major & Palmer (2001)
11. Hagen et al. (1996); Herreid (2005); Herreid, Schiller, Herreid, & Wright (2012)

12. Cunningham & Helms (1998); Schwartz et al. (2004); Wong & Hodson (2009, 2010)
13. Conant (1947); Irwin (2000); Solomon et al. (1992)
14. Dimopoulos & Koulaidis (2003); Wellington (1991); Wong et al. (2008)
15. Bell (2007); Crawford (2012)
16. Osborne et al. (2003)
17. Ault & Dodick (2010)
18. Clough (2006)
19. Friedman (2009); Hagen et al. (1996, v–vii); Minstrell & Kraus (2005)
20. Carter (2007)
21. Novak (2008)
22. Gangnon (2006)
23. Gabel (2005); Henke & Höttecke (2010); Jayakamur (2006)
24. Hagen et al. (1996)
25. Friedman (2010)
26. Leaf (2011)
27. Gros (2011)
28. Novak (2008)
29. Moran (2009)
30. Leland (2007); Remillard-Hagen (2010)
31. Höttecke & Silva (2011)

| References

Adelmann, H. B. (1966). *Marcello Malpighi and the evolution of embryology*. Ithaca, NY: Cornell University Press.

Aikenhead, G. S. (1991). *Logical reasoning in science & technology*. Toronto, ON: John Wiley & Sons Canada.

Aikenhead, G., & Michell, H. (2011). *Bridging cultures: Indigenous and scientific ways of knowing nature*. Don Mills, ON: Pearson Canada.

Akerson, V. L., Abd-El-Khalick, F., & Lederman, N. G. (2000). Influence of a reflective activity-based approach on elementary teachers' conceptions of nature of science. *Journal of Research in Science Teaching, 37*, 295–317.

Alcoze, T., & Barrios-Chacon, M. (1993). *Multiculturalism in mathematics, science, and technology: Readings and activities.* Menlo Park, CA: Addison-Wesley.

Allchin, D. (1992). Phlogiston after oxygen. *Ambix, 39*, 110–116.

Allchin, D. (1993). Of squid hearts and William Harvey. *Science Teacher, 60*(7), 26–33.

Allchin, D. (1994). James Hutton and phlogiston. *Annals of Science, 51*, 615–635.

Allchin, D. (1996). In the shadows of giants: Boyle's law?, Bunsen's burner?, Petri's dish, and the politics of scientific renown. Minneapolis, MN: SHiPS Resource Center. Retrieved from http://ships.umn.edu/updates/shadows.htm

Allchin, D. (1999a). Do we see through a social microscope?: Credibility as a vicarious selector. *Philosophy of Science 60*(Proceedings), S287–S298.

Allchin, D. (1999b). Values in science: An educational perspective. *Science & Education, 8*, 1–12.

Allchin, D. (2001). Error types. *Perspectives on Science, 9*, 38–59.

Allchin, D. (2005a). Genes "R" us? *American Biology Teacher, 67*, 244–246.

Allchin, D. (2005b). The Committee on Uranium, 1939. Minneapolis, MN: SHiPS Resource Center. Retrieved from http://ships.umn.edu/modules/phys/uranium

Allchin, D. (2005c). The dilemma of dominance. *Biology and Philosophy, 20*, 427–451.

Allchin, D. (2006a). Wallowing in the wastebin. *Science, 311*, 781–782.

Allchin, D. (2006b). Why respect for history—and historical error—matters. *Science & Education, 15*, 91–111.

Allchin, D. (2008). Naturalizing as an error-type in biology. *Filosofia e História da Biologia, 3*, 95–117.

Allchin, D. (2012a). Skepticism & the architecture of trust. *American Biology Teacher, 74*, 358–362.

Allchin, D. (2012b). Teaching the nature of science through scientific errors. *Science Education, 96*, 904–926.

American Association for the Advancement of Science, Project 2061. (1993).

Benchmarks for scientific literacy. New York, NY: Oxford University Press.

American Association for the Advancement of Science, Project 2061. (2009). *Benchmarks for scientific literacy*, Chap. 1. Retrieved from http://www.project2061. org/publications/bsl/online/index.php?chapter=1

American Chemical Society. (2006). *Chemistry in the community* (5th ed). New York, NY: W.H. Freeman.

Anand, P. (2002). Decision-making when science is ambiguous. *Science, 295*, 1839.

Andrews, T. (1869). On the continuity of the gaseous and liquid states of matter. *Proceedings of the Royal Society of London, 18*, 42–45.

Arora, A., & Kean, E. (1992). Perceptions of doing science: Science teachers' reflections. In S. Hills (Ed.), *The history and philosophy of science in science education* (Vol. 1, pp. 53–68). Kingston, ON: Mathematics, Science, Technology and Teacher Education Group and Faculty of Education, Queen's University.

Asimov, I. (1964). *A short history of biology*. Garden City, NY: Natural History Press.

Ault, C. R., Jr., & Dodick, J. (2010). Tracking the footprints puzzle: The problematic persistence of science-as-process in teaching the nature and culture of science. *Science Education, 94*, 1092–1122.

Barba, R. H., Pang, V. O., & Tran, M. T. (1992). Who *really* discovered aspirin? *Science Teacher, 59*(5), 26–27.

Barbacci, S., Bugini, A., Brenni, P., & Giatti, A. (2011). The discovery of dynamic electricity and the transformation of distance communications. Florence, Italy: Fondazione Scienza e Tecnica. Retrieved from http://hipstwiki.wetpaint.com/page/Case+Study+1

Barker, M. (1995). Esteeming prior knowledge: Historical views and students' intuitive ideas about plant nutrition. In F. Finley, D. Allchin, D. Rhees, & S. Fifield (Eds.), *Proceedings, Third International History, Philosophy and Science Teaching Conference* (pp. 97–102). Minneapolis: University of Minnesota Office of Continuing Education.

Barth, M. (1995). Snell's law/models of light: A historic approach to teaching science. In F. Finley, D. Allchin, D. Rhees, & S. Fifield (Eds.), *Proceedings, Third International History, Philosophy and Science Teaching Conference* (pp. 103–114). Minneapolis: University of Minnesota Office of Continuing Education.

Bauer, H. H. (1992). *Scientific literacy and the myth of the scientific method*. Urbana: University of Illinois Press.

Baumel, H. B., & Berger, J. J. (1973). *Biology: Its people and its papers*. Washington, DC: National Science Teachers Association.

Bazerman, C. (1988). *Shaping written knowledge: The genre and activity of the experimental article in science*. Madison: University of Wisconsin Press.

Bechtel, W. (2006). *Discovering cell mechanisms: The creation of modern cell biology*. New York, NY: Cambridge University Press.

Bechtel, W., & Richardson, R. C. (1993). *Discovering complexity: Decomposition and localization as strategies in scientific research*. Princeton, NJ: Princeton University Press.

Becker, B. J. (1992). Incorporating primary source material in secondary and college science curricula. In S. Hills (Ed.), *The history and philosophy of science in science education* (Vol. 1, pp. 69–76). Kingston, ON: Mathematics, Science, Technology and Teacher Education Group and Faculty of Education, Queen's University.

Bell, R. L. (2007). *Teaching the nature of science through process skills: Activities for grades 3–8*. Upper Saddle River, NJ: Pearson Education.

Besson, U. (2012). The history of the cooling law: When the search for simplicity can be an obstacle. *Science & Education, 21,* 1085–1110.

Beveridge, W. I. B. (1950). *The art of scientific investigation.* New York, NY: Random House.

Biagioli, M. (1993). *Galileo, courtier.* Chicago, IL: University of Chicago Press.

Bickhard, M. H. (1997). Constructivisms and relativisms: A shopper's guide. *Science and Education, 6,* 29–42.

Black, M. (1962). Models and archetypes. In *Models and metaphors* (pp. 219–243). Ithaca, NY: Cornell University Press.

Bodner, G. (2002). Thermite reaction. Lecture demonstration sheets. Purdue, IN: Purdue University Division of Chemical Education. Retrieved from http://chemed.chem.purdue.edu/genchem/demosheets/5.3.html

Bonwell, C. C., & Eison, J. A. (1991). *Active learning: Creating excitement in the classroom.* AEHE-ERIC Higher Education Report No.1. Washington, DC: Jossey-Bass.

Booth, V. (1993). *Communicating in science* (2nd ed.). Cambridge, England: Cambridge University Press.

Boring, E. G. (1954). The nature and history of experimental control. *American Journal of Psychology, 67,* 573–589.

Bourdieu, P. (1975). The specificity of the scientific field and the social conditions of the progress of reason (R. Nice, Trans.). *Social Science Information, 14*(6), 19–47.

Boyle, R. (1660). *New experiments physico-mechanicall, touching the spring of the air* […]. Oxford, England: H. Hall.

Boyle, R. (1661). *Certain physiological essays.* London, England: H. Herringman.

Boyle, R. (1662). *New experiments physico-mechanical, touching the air* […]. London, England: H. Hall.

Boyle, R. (1673). An essay about the origins & virtues of gems [1672]. In *Essays of the strange subtilty, great efficacy, determinate nature, of effluviums* […]. London, England: W.G. for M. Pitt.

Boyle, R. (1682). *A defence of the doctrine touching the spring and weight of the air.* London, England: Miles Flesher.

Boyle, R. (1699). *The works of the Honourable Robert Boyle, Esq., epitomized.* London, England: J. Phillips and J. Taylor.

Brannigan, A. (1981). *The social basis of scientific discoveries.* New York, NY: Cambridge University Press.

Broad, W., & Wade, N. (1982). *Betrayers of the truth: Fraud and deceit in the halls of science.* New York, NY: Simon and Schuster.

Brock, W. H. (1993). *The Norton history of chemistry.* New York, NY: W.W. Norton.

Brown, A. C. (1866). Note on phlogistic theory. *Proceedings of the Royal Society of Edinburgh, 5*(V), 328–330.

Brush, S. G. (1974). Should the history of science be rated X? *Science, 183,* 1164–1172.

Brush, S. G. (1995). Scientists as historians. *Osiris, 10,* 215–231.

Brush, S. G. (2002). Cautious revolutionaries: Maxwell, Planck, Hubble. *American Journal of Physics, 70,* 119–127.

Buller, D. J. (2005). *Adapting minds: Evolutionary psychology and the persistent quest for human nature.* Cambridge, MA: MIT Press.

Burdett, P. (1989). Adventures with N-rays: An approach to teaching about scientific theory and theory evaluation. In R. Millar (Ed.), *Doing science: Images of science in science education* (p. 180). Bristol, PA: Taylor & Francis.

Burke, J. (1978). *Connections.* London, England: Macmillan.

Burke, J. (1996). *The pinball effect.* London, England: Macmillan.

Burke, J. (1999). *The knowledge web.* New York, NY: Simon & Schuster.

Burke, J. (2003). *Twin tracks.* New York, NY: Simon & Schuster.

Buster, S. (1995). Using the historical approach to avoid misconceptions in students' understanding of electricity. In F. Finley, D. Allchin, D. Rhees, & S. Fifield (Eds.), *Proceedings, Third International History, Philosophy and Science Teaching Conference* (pp. 170–175). Minneapolis: University of Minnesota Office of Continuing Education.

Butterfield, H. H. (1959). *The Whig interpretation of history.* London, England: G. Bell and Sons.

Campbell, N. A., Reece, J. B., & Mitchell, L. G. (1999). *Biology* (5th ed.). Menlo Park, CA: Benjamin/Cummings.

Carrier, M. (1991). What is wrong with the miracle argument? *Studies in History and Philosophy of Science A, 22,* 23–36.

Carroll, J. W. (1994). *Laws of nature.* Cambridge, England: Cambridge University Press.

Carter, K. (2007). Picture perfect?: Making sense of the vast diversity of life on Earth. Minneapolis, MN: SHiPS Resource Center. Retrieved from http://ships. umn.edu/modules/biol/lincei.htm

Cartwright, N. (1983). *How the laws of physics lie.* Oxford, England: Clarendon Press.

Cartwright, N. (1999). *The dappled world: A study of the boundaries of science.* Cambridge, England: Cambridge University Press.

Carvalho, A. M. P. (1990). The influence of the history of momentum and its conservation on the teaching of mechanics in high schools. In D. Herget (Ed.), *More History and Philosophy of Science in Science Teaching: Proceedings of the First International Conference* (pp. 212–219). Tallahassee: Florida State University Departments of Science Education and Philosophy.

Chalmers, A. F. (1990). *Science and its fabrication.* Minneapolis: University of Minnesota Press.

Chambers, D. W. (1983). Stereotypic images of the scientist: The draw-a-scientist test. *Science Education, 67,* 255–265.

Chang, H. (2009). We have never been Whiggish (about phlogiston). *Centaurus, 51,* 239–264.

Clement, J. (1983). A conceptual model discussed by Galileo and used intuitively by physics students. In D. Gentner & A. L. Stevens (Eds.), *Mental models* (pp. 325–340). Hillsdale, NJ: Lawrence Erlbaum.

Cliff, W. H., & Nesbitt, L. M. (2005). An open or shut case? Contrasting approaches to case study design. *Journal of College Science Teaching, 34*(4), 14–17.

Clough, M. P. (2006). Learners' responses to the demands of conceptual change: Considerations for effective nature of science instruction. *Science & Education, 15,* 463–494.

Clough, M. P. (2007). Teaching the nature of science to secondary and post-secondary students: Questions rather than tenets. *Pantaneto Forum, 25.* Retrieved from http://www.pantaneto.co.uk/issue25/clough.htm

Clough, M. P. (2010). The story behind the science: Bringing science and scientists to life in post-secondary science education. *Science & Education, 20,* 701–717.

Clough, M. P., & Olson, J. K. (2008). Teaching and assessing the nature of science: An introduction. *Science & Education, 17,* 143–145.

Cohen, I. B. (1985). *Revolution in science.* Cambridge, MA: Harvard University Press.

Cole, S. (1992). *Making science: Between nature and society.* Cambridge, MA: Harvard University Press.

Collins, H. (1992). *Changing order: Replication and induction in scientific practice.* Chicago, IL: University of Chicago Press.

Collins, H. (2004). *Gravity's shadow: The search for gravitational waves.* Chicago, IL: University of Chicago Press.

Collins, H., & Pinch, T. (1993). *The golem: What everyone should know about science.* Cambridge, England: Cambridge University Press.

Collins, H., & Pinch, T. (2005). *Dr. Golem: How to think about medicine.* Chicago, IL: University of Chicago Press.

Collison, G. (1992). *Science and the cosmos* (Doctoral dissertation). University of Massachussetts, Amherst.

Comfort, N. (2001). *The tangled field: Barbara McClintock's search for the patterns of genetic control.* Cambridge, MA: Harvard University Press.

Committee on Conceptual Framework for the New K–12 Science Education Standards, National Research Council. (2012). *A framework for K–12 science education.* Washington, DC: National Academies Press. Retrieved from http://www.nap.edu/catalog.php?record_id=13165.

Committee on Science, Engineering, and Public Policy, National Academy of Sciences, National Academy of Engineering, and Institute of Medicine. (2009). *On being a scientist,* 3rd ed. Washington, DC: National Academy Press. Retrieved from http://www.nap.edu/catalog.php?record_id=12192

Committee on Undergraduate Science Education, National Research Council. (1997). Science teaching reconsidered: A handbook. Washington, DC: National Academy Press. Retrieved from http://www.nap.edu/openbook.php?isbn=0309054982

Conant, J. B. (1947). *On understanding science: An historical approach.* New Haven, CT: Yale University Press.

Conant, J. B. (1957). The overthrow of the phlogistic theory. In J. B. Conant & L. K. Nash (Eds.), *Harvard case histories in experimental science* (Vol. 1, pp. 65–115). Cambridge, MA: Harvard University Press.

Conant, J. B., & Nash, L. K. (Eds.). (1957). *Harvard case histories in experimental science* (Vols. 1 and 2). Cambridge, MA: Harvard University Press.

Confrey, J., & Smith, E. (1989). Alternative representations of ratio: The Greek concept of anthyphairesis and modern decimal notation. In D. Herget (Ed.), *The History and Philosophy of Science in Science Teaching: Proceedings of the First International Conference* (pp. 71–82). Tallahassee: Florida State University Departments of Science Education and Philosophy.

Copeland, M., Larson, D., & Morton, D. (1996). Polymers & serendipity: Case studies. Minneapolis, MN: SHiPS Resource Center. Retrieved from http://ships.umn.edu/modules/scimath/polymer1.htm

Corcos, A. F., & Monaghan, F. V. (1993). *Gregor Mendel's experiments on plant hybrids: A guided study.* New Brunswick, NJ: Rutgers University Press.

Cordingley, B. (2001). *In your face: What facial features reveal about the people you know and love.* East Rutherford, NJ: New Horizon Press.

Craven, J. A., Hand, B., & Vaughan, P. (2002). Assessing explicit and tacit conceptions of the nature of science among preservice elementary teachers. *International Journal of Science Education, 24,* 785–802.

Crawford, B. A. (2012). Moving the essence of inquiry into the classroom: Engaging teachers and students in authentic science. In K. C. D. Tan & M. Kim (Eds.), *Issues and challenges in science education research: Moving forward* (pp. 25–42). Dordrecht, The Netherlands: Springer.

Creager, A. N. H. (2002). *The life of a virus: Tobacco mosaic virus as an experimental model, 1930–1965.* Chicago, IL: University of Chicago Press.

Creager, A. N. H., Lunbeck, E., & Wise, M. N. (Eds.). (2007). *Science without laws: Model systems, cases, exemplary narratives.* Durham, NC: Duke University Press.

Crofton, I. (2010). *The totally useless history of science.* London, England: Quercus.

Cunningham, C. M., & Helms, J. V. (1998). Sociology of science as a means to a more authentic, inclusive science education. *Journal of Research in Science Teaching, 35,* 483–499.

Daempfle, P. A. (2013). *Good science, bad science, pseudoscience , and just plain bunk: How to tell the difference.* Lanham, MD: Rowman & Littlefield.

Darden, L. (1991). *Theory change in science: Strategies from Mendelian genetics.* New York, NY: Oxford University Press.

Daston, L., & Park, K. (1998). *Wonders and the order of nature.* New York, NY: Zone Books.

Davson-Galle, P. (2002). Science, values and objectivity. *Science & Education, 11,* 191–202.

de Berg, K. C. (1990). The historical development of the pressure-volume law for gases. *Australian Science Teachers Journal, 36,* 14–20.

de Berg, K. C. (1995). Revisiting the pressure-volume law in history—What can it teach us about the emergence of mathematical relationships in science? *Science & Education, 4,* 47–64.

De Mey, M. (1982). *The cognitive paradigm.* Chicago, IL: University of Chicago Press.

DeKosky, R., & Allchin, D. (Eds.). (2008). *An introduction to history of science in non-Western traditions* (2nd ed.). Seattle, WA: History of Science Society. (Reprinted online at http://www.hssonline.org/publications/NonWesternPub/introchapters.html)

DeMeo, S. (1992). An immaculate conception: Le Chatelier and equilibrium. In S. Hills (Ed.), *The history and philosophy of science in science education* (Vol. 1, pp. 227–237). Kingston, ON: Mathematics, Science, Technology and Teacher Education Group and Faculty of Education, Queen's University.

Derry, G. N. (1999). *What science is and how it works.* Princeton, NJ: Princeton University Press.

Di Trocchio, F. (1991). Mendel's experiments: A reinterpretation. *Journal of the History of Biology, 24,* 485–519.

Diamond, J. M. (1997). *Guns, germs, and steel: The fates of human societies.* New York, NY: W.W. Norton.

Dimopoulos, K., & Koulaidis, V. (2003). Science and technology education for citizenship: The potential role of the press. *Science Education, 87,* 241–256.

Dobbs, B. J. T. (1975). *The foundations of Newton's alchemy: Or, 'The hunting of the greene lyon.'* Cambridge, England: Cambridge University Press.

Dolby, R. G. A. (1996). A theory of pathological science. In *Uncertain knowledge* (pp. 227–244). Cambridge, England: Cambridge University Press.

Dolphin, G. (2009). Evolution of the theory of the earth: A contextualized approach for teaching the history of the theory of plate tectonics to ninth grade students. *Science & Education, 18,* 425–441.

Driscoll, M., Dunn, E., Siems, D., & Swanson, B. K. (2009). *Charles Darwin, the Copley Medal, and the rise of naturalism 1862–1864.* New York, NY: Pearson/Longman.

Ducheyne, S. (2005). Joan Baptista van Helmont and the question of experimental modernism. *Physis: Rivista Internazionale di Storia della Scienza, 42,* 305–332. Retrieved from http://logica.ugent.be/steffen/Physis%20-%20Van%20Helmont.pdf

Ducheyne, S. (2008). J. B. van Helmont. In A. Pilchak (Ed.), *New dictionary of scientific biography* (Vol. 3, pp. 277–281). Detroit, MI: Thomson Gale.

Dunbar, K. N. (2002). Understanding the role of cognition in science: The Science as Category framework. In P. Carruthers, S. Stich, & M. Siegal (Eds.), *The cognitive basis of science* (pp. 154–170). Cambridge, England: Cambridge University Press.

Duschl, R. (1987). Causes of earthquakes: An inquiry into the plausibility of competing explanations. *Science Activities, 24*(3), 8–14.

Duschl, R. (1990). *Restructuring science education: The importance of theories and their development.* New York, NY: Teachers College Press.

Duschl, R. (1993). Causes of earthquakes. In K. Kachman & C. Sutton (Eds.), *Curriculum reform in college science* (pp. 185–211). Pittsburgh, PA: University of Pittsburgh Department of Education.

Duschl, R., Schweingruber, H. A., & Shouse, A. W. (Eds.). (2007). *Taking science to school: Learning and teaching science in grades K–8.* Washington, DC: National Academies Press.

Eakin, R. M. (1975). *Great scientists speak again.* Berkeley: University of California Press.

Elby, A., & Hammer, D. (2001). On the substance of a sophisticated epistemology. *Science Education, 85,* 554–567.

Elkana, Y., & Goodfield, J. (1968). Harvey and the problem of the "capillaries." *Isis, 59,* 61–73.

Elliott, P. (2006). Reviewing newspaper articles as a technique for enhancing the scientific literacy of student-teachers. *International Journal of Science Education, 28,* 1245–1265.

Encyclopedia Britannica. (1771). Of the phlogiston. I, 68–69. Retrieved from books. google.com/books?id=xnW408_9_SMC&pg=PA68#v=onepage&q&f=false

Endersby, J. (2007). *A guinea pig's history of biology.* Cambridge, MA: Harvard University Press.

Endler, J. A. (1986). *Natural selection in the wild.* Princeton, NJ: Princeton University Press.

Falkenberg, L. (2009). Teacher brings Darwin to life in lessons. *Houston Chronicle,* 16 February. Retrieved from http://www.chron.com/news/falkenberg/article/Teacher-brings-Darwin-to-life-in-lessons-1721943.php

Faria, C., Pereira, G., & Chagas , I. (2011). A naturalist who became a pioneer of experimental marine oceanography in Portugal: Assets for science education. Lisbon, Portugal: University of Lisbon and The Aquarium Vasco da Gama. Retrieved from http://hipstwiki.wetpaint.com/page/King+D+Carlos%2C+a+n aturalist+oceanographer

Fazio, F. (1992). Using Robert Boyle's original data in physics and chemistry classrooms. *Journal of College Science Teaching, 21,* 363–365.

Feder, K. L. (1999). *Frauds, myths and mysteries: Science and pseudoscience in archaeology* (3rd ed.). Mountain View, CA: Mayfield.

Fee, E. (1979). Nineteenth-century craniology: The study of the female skull. *Bulletin of the History of Medicine, 53,* 415–433.

Feyerabend, P. (2000). *Conquest of abundance: A tale of abstraction versus the richness of being.* Chicago, IL: University of Chicago Press.

Finson, K. D. (2002). Drawing a scientist: What we do and do not know after fifty years of drawings. *School Science and Mathematics, 102,* 335–345.

Fischer, E. P., & Lipson, C. (1988). *Thinking about science: Max Delbrück and the origins of molecular biology.* New York, NY: W.W. Norton.

Flannery, M. (2001). The white coat: A symbol of science and medicine as a male pursuit. *Thyroid, 11,* 947–951.

Ford, M. (2008). 'Grasp of practice' as a reasoning resource for inquiry and nature of science understanding. *Science & Education, 17,* 147–177.

Franco, C. (1992). History of science and psychogenesis: A comparative study on Galileo's free fall law and ideas of speed in the child. In S. Hills (Ed.), *The history and philosophy of science in science education* (Vol. 1, pp. 323–330). Kingston, ON: Mathematics, Science, Technology and Teacher Education Group and Faculty of Education, Queen's University.

Frank, R. G., Jr. (1980). *Harvey and the Oxford physiologists: Scientific ideas and social interaction.* Berkeley: University of California Press.

Franklin, A. (1986). *The neglect of experiment.* Cambridge, England: Cambridge University Press.

Franklin, A. (1991). *Experiment, right or wrong.* Cambridge, England: Cambridge University Press.

Franklin, A. (1997). Calibration. *Perspectives on Science, 5,* 31–80.

Franklin, A., Edwards, A. W. F., Fairbanks, D. J., Hartl, D. L., & Seidenfeld, T. (2008). *Ending the Mendel-Fisher controversy.* Pittsburgh, PA: University of Pittsburgh Press.

Fredrickson, D. S. (2001). *The recombinant DNA controversy, a memoir: Science, politics, and the public interest, 1974–1981.* Washington, DC: ASM Press.

Freely, J. (2009). *Aladdin's lamp: How Greek science came to Europe through the Islamic world.* New York, NY: Vintage.

Friedman, A. (2009, June). *But what does it look like? Exploring the use of the history of science in one high school's biology classroom.* Paper presented at the 9th International History, Philosophy and Science Teaching Conference, Notre Dame, Indiana.

Friedman, A. (2010). *Alfred Russel Wallace & the origin of new species.* Minneapolis, MN: SHiPS Resource Center. Retrieved from http://ships.umn.edu/modules/biol/wallace.htm

Fritze, R. H. (2009). *Invented knowledge: False history, fake science and pseudo-religions.* London, England: Reaktion Books.

Gabel, K. (2005). The earliest microscopes. Minneapolis, MN: SHiPS Resource Center. Retrieved from http://ships.umn.edu/modules/biol/m-scope.htm

Galison, P. (1987). *How experiments end*. Chicago, IL: University of Chicago Press.

Gallagher, A. (2005). Freedom from decision: The psychology of B. F. Skinner. Minneapolis, MN: SHiPS Resource Center. Retrieved from http://ships.umn.edu/modules/biol/skinner.htm

Gangnon, D. (2006). The chemistry of fabric dyes: Chemical reactions of the king of colors: indigo. Minneapolis, MN: SHiPS Resource Center. Retrieved from http://ships.umn.edu/modules/chem/indigo.htm

Gaon, S., & Norris, S. P. (2001). The undecidable grounds of scientific expertise: Science education and the limits of intellectual independence. *Journal of Philosophy of Education, 35*, 187–201.

Gardner, M. (1957). *Fads and fallacies in the name of science*. New York, NY: Dover.

Gardner, R. (1990). *Famous experiments you can do*. New York, NY: Franklin Watts.

Garrison, J. (1995). An alternative to von Glasersfeld's subjectivism in science education: Deweyan social constructivism. In F. Finley, D. Allchin, D. Rhees, & S. Fifield (Eds.), *Proceedings, Third International History, Philosophy and Science Teaching Conference* (pp. 432–440). Minneapolis: University of Minnesota Office of Continuing Education.

Gauld, C. F. (1977). The role of history in the teaching of science. *Australian Science Teachers Journal, 23*(3), 47–52.

Gauld, C. F. (1991). History of science, individual development and science teaching. *Research in Science Education, 21*, 133–140.

Gauld, C. F. (1992). The historical anecdote as a "caricature": A case study. *Research in Science Education, 22*, 149–156.

Gauld, C. F. (1995). The Newtonian solution to the problem of impact in the 17th and 18th centuries and teaching Newton's third law today. In F. Finley, D. Allchin, D. Rhees, & S. Fifield (Eds.), *Proceedings, Third International History, Philosophy and Science Teaching Conference* (pp. 441–452). Minneapolis: University of Minnesota Office of Continuing Education.

Gay-Lussac, J.-L. (1802). The expansion of gases by heat. *Annales de Chimie, 43*, 137–175.

Geelan, D. R. (1997). Epistemological anarchy and the many forms of constructivism. *Science & Education, 6*, 15–28.

Giere, R. N. (1984). *Understanding scientific reasoning* (2nd ed.). New York, NY: Holt, Rinehart and Winston.

Giere, R. N. (1995). The skeptical perspective: Science without laws of nature. In F. Weinert (Ed.), *Laws of nature: Essays on the philosophical, scientific and historical dimensions* (pp. 120–138). Berlin, Germany: De Gruyter.

Giere, R. N. (1999). *Science without laws*. Chicago, IL: University of Chicago Press.

Gieryn, T. F. (1999). *Cultural boundaries of science: Credibility on the line*. Chicago, IL: University of Chicago Press.

Gilovich, T. (1991). *How we know what isn't so: The fallibility of human reason in everyday life*. New York, NY: Free Press.

Glen, W. (Ed.). (1994). *The mass-extinction debates: How science works in a crisis*. Palo Alto, CA: Stanford University Press.

Glynn, S. M., & Muth, K. D. (1994). Reading and writing to learn science: Achieving scientific literacy. *Journal of Research in Science Teaching, 9*, 1057–1069.

Goldman, A. I. (1986). *Epistemology and cognition.* Cambridge, MA: Harvard University Press.

Goldman, A. I. (1999). *Knowledge in a social world.* Oxford, England: Oxford University Press.

Goldman, A. I. (2002). *Pathways to knowledge: Private and public.* Oxford, England: Oxford University Press.

Goodrich, D. (2011, January 26). Parents have valid reasons to be wary of vaccinations. *Star Tribune,* p. A10.

Gould, S. J. (1977). On heroes and fools in science. In *Ever since Darwin: Reflections in natural history* (pp. 201–206). New York, NY: W.W. Norton.

Gould, S. J. (1980). *The panda's thumb: More reflections in natural history.* New York, NY: W.W. Norton.

Gould, S. J. (1981). *The mismeasure of man.* New York, NY: W.W. Norton.

Gould, S. J. (1983). Science and Jewish immigration. In *Hen's teeth and horse's toes: Further reflections in natural history* (pp. 291–302). New York, NY: W.W. Norton.

Gould, S. J. (1990). *Wonderful life: The Burgess Shale and the nature of history.* New York, NY: W.W. Norton.

Grandy, R. E. (1997). Constructivisms and objectivity: Disentangling metaphysics from pedagogy. *Science & Education, 6,* 43–53.

Grapí, P. (1992). Study of some conditions involved in the construction of the concept of "chemical reversibility." In S. Hills (Ed.), *The history and philosophy of science in science education* (Vol. 1, pp. 449–457). Kingston, ON: Mathematics, Science, Technology and Teacher Education Group and Faculty of Education, Queen's University.

Gregory, A. (2001). *Harvey's heart: The discovery of blood circulation.* Cambridge, England: Icon Books.

Gros, P. P. (2011). Carleton Gajdusek & kuru. Minneapolis, MN: SHiPS Resource Center. Retrieved from http://ships.umn.edu/modules/biol/gajdusek.htm

Guinta, C. J. (2001). Using history to teach scientific method: The role of errors. *Journal of Chemical Education, 78,* 623–627.

Habben, D., Mehrle, U., Heering, P., Meya, J., Reiß, F., Rohlfs, G., & Sibum, H. O. (1994). *Vom Bernstein zur Volta-Säule: Geschichte der Elektrizität im Unterricht* [*From amber to the Voltaic pile: History of electricity in the classroom*]. Marburg, Germany: Redaktions-Gemeinshaft Soznat.

Hacking, I. (1983). *Representing and intervening: Introductory topics in the philosophy of natural science.* Cambridge, England: Cambridge University Press.

Hacking, I. (1990). *The taming of chance.* Cambridge, England: Cambridge University Press.

Hagen, J. B. (1993). Kettlewell and the peppered moths reconsidered. *BioScene, 19*(3), 3–9.

Hagen, J. B. (1999). Retelling experiments: H.B.D. Kettlewell's studies of industrial melanism in peppered moths. *Biology & Philosophy, 14,* 39–54.

Hagen, J. [B.], Allchin, D., & Singer, F. (1996). *Doing biology.* Glenview, IL: HarperCollins.

Haggerty, S. M. (1992). Student teachers' perceptions of science and science teaching. In S. Hills (Ed.), *The history and philosophy of science in science education* (Vol. 1, pp. 483–494). Kingston, ON: Mathematics, Science, Technology and Teacher Education Group and Faculty of Education, Queen's University.

Handschy, M. (1982). Re-examination of the 1887 Michelson-Morley experiment. *American Journal of Physics, 50,* 987–990.

Haraway, D. (1989). *Primate visions: Gender, race, and nature in the world of modern science.* New York, NY: Routledge.

Harding, S. (1991). *Whose science? Whose knowledge? Thinking from women's lives.* Ithaca, NY: Cornell University Press.

Harding, S. (1998). *Is science multicultural? Postcolonialisms, feminisms, and epistemologies.* Bloomington: Indiana University Press.

Hardwig, J. (1991). The role of trust in knowledge. *Journal of Philosophy, 88,* 693–708.

Hartl, D. L., & Orel, V. 1992. What did Gregor Mendel think he discovered? *Genetics, 131,* 245–253. (Reprint retrieved from http://www.mendelweb.org/MWhartl.intro.html)

Harvey, W. (1628/1952). *On the motion of the heart and blood in animals* (R. Willis, Trans.). (Reprinted in *Great books of the Western World,* Vol. 28, 1952, Chicago, IL: Encyclopedia Britannica.)

Harvey, W. (1649/1952). *Anatomical disquisition on the circulation of the blood, to Jean Riolan* (R. Willis, Trans.). (Reprinted in *Great books of the Western World,* Vol. 28, 1952, Chicago, IL: Encyclopedia Britannica.)

Heering, P. (1992a). On Coulomb's inverse square law. *American Journal of Physics, 60,* 988–996.

Heering, P. (1992b). On J. P. Joule's determination of the mechanical equivalent of heat. In S. Hills (Ed.), *The history and philosophy of science in science education* (Vol. 1, pp. 495–505). Kingston, ON: Mathematics, Science, Technology and Teacher Education Group and Faculty of Education, Queen's University.

Helmont, J. B. van. (1664). *Works.* London: Lodowick Lloyd.

Hempel, C. (1966). *Philosophy of natural science.* Englewood Cliffs, NJ: Prentice-Hall.

Henke, A., & Höttecke, D. (2010). History of electricity (episodes 1–5). Retrieved from http://hipstwiki.wetpaint.com/page/history+of+electricity

Henry, J. (1997). *The Scientific Revolution and the origins of modern science.* New York, NY: St. Martin's Press.

Herreid, C. F. (2005). The interrupted case method. *Journal of College Science Teaching, 35*(2), 4–5.

Herreid, C. F., Schiller, N. A., Herreid, K. F., & Wright, C. (2012). My favorite case and what makes it so. *Journal of College Science Teaching, 42,* 70–75.

Herschel, J. F. W. (1830/1987). *A preliminary discourse on the study of natural philosophy.* Reprinted with a foreword by A. Fine, 1987, Chicago, IL: University of Chicago Press.

Hershey, D. (1991). Digging deeper into Helmont's famous willow tree experiment. *American Biology Teacher, 53*(8), 458–460.

Hershey, D. (2003). Misconceptions about Helmont's willow experiment. *Plant Science Bulletin, 49,* 78–83.

Hess, D. J. (1997). *Science studies: An advanced introduction.* New York: New York University Press.

Hindmarsh, R., & Gottweis, H. (Eds.). (2005). *Recombinant regulation: The Asilomar legacy 30 years on. Science as Culture, 14,* 299–307.

History and Philosophy in Science Teaching Consortium [HIPST]. (2010). HIPST developed cases. Retrieved from http://hipstwiki.wetpaint.com/page/

hipst+developed+cases

Ho, D. (1999, March). Bacteriologist Alexander Fleming. *Time 100.* Retrieved from http://www.time.com/time/time100/scientist/profile/fleming.html

Hodson, D. (2008). *Towards scientific literacy.* Rotterdam, The Netherlands: Sense.

Holmes, F. L. (2001). *Meselson, Stahl, and the replication of DNA: A history of 'the most beautiful experiment in biology.'* New Haven, CT: Yale University Press.

Holton, G. (1978a). On the psychology of scientists, and their social concerns. In *The scientific imagination: Case studies* (pp. 229–252). Cambridge, England: Cambridge University Press.

Holton, G. (1978b). Subelectrons, presuppositions and the Millikan-Ehrenhaft dispute. *Historical Studies in the Physical Sciences, 9,* 166–224. (Reprinted in *The scientific imagination: Case studies* (pp. 25–83), 1978, Cambridge, England: Cambridge University Press.)

Holton, G. (1988). Einstein, Michelson and the 'crucial' experiment. In *Thematic origins of scientific thought, Kepler to Einstein* (2nd ed., pp. 279–370). Cambridge, MA: Harvard University Press.

Honey, J. M. (1992). Progression in understanding history: Its effects upon understanding the nature of science. In S. Hills (Ed.), *The history and philosophy of science in science education* (Vol. 1, pp. 507–516). Kingston, ON: Mathematics, Science, Technology and Teacher Education Group and Faculty of Education, Queen's University.

Horibe, S. (2010). Robert Hooke, Hooke's law & the watch spring. Minneapolis, MN: SHiPS Resource Center. Retrieved from http://ships.umn.edu/modules/phys/hooke.htm

Höttecke, D., & Silva, C. C. (2011). Why implementing history and philosophy in school science education is a challenge: An analysis of obstacles. *Science & Education, 20,* 293–316.

Howe, E. M. (2009). Henry David Thoreau, forest succession & the nature of science. *American Biology Teacher, 71,* 397–404.

Howe, E. [M.] (2010). Teaching with the history of science: Understanding sickle-cell anemia and the nature of science. Retrieved from http://www1.assumption.edu/users/emhowe/Sickle Case/start.htm

Hull, D. (1988). *Science as a process.* Chicago, IL: University of Chicago Press.

Hunter, M. (2009). *Boyle: Between God and science.* New Haven, CT: Yale University Press.

Ihde, D. (1997). Why not science critics? *International Studies in Philosophy, 29,* 45–54.

Irwin, A. R. (2000). Historical case studies: Teaching the nature of science in context. *Science Education, 84,* 5–26.

Irzik, G., & Irzik, S. (2002). Which multiculturalism? *Science & Education, 11,* 393–403.

Jackson, M., & Mendoza, K. (Producers). (1979). *The search for solutions* [video series]. Bartlesville, OK: Phillips Petroleum/Karol Media.

Jardine, L. (1999). *Ingenious pursuits: Building the Scientific Revolution.* New York, NY: Doubleday.

Jardine, N. (1994). A trial of Galileos. *Isis, 85,* 279–283.

Jayakumar, S. (2006). Splendor of the spectrum: Bunsen, Kirchoff and the origin of spectroscopy. Minneapolis, MN: SHiPS Resource Center. Retrieved from http://ships.umn.edu/modules/chem/spectro.htm

Jensen, M. S., & Finley, F. N. (1995). Changes in students' understanding of evolution by natural selection when using a historically rich curriculum. In F. Finley, D. Allchin, D. Rhees, & S. Fifield (Eds.), *Proceedings, Third International History, Philosophy and Science Teaching Conference* (pp. 564–571). Minneapolis: University of Minnesota Office of Continuing Education.

Johnson, D. W., Johnson, R. T., & Holubec, E. J. (1986). *Circles of learning: Cooperation in the classroom.* Edina, MN: Interaction Book Company.

Johnson, S. (2008). *The invention of air.* New York, NY: Riverhead Books.

Johnson, S., & Stewart, J. (1990). Using philosophy of science in curriculum development: An example from high school genetics. *International Journal of Science Education, 12,* 297–307.

Jones, J. H. (1981). *Bad blood: The Tuskegee syphilis experiment.* New York, NY: Free Press.

Judson, H. F. (1980). *The search for solutions.* New York, NY: Holt, Rinehart, and Winston.

Judson, H. F. (2004). *The great betrayal: Fraud in science.* Orlando, FL: Houghton Mifflin Harcourt.

Juel, R. (2011, August 27). Don't be so certain on scientific assumptions. *Star Tribune,* p. A14.

Jungck, J. (1996). Ignorance, error and chaos: Local learning/global research [in Japanese]. *Japanese Journal of Contemporary Philosophy or Modern Thought, 24*(11), 363–376.

Kahneman, D. (2011). *Thinking, fast and slow.* New York, NY: Farrar, Straus and Giroux.

Kamin, L. J. (1974). *The science and politics of IQ.* Potomac, MD: L. Erlbaum Associates.

Keiny, S. & Gorodetsky, M. (1995). Beyond collaboration: Creating a community of learners. In F. Finley, D. Allchin, D. Rhees, & S. Fifield (Eds.), *Proceedings, Third International History, Philosophy and Science Teaching Conference* (pp. 589–600). Minneapolis: University of Minnesota Office of Continuing Education.

Kelly, G. J., Carlsen, W. S., & Cunningham, C. M. (1993). Science education in sociocultural context. *Science Education, 77,* 207–220.

Kelly, G. J., Chen, C., & Crawford, T. (1998). Methodological considerations for studying science-in-the-making in educational settings. *Research in Science Education, 28,* 23–49.

Kettlewell, H. B. D. (1955). Selection experiments on industrial melanism in the Lepidoptera. *Heredity, 9,* 323–342.

Kettlewell, H. B. D. (1956). Further selection experiments on industrial melanism in the Lepidoptera. *Heredity, 10,* 287–301.

Kettlewell, H. B. D. (1959, March). Darwin's missing evidence. *Scientific American, 200*(3), 48–53.

Kettlewell, [H.] B. [D.] (1973). *The evolution of melanism: The study of recurring necessity; with special reference to industrial melanism in the Lepidoptera.* Oxford, UK: Clarendon Press.

Khishfe, R., & Abd-El-Khalick, F. (2002). Influence of explicit and reflective versus implicit inquiry-oriented instruction on sixth graders' views of nature of science. *Journal of Research in Science Teaching, 39,* 551–578.

Kim, M. G. (2008). The 'instrumental' reality of phlogiston. *HYLE, 14*, 27–51.

Kintisch, E. (2010). Critics are far less prominent than supporters. *Science, 328*, 1622.

Kipling, R. (1902). *Just so stories*. London, England: Macmillan.

Kipnis, N. (1993). *Rediscovering optics*. Minneapolis, MN: Bena Press.

Klassen, S. (2007). The application of historical narrative in science learning: The Atlantic Cable story. *Science & Education, 16*, 335–352.

Klein, M. J. (1972). The use and abuse of historical teaching in physics. In S. G. Brush & A. L. King (Eds.), *History in the teaching of physics* (pp. 12–18). Hanover, NH: University Press of New England.

Klopfer, L. E. (1964–1966). *History of science cases*. Chicago, IL: Science Research Associates.

Knorr-Cetina, K. D. (1981). *The manufacture of knowledge: An essay on the constructivist and contextual nature of science*. Oxford, England: Pergamom Press.

Kohler, R. E. (1994). *Lords of the fly:* Drosophila *genetics and the experimental life*. Chicago, IL: University of Chicago Press.

Kohn, A. (1989). *Fortune or failure: Missed opportunities and chance discoveries*. Oxford, England: Basil Blackwell.

Kolstø, S. D. (2001). Scientific literacy for citizenship: Tools for dealing with the science dimension of controversial socioscientific issues. *Science Education, 85*, 291–300.

Korpan, C. A., Bisanz, G. L., Bisanz, J., & Henderson, J. M. (1997). Assessing literacy in science: Evaluation of scientific news briefs. *Science Education, 81*, 515–532.

Kosso, P. (1992). *Reading the book of nature: An introduction to the philosophy of science*. New York, NY: Cambridge University Press.

Krajcik, J. S., & Sutherland, L. M. (2010). Supporting students in developing literacy in science. *Science, 328*, 456–459.

Kraut, A. M. (2003). *Goldberger's war: The life and work of a public health crusader*. New York, NY: Hill & Wang.

Kroeber, E., Wolff, W. H., & Weaver, R. L. (1969). *Biology* (2nd ed.). Lexington, MA: D.C. Heath.

Krüger, J., Ruhrig, J., & Höttecke, D. (2013, in press). Lehrerperspektiven auf unsichere Evidenz II: Ergebnisse einer Gruppendiskussionsstudie. In S. Bernholt (Ed.), *Zur Didaktik der Chemie und Physik, GDCP-Jahrestagung in Hannover 2012*.

Kuhn, T. S. (1970). *The structure of scientific revolutions* (2nd ed.). Chicago, IL: University of Chicago Press.

Lakatos, I. (1978). *The methodology of scientific research programmes: Philosophical papers* (Vol. 1). Cambridge, England: Cambridge University Press.

Lakoff, G., & Johnson, M. (1980). *Metaphors we live by*. Chicago, IL: University of Chicago Press.

Latour, B. (1987). *Science in action: How to follow scientists and engineers through society*. Cambridge, MA: Harvard University Press.

Latour, B. (1988). *The pasteurization of France*. Cambridge, MA: Harvard University Press.

Latour, B., & Woolgar, S. (1979). *Laboratory life: The construction of scientific facts*. Princeton, NJ: Princeton University Press.

Laudan, L. (1984). *Science and values*. Berkeley: University of California Press.

Lawson, A. E. (2000). The generality of the hypothetico-deductive reasoning: Making scientific thinking explicit. *American Biology Teacher, 62*, 482–495.

Lawson, A. E. (2002). What does Galileo's discovery of Jupiter's moons tell us about the process of scientific discovery? *Science & Education, 11*, 1–24.

Lawson, A. E. (2003). Allchin's shoehorn, or why science is hypothetico-deductive. *Science & Education, 12*, 331–337.

Leaf, J. (2011). Charles Keeling & measuring atmospheric CO_2. Minneapolis, MN: SHiPS Resource Center. Retrieved from http://ships.umn.edu/modules/earth/keeling.htm

Lederman, N. G., Wade, P., & Bell, R. L. (1998). Assessing the nature of science: What is the nature of our assessments? *Science & Education, 7*, 595–615.

Leinhard, J. H. (1997). William Harvey. Engines of our ingenuity, No. 336 [Radio episode]. Houston, TX: KUHF-FM. Retrieved from http://www.uh.edu/engines/epi336.htm

Leland, T. (2007). Interpreting Native American herbal remedies. Minneapolis, MN: SHiPS Resource Center. Retrieved from http://ships.umn.edu/modules/biol/native-herb.htm

Leonard, W. H., Penick, J. E., & Speziale, B. J. (2008). *BioComm: Biology in a community context*. Armonk, NY: It's About Time, Herff Jones Education.

Levere, T. H. (2001). *Transforming matter: A history of chemistry from alchemy to the buckyball*. Baltimore, MD: Johns Hopkins University Press.

Levins, R. (1966). The strategy of model building in population biology. *American Scientist, 54*, 421–431. (Reprinted in E. Sober [Ed.], *Conceptual issues in evolutionary biology* (pp. 18–27), 1984, Cambridge, MA: MIT Press.)

Lewis, R. W. (1988). Biology: A hypothetico-deductive science. *American Biology Teacher, 50*, 362–366.

Lewontin, R. C., Rose, S., & Kamin, L. J. (1984). *Not in our genes: Biology, ideology, and human nature*. New York, NY: Pantheon.

Lockhead, J., & Dufresne, R. (1989). Helping students understand difficult science concepts through the use of dialogues with history. In D. Herget (Ed.), *The History and Philosophy of Science in Science Teaching: Proceedings of the First International Conference* (pp. 221–229). Tallahassee: Florida State University Departments of Science Education and Philosophy.

Longino, H. E. (1990). *Science as social knowledge*. Princeton, NJ: Princeton University Press.

Longino, H. E. (2001). *The fate of knowledge*. Princeton, NJ: Princeton University Press.

Losee, J. (1972). *A historical introduction to the philosophy of science*. Oxford, England: Oxford University Press.

Losee, J. (2005). *Theories on the scrap heap: Scientists and philosophers on the falsification, rejection, and replacement of theories*. Pittsburgh, PA: University of Pittsburgh Press.

Lühl, J. (1990). The history of atomic theory with its societal and philosophical implications in chemistry classes. In D. Herget (Ed.), *More History and Philosophy of Science in Science Teaching: Proceedings of the First International Conference* (pp. 266–273). Tallahassee: Florida State University Departments of Science Education and Philosophy.

Lundberg, M. A., Levin, B. B., & Harrington, H. L. (Eds.). (1999). *Who learns what from cases and how?* Mahwah, NJ: Lawrence Erlbaum Associates.

Lyons, S. [L.] (1998). Science or pseudoscience: Phrenology as a cautionary tale for evolutionary psychology. *Perspectives in Biology and Medicine, 41*, 491–503.

Lyons, S. L. (2009). *Species, serpents, spirits and skulls: Science at the margins in the Victorian age.* Albany: State University of New York Press.

MacCormac, E. R. (1976). *Metaphor and myth in science and religion.* Durham, NC: Duke University Press.

Macfarlane, G. (1985). *Alexander Fleming: The man and the myth.* Oxford, England: Oxford University Press.

Machold, D. K. (1990). The historical objections to the special theory of relativity and a method of instruction for overcoming these difficulties. In D. Herget (Ed.), *More History and Philosophy of Science in Science Teaching: Proceedings of the First International Conference* (pp. 258–265). Tallahassee: Florida State University Departments of Science Education and Philosophy.

Magnusson, S., & Templin, M. (1995). Scientific practice and science learning: The community basis of scientific literacy. In F. Finley, D. Allchin, D. Rhees, & S. Fifield (Eds.), *Proceedings, Third International History, Philosophy and Science Teaching Conference* (pp. 687–698). Minneapolis: University of Minnesota Office of Continuing Education.

Major, C. H., & Palmer, B. (2001). Assessing the effectiveness of problem-based learning in higher education: Lessons from the literature. *Academic Exchange Quarterly, 5*(1). Retrieved from http://www.rapidintellect.com/AEQweb/mop4spr01.htm

Malpighi, M. (1661/1929). About the lungs. J. Young. *Proceedings of the Royal Society of Medicine, 23*, 1–11.

Mamlok-Naaman, R., Ben-Zvi, R., Hofstein, A., Menis, J., & Erduran, S. (2005). Learning science through a historical approach: Does it affect the attitudes of non-science-oriented students towards science? *International Journal of Science and Mathematics Education, 3*, 485–507.

Martin, B. (1991). *Scientific knowledge in controversy: The social dynamics of the fluoridation debate.* Albany: State University of New York Press.

Martin, B. E., & Brouwer, W. (1991). The sharing of personal science and the narrative element in science education. *Science Education, 75*, 707–722.

Martin, J. R. (1989). What should science educators do about the gender bias in science? In D. Herget (Ed.), *The History and Philosophy of Science in Science Teaching: Proceedings of the First International Conference* (pp. 242–255). Tallahassee: Florida State University Departments of Science Education and Philosophy.

Mas, C. J. F., Perez, J. H., & Harris, H. H. (1968). Parallels between adolescents' conceptions of gases and the history of chemistry. *Journal of Chemical Education, 64*, 616–618.

Matthews, M. R. (1994). *Science teaching: The role of history and philosophy of science.* New York, NY: Routledge.

Matthews, M. R. (1999). *Time for science education.* New York, NY: Kluwer.

Matthews, M. R. (2009b). Book review: Steven Johnson, *The invention of air. Science & Education, 20*, 373–380.

Matthews, M. R. (2009a). Science and worldviews in the classroom: Joseph Priestley and photosynthesis. *Science & Education, 18*, 929–960. (Reprinted in

M. R. Matthews [Ed.], *Science, Worldviews and Education* (pp. 271–302), 2009, Dordrecht, The Netherlands: Springer.)

Mayer, R. E. (2004). Should there be a three-strikes rule against pure discovery learning? The case for guided methods of instruction. *American Psychologist, 59*, 14–19.

McCann, H. G. (1978). *Chemistry transformed: The Paradigmatic Shift from Phlogiston to Oxygen.* Norwood, NJ: Ablex.

McClellan, J. E., III, & Dorn, H. (1999). *Science and technology in world history: An introduction.* Baltimore, MD: Johns Hopkins University Press.

McClune, B., & Jarman, R. (2010). Critical reading of science-based news reports: Establishing a knowledge, skills and attitudes framework. *International Journal of Science Education, 32,* 727–752.

McComas, W. F. (1996). Ten myths of science: Reexamining what we think we know. *School Science & Mathematics, 96,* 10–16.

McComas, W. F. (1998). The principal elements of the nature of science: Dispelling the myths. In W. F. McComas (Ed.), *The nature of science in science education: Rationales and strategies* (pp. 53–70). Dordrecht, The Netherlands: Kluwer.

McComas, W. F., & Olson, J. K. (1998). The nature of science in international science education standards documents. In W. F. McComas (Ed.), *The nature of science in science education: Rationales and strategies* (pp. 41–52). Dordrecht, The Netherlands: Kluwer.

McEvoy, J. G. (1987). Causes and laws, powers and principles: The metaphysical foundations of Priestley's concept of phlogiston. In R. G. W. Anderson & C. Lawrence (Eds.), *Science, medicine and dissent: Joseph Priestley (1733–1804)* (pp. 55–71). London, England: Wellcome Trust/Science Museum.

Medawar, P. B. (1964). Is the scientific paper fraudulent? Yes; it misrepresents scientific thought. *Saturday Review, 47*(1 August), 42–43.

Melhado, E. M. (1989). Toward an understanding of the Chemical Revolution. *Knowledge in Society, 8,* 123–127.

Mendel, G. (1866/1966). Versuche über Pflanzenhybriden [Experiments on plant hybrids]. Translated in C. Stern & E. Sherwood (Eds.), *The origin of genetics: A Mendel source book* (pp. 1–48), 1966, San Francisco, CA: W.H. Freeman. Translation by C. T. Druery (1901) retrieved from http://www.mendelweb.org/MWpaptoc.html

Mendel, G. (1869/1966). On Hieracium-hybrids obtained by artificial fertilization. Translated in C. Stern & E. Sherwood (Eds.), *The origin of genetics: A Mendel source book* (pp. 49–50), 1966, San Francisco, CA: W.H. Freeman.

Mermelstein, E., & Young, K. C. (1995). The nature of error in the process of science and its implications for the teaching of science. In F. Finley, D. Allchin, D. Rhees, & S. Fifield (Eds.), *Proceedings, Third International History, Philosophy and Science Teaching Conference* (pp. 768–775). Minneapolis: University of Minnesota Office of Continuing Education.

Merton, R. K. (1973). *The sociology of science.* Chicago, IL: University of Chicago Press.

Merton, R. K., & Barber, E. (2003). *The travels and adventures of serendipity.* Princeton, NJ: Princeton University Press.

Micale, M. S. (1995). *Approaching hysteria: Disease and its interpretations.* Princeton, NJ: Princeton University Press.

Michael, J. (2006). Where's the evidence that active learning works? *Advances in Physiology Education, 30,* 159–167.

Michaels, D. (2008). *Doubt is their product: How industry's assault on science threatens your health.* Oxford, England: Oxford University Press.

Midgley, M. (1992). *Science as salvation: A modern myth and its meaning.* London, England: Routledge.

Millar, R. (2000). Science for public understanding: Developing a new course for 16–18 year old students. *Critical Studies in Education, 41,* 201–214.

Millar, R., & Osborne, J. (1998). *Beyond 2000: Science education for the future.* London, England: King's College.

Milne, C. (1998). Philosophically correct science stories? Examining the implications of heroic science stories for school science. *Journal of Research in Science Teaching, 35,* 175–187.

Milton, J. R. (1981). The origin and development of the concept of the 'laws of nature.' *European Journal of Sociology, 22,* 173–195.

Minstrell, J., & Kraus, P. (2005). Guided inquiry in the science classroom. In M. S. Donovan & J. D. Bransford (Eds.), *How students learn: History, mathematics, and science in the classroom* (pp. 475–513). Washington, DC: National Academies Press.

Mix, M. C., Farber, P., & King, K. I. (1996). *Biology: The network of life* (2nd ed.). Glenview, IL: HarperCollins.

Monaghan, F., & Corcos, A. F. (1990). The real objective of Mendel's paper. *Biology & Philosophy, 5,* 267–292.

Monk, M., & Osborne, J. (1997). Placing the history and philosophy of science on the curriculum: A model for the development of pedagogy. *Science Education, 81,* 405–424.

Monmonier, M. (1991). *How to lie with maps.* Chicago, IL: University of Chicago Press.

Montgomery, K. (2010). *Debating glacial theory, 1800–1870.* Minneapolis, MN: SHiPS Resource Center. Retrieved from http://glacialtheory.net

Moran, B. T. (2005). *Distilling knowledge: Alchemy, chemistry, and the Scientific Revolution.* Cambridge, MA: Harvard University Press.

Moran, E. (2009). Richard Lower and the "life force" of the body. Minneapolis, MN: SHiPS Resource Center. Retrieved from http://ships.umn.edu/modules/biol/lower.htm

Moss, J. D. (1993). *Novelties in the heavens: Rhetoric and science in the Copernican controversy.* Chicago, IL: University of Chicago Press.

Moss, L. (2002). *What genes can't do.* Cambridge, MA: MIT Press.

Murcia, K., & Schibeci, R. (1999). Primary student teachers' conceptions of the nature of science. *International Journal of Science Education, 21,* 1123–1140.

Musgrave, A. (1976). Why did oxygen supplant phlogiston? Research programmes in the chemical revolution. In C. Howson (Ed.), *Method and appraisal in the physical sciences: The critical background to modern science, 1800–1905* (pp. 181–210). Cambridge, England: Cambridge University Press.

Nash, L. K. (1951). An historical approach to the teaching of science. *Journal of Chemical Education, 28,* 146–151.

Nash, L. K. (1957). Plants and the atmosphere. In J. B. Conant & L. K. Nash (Eds.), *Harvard case histories in experimental science* (Vol. 2, pp. 323–426). Cambridge,

MA: Harvard University Press.

Nason, A., & Goldstein, P. (1969). *Biology: Introduction to life.* Menlo Park, CA: Addison Wesley.

National Research Council. (1996). *National science education standards.* Washington, DC: National Academy Press.

National Science Teachers Association. (2000). NSTA position statement: The nature of science. Retrieved from http://www.nsta.org/about/positions/natureofscience.aspx

Nendick, J., Scrancher, D., & Usher, O. (2007). Chlorine and Prout's hypothesis. In H. Chang & C. Jackson (Eds.), *An element of controversy: The life of chlorine in science, medicine, technology and war* (pp. 73–104). London, England: British Society for the History of Science.

Nersessian, N. J. (1989). Conceptual change in science and science education. *Synthese, 80,* 163–183.

Nersessian, N. J. (2002). The cognitive basis of model-based reasoning in science. In P. Carruthers, S. Stich & M. Siegal (Eds.), *The cognitive basis of science* (pp. 133–153). Cambridge, England: Cambridge University Press.

Nickerson, R. S. (1998). Confirmation bias: A ubiquitous phenomenon in many guises. *Review of General Psychology, 2,* 175–220.

Nogler, G. A. (2006). The lesser-known Mendel: His experiments on *Hieracium. Genetics, 172,* 1–6.

Norris, S. P. (1995). Learning to live with scientific expertise: Toward a theory of intellectual communalism for guiding science teaching. *Science Education, 79,* 201–217.

Norris, S. P. (1997). Intellectual independence for nonscientists and other content-transcendent goals of science education. *Science Education, 81,* 239–258.

Norris, S. P., & Phillips, L. M. (1994). Interpreting pragmatic meaning when reading popular reports of science. *Journal of Research in Science Teaching, 31,* 947–967.

Norris, S. P., Phillips, L. M., & Korpan, C. A. (2003). University students' interpretation of media reports of science and its relationship to background knowledge, interest, and reading difficulty. *Public Understanding of Science, 12,* 123–145.

Nott, M., & Wellington, J. (1998). Eliciting, interpreting and developing teachers' understandings of the nature of science. *Science & Education, 7,* 579–594.

Novak, L. (2008). Determining atomic weights: The role of Avogadro's hypothesis. Minneapolis, MN: SHiPS Resource Center. Retrieved from http://ships.umn.edu/modules/chem/avogadro.htm

Odling, W. (1871a). On the revived theory of phlogiston. *Proceedings of the Royal Institution of Great Britain, 6,* 315–325.

Odling, W. (1871b). *Pharmaceutical Journal* (3rd series), *1,* 977–981.

Olby, R. C. (1985). *Origins of Mendelism* (2nd ed.). Chicago, IL: University of Chicago Press.

Olby, R. C. (1997). Mendel, Mendelism and genetics. MendelWeb. Retrieved from http://www.mendelweb.org/MWolby.intro.html

O'Rafferty, M. (1995). Developing sociological insights on scientific norms using case studies on scientific misconduct. In F. Finley, D. Allchin, D. Rhees, & S. Fifield (Eds.), *Proceedings, Third International History, Philosophy and Science Teaching Conference* (pp. 905–913). Minneapolis: University of Minnesota

Office of Continuing Education.

Oreskes, N., & Conway, E. M. (2010). *Merchants of doubt: How a handful of scientists obscured the truth on issues from tobacco smoke to global warming.* New York, NY: Bloomsbury.

Organization for Economic Cooperation and Development. (2009). PISA 2009 assessment framework. Retrieved from http://www.oecd.org/pisa/pisaproducts/44455820.pdf

Osborne, J. (2007). Science education for the twenty-first century. *Eurasia Journal of Mathematics, Science and Technology Education, 3,* 173–184.

Osborne, J., Collins, S., Ratcliffe, M., Millar, R., & Duschl, R. (2003). What "ideas-about-science" should be taught in school science? A Delphi study of the expert community. *Journal of Research in Science Teaching, 40,* 692–720.

Pagel, W. (1967). *William Harvey's biological ideas.* New York, NY: Hafner.

Pagel, W. (1976). *New light on William Harvey.* Basel, Switzerland: S. Karger.

Pagel, W. (1982). *Joan Baptista van Helmont: Reformer of science and medicine.* Cambridge, England: Cambridge University Press.

Park, R. (2000). *Voodoo science: The road from foolishness to fraud.* Oxford, England: Oxford University Press.

Partington, J. R. (1961–1964). *A history of chemistry* (4 vols.). London, England: Macmillan.

Partington, J. R., & McKie, D. (1937–1939). Historical studies on the phlogiston theory, I–IV. *Annals of Science, 2,* 361–404; *3,* 1–38, 336–371; *4,* 113–149.

Pellegrino, J. W., Chudowsky, N., & Glaser, R. (Eds.). (2001). *Knowing what students know: The science and design of educational assessment.* Washington, DC: National Academies Press.

Phillips, L. M., & Norris, S. P. (1999). Interpreting popular reports of science: What happens when the reader's world meets the world on paper? *International Journal of Science Education, 21,* 317–327.

Pi Suñer, A. (1955). *Classics of biology* (C. M. Stern, Trans.). New York, NY: Philosophical Library.

Pickering, A. (1984). *Constructing quarks: A sociological history of particle physics.* Chicago, IL: University of Chicago Press.

Pickering, A. (1992). *Science as practice and culture.* Chicago, IL: University of Chicago Press.

Pickering, A. (1995). *The mangle of practice: Time, agency, and science.* Chicago, IL: University of Chicago Press.

Pigliucci, M. (2010). *Nonsense on stilts: How to tell science from bunk.* Chicago, IL: University of Chicago Press.

Pitt, J. C. (2000). *Thinking about technology: Foundations of the philosophy of technology.* New York, NY: Seven Bridges Press.

Poincaré, H. (1902). *La science et l'hypothèse.* Paris, France: E. Flammarion.

Porter, T. M. (1986). *The rise of statistical thinking, 1820–1900.* Princeton, NJ: Princeton University Press.

Posner, G. J., Strike, K. A., Hewson, P. W., & Gertzog, W. A. (1982). Accommodation of a scientific conception: Toward a theory of conceptual change. *Science Education, 66,* 211–227.

Postlethwait, J. H., & Hopson, J. L. (2003). *Explore life.* Pacific Grove, CA: Brooks/Cole.

Powell, J. L. (1998). *Night comes to the Cretaceous: Comets, craters, controversy, and the last days of the dinosaurs.* New York, NY: W.H. Freeman.

Price, D. J. de Solla (1963). *Little science, big science.* New York, NY: Columbia University Press.

Principe, L. M. (1998). *The aspiring adept: Robert Boyle and his alchemical quest.* Princeton, NJ: Princeton University Press.

Pyenson, L., & Sheets-Pyenson, S. (1999). *Servants of nature: A history of scientific institutions, enterprises, and sensibilities.* New York, NY: W.W. Norton.

Rader, K. (2004). *Making mice: Standardizing animals for American biomedical research, 1900–1955.* Princeton, NJ: Princeton University Press.

Randak, S. (1990). Role-playing in the classroom. *American Biology Teacher, 52,* 439–442.

Raup, D. M. (1992). *Extinction: Bad genes or bad luck?* New York, NY: W.W. Norton.

Regnault, H. V. (1847). Relations des expériences entreprises par ordre de Monsieur le Ministre des Travaux Publics, et sur la proposition de la Commission Centrale des Machines a Vapeur, pour déterminer les principales lois et les données numériques qui entrent dans le calcul des machines a vapeur. *Mémoires de l'Académie Royale des Sciences de l'Institut de France, Vol. 21.*

Reiß, F. (1995). Teaching science and the history of science by redoing historical experiments. In F. Finley, D. Allchin, D. Rhees, & S. Fifield (Eds.), *Proceedings, Third International History, Philosophy and Science Teaching Conference* (pp. 958–966). Minneapolis: University of Minnesota Office of Continuing Education.

Remillard-Hagen, E. (2010). Lady Mary Wortley Montagu & smallpox variolation in 18th-century England. Minneapolis, MN: SHiPS Resource Center. Retrieved from http://ships.umn.edu/modules/biol/smallpox.htm

Rheinberger, H.-J. (1997). *Toward a history of epistemic things: Synthesizing proteins in the test tube.* Palo Alto, CA: Stanford University Press.

Rheinberger, H.-J. (2009). Experimental reorientations. In G. Hon, J. Schickore, & F. Steinle (Eds.), *Going amiss in experimental research* (pp. 75–90). Dordrecht, The Netherlands: Springer.

Richardson, R. C. (2007). *Evolutionary psychology as maladapted psychology.* Cambridge, England: Cambridge University Press.

Rivers, I., & Wykes, D. L. (Eds.). (2008). *Joseph Priestley: Scientist, philosopher, and theologian.* Oxford, England: Oxford University Press.

Roberts, R. M. (1989). *Serendipity: Accidental discoveries in science.* New York, NY: John Wiley.

Robin, H. (1992). *The scientific image: From cave to computer.* New York, NY: H.N. Abrams.

Robin, N., & Ohlsson, S. (1989). Impetus then and now: A detailed comparison between Jean Buridan and a single contemporary subject. In D. Herget (Ed.), *The History and Philosophy of Science in Science Teaching: Proceedings of the First International Conference* (pp. 292–305). Tallahassee: Florida State University Departments of Science Education and Philosophy.

Ronan, C. A. (1985). *Science: Its history and development among the world's cultures.* New York, NY: Facts-on-File.

Rose, S. (1997). *Lifelines: Life beyond the gene.* New York, NY: Oxford University Press.

Roth, W.-M. (1997). From everyday science to science education: How science and technology studies inspired curriculum design and classroom research. *Science & Education, 6*, 373–396.

Rothbart, D., & Slayden, S. W. (1994). The epistemology of a spectrometer. *Philosophy of Science, 61*, 25–38.

Rousseau, D. L. (1992). Case studies in pathological science. *American Scientist, 80*, 54–63.

Rubin, J. (2011). Following the path of discovery: Repeat famous experiments and inventions. Retrieved from http://www.juliantrubin.com/bigten/pathdiscovery.html

Rudge, D. W. (1999). Taking the peppered moth with a grain of salt. *Biology & Philosophy, 14*, 9–37.

Rudolph, J. L. (2000). Reconsidering the 'nature of science' as a curriculum component. *Journal of Curriculum Studies, 32*, 403–419.

Rudwick, M. J. S. (1985). *The great Devonian controversy*. Chicago, IL: University of Chicago Press.

Ruhrig, J., Ohlsen, M., & Höttecke, D. (2013, in press). Lehrerperspektiven auf unsichere Evidenz I: Projektziele, -design und Erhebungsinstrumente. In S. Bernholt (Ed.), *Zur Didaktik der Chemie und Physik, GDCP-Jahrestagung in Hannover 2012*.

Rutherford, F. J., & Ahlgren, A. (1990). *Science for all Americans*. Oxford, England: Oxford University Press.

Ryder, J. (2001). Identifying science understanding for functional scientific literacy. *Studies in Science Education, 36*, 1–44.

Sadler, T. D., Chambers, F. W., & Zeidler, D. L. (2004). Student conceptualizations of the nature of science in response to a socioscientific issue. *International Journal of Science Education, 26*, 387–409.

Sagoff, M. (2003). The plaza and the pendulum: Two concepts of ecological science. *Biology & Philosophy, 18*, 529–552.

Sapp, J. (1990). The nine lives of Gregor Mendel. In H. E. Le Grand (Ed.), *Experimental inquiries* (pp. 137–166). Dordrecht, The Netherlands: Kluwer Academic. Retrieved from http://www.mendelweb.org/MWsapp.html

Sargent, R.-M. (1995). *The diffident naturalist*. Chicago, IL: University of Chicago Press.

Savan, B. (1988). *Science under siege: The myth of objectivity in scientific research*. Montreal, Canada: CBC Enterprises.

Scerri, E. R. (2007). *The periodic table: Its story and its significance*. New York, NY: Oxford University Press.

Schaffer, S. (1989). Glass works: Newton's prisms and the uses of experiment. In D. Gooding, T. Pinch, & S. Schaffer (Eds.), *The uses of experiment* (pp. 67–104). Cambridge, England: Cambridge University Press.

Scharmann, L. C., Smith, M. U., James, M. C., & Jensen, M. (2005). Explicit reflective nature of science instruction: Evolution, intelligent design, and umbrellaology. *Journal of Science Teacher Education, 16*, 27–41.

Schiebinger, L. (1993). *Nature's body: Gender in the making of modern science*. Boston, MA: Beacon Press.

Schofield, R. (2004). *The enlightened Joseph Priestley: A study of his life and work from 1773 to 1804*. University Park: Pennsylvania State University Press.

Schwartz, A. T., Bunce, D. M., Silberman, R. G., Stanitski, C. L., Stratton, W. J., &

Zipp, A. P. (1997). *Chemistry in context: Applying chemistry to society* (2nd ed.). Dubuque, IA: William C. Brown/American Chemical Society.

Schwartz, R. S., Lederman, N. G., & Crawford, B. A. (2004). Developing views of nature of science in an authentic context: An explicit approach to bridging the gap between nature of science and scientific inquiry. *Science Education, 88,* 610–645.

Scott, J. H. (1958). Qualitative adequacy of phlogiston. *Journal of Chemical Education, 29,* 360–363.

Seker, H., & Welsh, L. C. (2005). *The comparison of explicit and implicit ways of using history of science for students understanding of the nature of science.* Paper presented at the 8th International History, Philosophy, Sociology & Science Teaching Conference, Leeds, England.

Selin, H. (Ed.). (1997). *Encyclopaedia of the history of science, technology, and medicine in non-Western cultures.* Dordrecht, The Netherlands: Kluwer.

Selinger, E., & Crease, R. P. (2006). *The philosophy of expertise.* New York, NY: Columbia University Press.

Service, R. F. (2002). Bell Labs fires star physicist found guilty of forging data. *Science, 298,* 30–31.

Service, R. F. (2003). More of Bell Labs physicist's papers retracted. *Science, 299,* 31.

Shaffer, P. (1990). *Lettice and lovage.* New York, NY: Harper and Row.

Shahn, E. (1990). Foundations of science: A lab course for nonscience majors. In D. Herget (Ed.), *More History and Philosophy of Science in Science Teaching: Proceedings of the First International Conference* (pp. 311–352). Tallahassee: Florida State University Departments of Science Education and Philosophy.

Shamos, M. A. (1987). *Great experiments in physics: Firsthand accounts from Galileo to Einstein.* New York, NY: Dover.

Shapin, S. (1979). The politics of observation: Cerebral anatomy and social interests in the Edinburgh phrenology disputes. *Sociological Review Monographs, 27,* 139–178.

Shapin, S. (1988). The house of experiment in seventeenth-century England. *Isis, 79,* 373–404.

Shapin, S. (1989). The invisible technician. *American Scientist, 77,* 554–563.

Shapin, S. (1994). *A social history of truth: Civility and science in seventeenth-century England.* Chicago, IL: University of Chicago Press.

Shapin, S. (1996). *The Scientific Revolution.* Chicago, IL: University of Chicago Press.

Shapin, S., & Schaffer, S. (1985). *Leviathan and the air-pump: Hobbes, Boyle, and the experimental life.* Princeton, NJ: Princeton University Press.

Shapiro, A. E. (1996). The gradual acceptance of Newton's theory of light and color, 1672–1727. *Perspectives on Science, 4,* 59–140.

Shea, W. (1998). Galileo's Copernicanism: The science and the rhetoric. In P. Machamer (Ed.), *The Cambridge companion to Galileo* (pp. 211–243). Cambridge, England: Cambridge University Press.

Shulman, B. (1995). Not just a spice: The historical perspective as an essential ingredient in every math class. In F. Finley, D. Allchin, D. Rhees, & S. Fifield (Eds.), *Proceedings, Third International History, Philosophy and Science Teaching Conference* (pp. 1042–1049). Minneapolis: University of Minnesota Office of Continuing Education.

Sibum, H. O. (1995). Reworking the mechanical value of heat: Instruments of precision and gestures of accuracy in early Victorian England. *Studies in History and Philosophy of Science, 26,* 73–106.

Siegfried, R. (1964). The phlogistic conjectures of Humphry Davy. *Chymia, 9,* 117–124.

Silva, C., Silva, P., Passos, P., Morais, A. M., & Neves, I. P. (1995). Fraud in science: The sociology of science in the science classroom. In F. Finley, D. Allchin, D. Rhees, & S. Fifield (Eds.), *Proceedings, Third International History, Philosophy and Science Teaching Conference* (pp. 1062–1067). Minneapolis: University of Minnesota Office of Continuing Education.

Silverman, M. P. (1992). Raising questions: Philosophical significance of controversy in science. *Science & Education, 1,* 163–179.

Silverstein, A., Silverstein, V., & Silverstein Nunn, L. (2008). *Photosynthesis.* Minneapolis, MN: Twenty-First Century Books.

Sismondo, S. (1996). *Science without myth: On constructions, reality, and social knowledge.* Albany: State University of New York Press.

Sismondo, S., & Chrisman, N. (2001). Deflationary metaphysics and the nature of maps. *Philosophy of Science, 68*(Proceedings), S38–S49.

Slezak, P. (1994). Sociology of scientific knowledge and scientific education. *Science & Education, 3,* 265–294, 329–355.

Solomon, J. (1987). Social influences on the construction of pupils' understanding of science. *Studies in Science Education, 14,* 63–82.

Solomon, J., Duveen, J., Scot, L., & McCarthy, S. (1992). Teaching about the nature of science through history: Action research in the classroom. *Journal of Research in Science Teaching, 29,* 409–421.

Solomon, M. (2001). *Social empiricism.* Cambridge, MA: MIT Press.

Stanley, M. (2007). "The soul made flesh": An introduction to the nervous system. Minneapolis, MN: SHiPS Resource Center. Retrieved from http://ships.umn.edu/modules/biol/willis.htm

Starr, C. (1994). *Biology: Concepts and applications* (2nd ed.). Belmont, CA: Wadsworth.

Steinbach, S. E. (1998, December 4). How frivolous litigation threatens good science. *Chronicle of Higher Education,* p. A56.

Steinberg, M. S. (1992). What is electric potential?: Connecting Alessandro Volta and contemporary students. In S. Hills (Ed.), *The history and philosophy of science in science education* (Vol. 2, pp. 472–480). Kingston, ON: Mathematics, Science, Technology and Teacher Education Group and Faculty of Education, Queen's University.

Steinle, F. (1995). The amalgamation of a concept—Laws of nature in the new sciences. In F. Weinert (Ed.), *Laws of nature: Essays on the philosophical, scientific and historical dimensions* (pp. 316–368). Berlin, Germany: Walter de Gruyter.

Steinle, F. (2002). Negotiating experiment, reason and theology: The concept of laws of nature in the early Royal Society. In W. Detel & C. Zittel (Eds.), *Ideals and cultures of knowledge in early modern Europe* (pp. 197–212). Berlin, Germany: Akademie Verlag.

Stephens, T., & Brynner, R. (2001). *Dark remedy: The impact of thalidomide and its revival as a vital medicine.* New York, NY: Basic Books.

Sterling, J. (1994). The world according to Disney: Trading in fantasies. *Earth Island Journal, 9*(3), 32–33.

Stocklmayer, S. M., & Treagust, D. F. (1994). A historical analysis of electric currents in textbooks: A century of influence on physics education. *Science & Education, 3*, 131–154.

Stork, H. (1995). Using history and philosophy of science in science teaching. In F. Finley, D. Allchin, D. Rhees, & S. Fifield (Eds.), *Proceedings, Third International History, Philosophy and Science Teaching Conference* (pp. 1135–1145). Minneapolis: University of Minnesota Office of Continuing Education.

Sudduth, W. M. (1978). Eighteenth-century identifications of electricity with phlogiston. *Ambix, 25*, 131–147.

Sutherland, S. (1992). *Irrationality: Why we don't think straight!* New Brunswick, NJ: Rutgers University Press.

Swanson, R. P. (1995a). Science education recapitulates science history: How can history guide modern science curriculum? In F. Finley, D. Allchin, D. Rhees, & S. Fifield (Eds.), *Proceedings, Third International History, Philosophy and Science Teaching Conference* (pp. 1192–1194). Minneapolis: University of Minnesota Office of Continuing Education.

Swanson, R. P. (1995b). The electrical atom: A series of lessons for high school teachers. In F. Finley, D. Allchin, D. Rhees, & S. Fifield (Eds.), *Proceedings, Third International History, Philosophy and Science Teaching Conference* (pp. 1181–1191). Minneapolis: University of Minnesota Office of Continuing Education.

Swenson, L. S., Jr. (1970). The Michelson-Morley-Miller experiment before and after 1905. *Journal for the History of Astronomy, 1*, 56–78.

Taton, R. (1962). *Reason and chance in scientific discovery* (A. J. Pomerans, Trans.). New York, NY: Science Editions.

Taylor, P. J. (2005). *Unruly complexity: Ecology, interpretation, engagement.* Chicago, IL: University of Chicago Press.

Teresi, D. (2002). *Lost discoveries: The ancient roots of modern science—from the Babylonians to the Maya.* New York, NY: Simon & Schuster.

Thagard, P. (1990). The conceptual structure of the chemical revolution. *Philosophy of Science, 57*, 183–209.

Thomas, J. (2000). Using current controversies in the classroom: Opportunities and concerns. *Melbourne Studies in Education, 41*, 133–144.

Thomas, L. (1981). Introduction. In H. F. Judson, *The search for solutions* (pp. ix–x). New York, NY: Holt, Rinehart, Winston.

Tobin, A. J., & Dusheck, J. (2005). *Asking about life* (3rd ed.). Thomson Brooks/Cole.

Tobin, K., & Roth, W.-M. (1995). Briding the great divide: Teaching from the perspective of one who knows and learning from the perspective of one who does not know. In F. Finley, D. Allchin, D. Rhees, & S. Fifield (Eds.), *Proceedings, Third International History, Philosophy and Science Teaching Conference* (pp. 1204–1216). Minneapolis: University of Minnesota Office of Continuing Education.

Toulmin, S. (1960). *The philosophy of science.* New York, NY: Harper and Row.

Toumey, C. P. (1996). *Conjuring science: Scientific symbols and cultural meanings in American life.* New Brunswick, NJ: Rutgers University Press.

Traweek, S. (1988). *Beamtimes and lifetimes: The world of high energy physics.* Cambridge, MA: Harvard University Press.

Tufte, E. R. (1997). *Visual explanations: Images and quantities, evidence and narrative.* Cheshire, CT: Graphics Press.

Turnbull, D. (1993). *Maps are territories: Science is an atlas.* Chicago, IL: University of

Chicago Press. Also online at http://territories.indigenousknowledge.org

Turnbull, D. (2000). *Masons, tricksters and cartographers: Comparative studies in the sociology of scientific and indigenous knowledge.* Amsterdam, Netherlands: Harwood Academic.

Tyson, L. M., Venville, G. J., Harrison, A. G., & Treagust, D. F. (1997). A multidimensional framework for interpreting conceptual change events in the classroom. *Science Education, 81,* 387–404.

Uglow, J. (2003). *The Lunar Men: Five friends whose curiosity changed the world.* New York, NY: Farrar, Straus and Giroux.

van der Waals, J. D. (1910). The equation of state for gases and liquids. (Reprinted in *Nobel Lectures: Physics 1901–1921,* 1967, Amsterdam, The Netherlands: Elsevier.) Retrieved from http://www.nobelprize.org/nobel_prizes/physics/laureates/1910/waals-lecture.html

van Klooster, H. S. (1947). Jan Baptist van Helmont. *Journal of Chemical Education, 24,* 319.

Vogel, G., Proffitt, F., & Stone, R. (2004). Ecologists roiled by misconduct case. *Science, 303,* 606–609.

Wallace, M. (1996). *Mickey Mouse history and other essays on American memory.* Philadelphia, PA: Temple University Press.

Walvig, S. (2010). Contested currents: The race to electrify America. Minneapolis, MN: SHiPS Resource Center. Retrieved from http://ships.umn.edu/modules/phys/currents/pages/intro.htm

Wandersee, J. H. (1986). Can the history of science help science educators anticipate students' misconceptions? *Journal of Research in Science Teaching, 23,* 581–597.

Wandersee, J. H. (1992). The historicality of cognition: Implications for science education research. *Journal of Research in Science Teaching, 29,* 423–434.

Watson, J. D. (1968). *The double helix.* New York, NY: Atheneum.

Webster, C. (1965). The discovery of Boyle's law, and the concept of the elasticity of air in the seventeenth century. *Archive for History of the Exact Sciences, 2,* 441–502.

Weck, M. (1995). Are today's models tomorrow's misconceptions? In F. Finley, D. Allchin, D. Rhees, & S. Fifield (Eds.), *Proceedings, Third International History, Philosophy and Science Teaching Conference* (pp. 1286–1294). Minneapolis: University of Minnesota Office of Continuing Education.

Wellington, J. (1991). Newspaper science, school science: Friends or enemies? *International Journal of Science Education, 13,* 363–372.

Westerlund, J. F., & Fairbanks, D. J. (2010). Gregor Mendel's classic paper and the nature of science in genetics courses. *Heritas, 147,* 293–303.

WGBH. (1998). Fleming discovers penicillin, 1928–1945. Boston, MA: WGBH-TV. Retrieved from http://www.pbs.org/wgbh/aso/databank/entries/dm28pe.html

Whitaker, M. A. B. (1979). History and quasi-history in physics education—part 2. *Physics Education, 14, 108,* 239–242.

White, H. (1987). *The content of the form: Narrative discourse and historical representation.* Baltimore, MD: Johns Hopkins University Press.

White, M. (1997). *Isaac Newton: The last sorcerer.* Reading, MA: Perseus Books.

Wimsatt, W. C. (2007). *Re-engineering philosophy for limited beings: Piecewise approximations to reality.* Cambridge, MA: Harvard University Press.

Wong, S. L., & Hodson, D. (2009). From the horse's mouth: What scientists say about scientific investigation and scientific knowledge. *Science Education, 93,* 109–130.

Wong, S. L., & Hodson, D. (2010). More from the horse's mouth: What scientists say about science as a social practice. *International Journal of Science Education, 32,* 1431–1463.

Wong, S. L., Hodson, D., Kwan J., & Yung, B. H. W. (2008). Turning crisis into opportunity: Enhancing student-teachers' understanding of nature of science and scientific inquiry through a case study of the scientific research in severe acute respiratory syndrome. *International Journal of Science Education, 30,* 1417–1439.

Woodward, J. (2003). *Making things happen: A theory of causal explanation.* Oxford, England: Oxford University Press.

Young, J. (1929). Malpighi's "De pulmonibus" [introduction]. *Proceedings of the Royal Society of Medicine, 23,* 1.

Ziman, J. (1976). *The force of knowledge: The scientific dimension of society.* Cambridge, England: Cambridge University Press.

Ziman, J. (1978). *Reliable knowledge: An exploration of the grounds for belief in science.* Cambridge, England: Cambridge University Press.

| Acknowledgments

My debts are numerous and the occasion for much warm appreciation.

My first thanks go to my students at Georgetown Day High School in Washington, DC, where many of the perspectives that shaped this book originated in the early 1980s. Also to my students at the University of Minnesota, who participated in the historical simulations presented in Chapter 12 and 13 in various stages of their development, or who worked with me to assemble their own historical case studies for teaching science (and whose struggles contributed to the advice shared in Chapter 14).

I am especially grateful to Peter Heering, Dietmar Höttecke, Cibelle Celestino-Silva, Falk Reiß, Art Stinner, Fabio Bevilaqua, John Jungck, and Rick Duschl, whose generous invitations and support provided occasion for me to consolidate my emerging thoughts and present them for scholarly discussion (now found in Chapters 2, 3, 4, 7, 8, and 9). I send similar thanks to Joel Hagen, for engaging me in developing a set of historical case studies and promoting my work at a critical stage (as reflected in Chapter 10), and to Rick Swanson, for recruiting me to collaborate in his classroom (as described in Chapter 11). I hope the work here honors these gestures.

I am particularly thankful to Glenn Dolphin, Ami Friedman, and Greg Kelly, who each reviewed the whole assembled manuscript and whose comments proved invaluable.

Nearly every perspective presented here benefited from collegial comments, both affirming and critical. In many cases I can recognizably trace contributions to particular conversations or to fruitful ongoing dialogue. Among those I wish to acknowledge (in addition to those mentioned above) are Roberto de Martins Andrade, Jim Andrews, Libby Anthony, Agustín Adúriz-Bravo, Nancy Brickhouse, Jack Bristol, Michael Clough, Pamela Goss Cook, Alan Dean, Robert Dennison, Steve Fifield, Fred Finley, Maura Flannery, Grahame Gooday, Alan Gross, Piers Hale, Andreas Henke, Clyde Herreid, Ted Hodapp, Bruce King, Cathrine Klassen, Carl Lieb, Sherrie Lyons, Ralph Mason, Roberta Millstein, Maria Elice Prestes, Patti Ross, David Rudge, Alan Shapiro, Patti Soderberg, Ethel Stanley, Peter Taylor, Nadine Weidman, Alex Werth, Alison Wylie, and Michael Ziomko. I have also profited from innumerable others who participated in workshops where the ideas here were discussed or who anonymously reviewed earlier publications.

Finally, my gratitude to Richard Earles, Deb Sampson, and Theanna Hegge, whose professional guidance helped shepherd the book to publication.

In a world where argumentative bitterness and castigation abound, I am grateful for finding encouragement, supportive criticism, and assistance among so many colleagues.

In addition, I am indebted to the respective publishers for publishing the original papers and gratefully acknowledge permission to republish them in an adapted form. Chapter 2 is adapted in part from "The Power of History as a Tool for Teaching Science," in A. Dally, T. Nielsen, & F. Reiß (Eds.), *History and Philosophy of Science: A Means to Better Scientific Literacy?*, 1997, Loccum: Evangelische Akadamie Loccum, pp. 70–98. Chapter 3 is adapted in part from "Scientific Myth-Conceptions," 2008, *Science Education*, 87, pp. 329–351. Chapter 4 is adapted from "How *Not* to Teach History in Science," 2000, *Journal of College Science Teaching*, 30, pp. 33–37. Chapter 5 is adapted from "Pseudohistory and Pseudoscience," 2004, *Science & Education, 13,* pp. 179–195; and "Lawson's Shoehorn, or Should the Philosophy of Science be Rated 'X'?", 2003, *Science & Education*, 12, pp. 315–329. Chapter 6 is adapted from "Should the Sociology of Science be Rated 'X'?", 2004, *Science Education*, 88, pp. 934–946. Chapter 7 is adapted in part from "Kettlewell's Missing Evidence, A Study in Black and White," 2001, *Journal of College Science Teaching*, 31, pp. 240–245. Chapter 8 is adapted from "Teaching Science Lawlessly," in P. Heering & D. Osewold (Eds.), *Constructing Scientific Understanding through Contextual Teaching*, 2007, Berlin: Frank & Timme, pp. 13–31; and "How School Science Lies," 1999, paper presented at the 5th International Conference on History, Philosophy and Science Teaching, Pavia: University of Pavia. Chapter 9 is adapted in part from "Evaluating Knowledge of the Nature of (Whole) Science," 2011, *Science Education*, 95, pp. 518–542. Chapter 10 is adapted from "Christiaan Eijkman & the Cause of Beriberi," in J. Hagen, D. Allchin, & F. Singer, *Doing Biology*, 1996, Glenview, IL: Harper Collins, pp. 116–127. Chapter 11 is adapted from "Rekindling Phlogiston: From Classroom Case Study to Interdisciplinary Relationships," 1997, *Science & Education*, 6, pp. 473–509. Chapter 14 is adapted in part from "The Minnesota Case Study Collection: New Historical Inquiry Cases for Nature of Science Education," 2012, *Science & Education*, 21, pp. 1263–1282.

| Index